Y0-AAY-433

Methods in Enzymology

Volume 242

NEOGLYCOCONJUGATES

Part A

Synthesis

METHODS IN ENZYMOLOGY

EDITORS-IN-CHIEF

John N. Abelson   Melvin I. Simon

DIVISION OF BIOLOGY
CALIFORNIA INSTITUTE OF TECHNOLOGY
PASADENA, CALIFORNIA

FOUNDING EDITORS

Sidney P. Colowick and Nathan O. Kaplan

Methods in Enzymology

Volume 242

Neoglycoconjugates

Part A

Synthesis

EDITED BY

Y. C. Lee

Reiko T. Lee

BIOLOGY DEPARTMENT
THE JOHNS HOPKINS UNIVERSITY
BALTIMORE, MARYLAND

QP601
C71
v.242
1994

ACADEMIC PRESS
San Diego New York Boston London Sydney Tokyo Toronto

This book is printed on acid-free paper. ∞

Copyright © 1994 by ACADEMIC PRESS, INC.
All Rights Reserved.
No part of this publication may be reproduced or transmitted in any form or by any means, electronic or mechanical, including photocopy, recording, or any information storage and retrieval system, without permission in writing from the publisher.

Academic Press, Inc.
A Division of Harcourt Brace & Company
525 B Street, Suite 1900, San Diego, California 92101-4495

United Kingdom Edition published by
Academic Press Limited
24-28 Oval Road, London NW1 7DX

International Standard Serial Number: 0076-6879

International Standard Book Number: 0-12-182143-9

PRINTED IN THE UNITED STATES OF AMERICA
94 95 96 97 98 99 EB 9 8 7 6 5 4 3 2 1

Table of Contents

Section I. Neoglycoproteins

Section II. Neoglycolipids

Section III. Synthetic Polymers

Contributors to Volume 242

Article numbers are in parentheses following the names of contributors.
Affiliations listed are current.

TOSHIHIRO AKAIKE (21), *Faculty of Bioscience and Technology, Tokyo Institute of Technology, Yokohama 227, Japan*

F. O. ANDERSSON (18, 25), *Department of Chemistry, University of Ottawa, Ottawa, Ontario, Canada K1N 6N5*

SABINE ANDRÉ (4), *Institut für Physiologische Chemie der Universität, D-80539 München, Germany*

SIVASUBRAMANIAN ARAVIND (27), *Department of Chemistry, University of Ottawa, Ottawa, Ontario, Canada K1N 6N5*

JOHN W. BAYNES (1), *Department of Chemistry and Biochemistry, School of Medicine, University of South Carolina, Columbia, South Carolina 29208*

PATRICK J. BRENNAN (3), *Department of Microbiology, Colorado State University, Fort Collins, Colorado 80523*

DELPHI CHATTERJEE (3), *Department of Microbiology, Colorado State University, Fort Collins, Colorado 80523*

R. A. CHILDS (19), *Glycoconjugates Section, MRC Clinical Research Centre, Harrow, Middlesex HA1 3UK, United Kingdom*

SANG-NAE CHO (3), *Department of Microbiology, Yonsei University College of Medicine, Seoul, Korea*

ANDRÉ DANGUY (4), *Laboratoire de Biologie Animale et Histologie Comparée, Universite Libre de Bruxelles, D-1050 Bruxelles, Belgium*

STEPHEN P. DOUGLAS (26), *Department of Molecular and Medical Genetics, Protein Engineering Network of Centres of Excellence, University of Toronto, Toronto, Ontario, Canada M58 1A8*

HEINZ EGGE (13), *Institut für Physiologische Chemie der Universität Bonn, D-53115 Bonn, Germany*

T. FEIZI (19), *Glycoconjugates Section, MRC Clinical Research Centre, Harrow, Middlesex HA1 3UK, United Kingdom*

TSUYOSHI FUJIWARA (3), *Institute for Natural Science, Nara, Japan*

TETSUYA FURUIKE (22), *Division of Ecological Science, Graduate School of Environmental Earth Science, Hokkaido University, Sapporo 060, Japan*

HANS-JOACHIM GABIUS (4, 6), *Institut für Physiologische Chemie, Ludwig-Maximilians-Universität, D-80539 München, Germany*

SIGRUN GABIUS (4), *Institut für Physiologische Chemie der Universität, D-80539 München, Germany*

IRWIN J. GOLDSTEIN (10, 11), *Department of Biological Chemistry, University of Michigan, Ann Arbor, Michigan 48109*

AKIRA HASEGAWA (15, 16, 17), *Department of Applied Bioorganic Chemistry, Gifu University, Gifu 501-11, Japan*

KENICHI HATANAKA (23), *Faculty of Bioscience and Biotechnology, Tokyo Institute of Technology, Yokohama 227, Japan*

COLLEEN E. HAYES (10), *Department of Biochemistry, University of Wisconsin–Madison, Madison Wisconsin 53706*

MISAO HIROTO (7), *Human Science and Technology Center, Department of Materials Science and Technology, Toin University of Yokohama, Yokohama 225, Japan*

YUJI INADA (7), *Human Science and Technology Center, Department of Materials Science and Technology, Toin University of Yokohama, Yokohama 225, Japan*

HIDEHARU ISHIDA (17), *Department of Applied Bioorganic Chemistry, Gifu University, Gifu 501-11, Japan*

ELISABET KALLIN (12, 20), *IsoSep AB, S-146 36 Tullinge, Sweden*

KLAUS KAYSER (4), *Abteilung Pathologie, Thoraxklinik, D-69126 Heidelberg, Germany*

MAKOTO KISO (15, 16, 17), *Department of Applied Bioorganic Chemistry, Gifu University, Gifu 501-11, Japan*

KAZUKIYO KOBAYASHI (21), *Faculty of Agriculture, Nogaya University, Chikusa, Nagoya 464-01, Japan*

MAKOTO KODAMA (6), *National Institute for Advanced Interdisciplinary Research, Bionic Design Research Group, Ibaraki 305, Japan*

YOH KODERA (7), *Human Science and Technology Center, Department of Materials Science and Technology, Toin University of Yokohama, Yokohama 225, Japan*

JIRI J. KREPINSKY (26), *Department of Molecular and Medical Genetics, Protein Engineering Network of Centres of Excellence, University of Toronto, Toronto, Ontario, Canada M58 1A8*

C. A. LAFERRIÈRE (8, 9, 25), *Department of Chemistry, University of Ottawa, Ottawa, Ontario, Canada K1N 6N5*

SU-CHEN LI (14), *Department of Biochemistry, Tulane University, School of Medicine, New Orleans, Lousiana 70112*

YU-TEH LI (14), *Department of Biochemistry, Tulane University, School of Medicine, New Orleans, Louisiana 70112*

T. JUHANI LINNA (5), *Syntex Development Research, Syntex (USA) Inc., Palo Alto, California 94303*

JÖRGEN LÖNNGREN (11), *Pro Gene AB, Uppsala, Sweden*

JAMES A. MAHONEY (2), *Departments of Pharmacology and Molecular Sciences, The Johns Hopkins University School of Medicine, Baltimore, Maryland 21205*

KOJI MATSUOKA (22), *Division of Biological Science, Graduate School of Science, Hokkaido University, Sapporo 060, Japan*

AYAKO MATSUSHIMA (8), *Human Science and Technology Center, Department of Materials Science and Technology, Toin University of Yokohama, Yokohama 225, Japan*

S. J. MEUNIER (8), *Department of Chemistry, University of Ottawa, Ottawa, Ontario, Canada K1N 6N5*

HIROYUKI NISHIMURA (7), *Human Science and Technology Center, Department of Materials Science and Technology, Toin University of Yokohama, Yokohama 225, Japan*

SHIN-ICHIRO NISHIMURA (22), *Division of Biological Science, Graduate School of Science, Hokkaido University, Sapporo 060, Japan*

WILLIAM K. C. PARK (27), *Department of Chemistry, University of Ottawa, Ottawa, Ontario, Canada K1N 6N5*

GOTTFRIED POHLENTZ (13), *Institut für Physiologische Chemie, der Universität Bonn, D-53113 Bonn, Germany*

CHERYL M. REICHERT (10), *Department of Biological Chemistry, University of Michigan, Ann Arbor, Michigan 48109*

JOCELYN RENAUD (27), *Department of Chemistry, University of Ottawa, Ottawa, Ontario, Canada K1N 6N5*

A. ROMANOWSKA (8, 18, 24, 27), *Department of Chemistry, University of Ottawa, Ottawa, Ontario, Canada K1N 6N5*

R. ROY (8, 9, 18, 24, 25, 27), *Department of Chemistry, University of Ottawa, Ottawa, Ontario, Canada K1N 6N5*

SUBRAMANIAM SABESAN (5), *Central Science and Engineering, DuPont Company, Wilmington, Delaware 19880*

RONALD L. SCHNAAR (2), *Departments of Pharmacology and Neuroscience, The Johns Hopkins University School of Medicine, Baltimore, Maryland 21205*

SUZANNE R. THORPE (1), *Department of Chemistry and Biochemistry, University of South Carolina, Columbia, South Carolina 29208*

F. D. TROPPER (8, 24), *Department of Chemistry, University of Ottawa, Ottawa, Ontario, Canada K1N 6N5*

TAICHI USUI (21), *Faculty of Agriculture, Shizuoka University, Ohya, Shizuoka 836, Japan*

DENNIS M. WHITFIELD (26), *Department of Molecular and Medical Genetics, Protein Engineering Network of Centres of Excellence, University of Toronto, Toronto, Ontario, Canada M58 1A8*

NOBORU YAMAZAKI (6), *National Institute of Materials and Chemical Research, Functional Molecules Laboratory, Ibaraki 305, Japan*

Preface

Rising interest in glycobiology in recent years has led to similar interest in neoglycoconjugates, a term given to carbohydrates conjugated to a variety of materials, including proteins, lipids, and synthetic polymers. Carbohydrates to be conjugated can be synthetically obtained or isolated from natural products. Such neoglycoconjugates are useful tools in a wide area of disciplines, ranging from clinical medicine to synthetic chemistry. They serve as a chemomimetic of natural glycoconjugates with well-defined structure and provide certain advantages over natural glycoconjugates.

Methods in Enzymology Volumes 242 and 247 contain many practical methods on how to prepare and use neoglycoconjugates, which have been contributed by a group of international experts. This Volume (242) deals with synthesis and Volume 247 with biomedical applications. Readers should peruse all sections or utilize the indexes, since many chapters contain both preparation and application.

We were pleased to be able to develop these volumes in the *Methods in Enzymology* series which complement our treatise "Neoglycoconjugates: Preparation and Applications" (1994, Academic Press).

Y. C. Lee
Reiko T. Lee

METHODS IN ENZYMOLOGY

VOLUME I. Preparation and Assay of Enzymes
Edited by SIDNEY P. COLOWICK AND NATHAN O. KAPLAN

VOLUME II. Preparation and Assay of Enzymes
Edited by SIDNEY P. COLOWICK AND NATHAN O. KAPLAN

VOLUME III. Preparation and Assay of Substrates
Edited by SIDNEY P. COLOWICK AND NATHAN O. KAPLAN

VOLUME IV. Special Techniques for the Enzymologist
Edited by SIDNEY P. COLOWICK AND NATHAN O. KAPLAN

VOLUME V. Preparation and Assay of Enzymes
Edited by SIDNEY P. COLOWICK AND NATHAN O. KAPLAN

VOLUME VI. Preparation and Assay of Enzymes (*Continued*)
Preparation and Assay of Substrates
Special Techniques
Edited by SIDNEY P. COLOWICK AND NATHAN O. KAPLAN

VOLUME VII. Cumulative Subject Index
Edited by SIDNEY P. COLOWICK AND NATHAN O. KAPLAN

VOLUME VIII. Complex Carbohydrates
Edited by ELIZABETH F. NEUFELD AND VICTOR GINSBURG

VOLUME IX. Carbohydrate Metabolism
Edited by WILLIS A. WOOD

VOLUME X. Oxidation and Phosphorylation
Edited by RONALD W. ESTABROOK AND MAYNARD E. PULLMAN

VOLUME XI. Enzyme Structure
Edited by C. H. W. HIRS

VOLUME XII. Nucleic Acids (Parts A and B)
Edited by LAWRENCE GROSSMAN AND KIVIE MOLDAVE

VOLUME XIII. Citric Acid Cycle
Edited by J. M. LOWENSTEIN

VOLUME XIV. Lipids
Edited by J. M. LOWENSTEIN

VOLUME XV. Steroids and Terpenoids
Edited by RAYMOND B. CLAYTON

VOLUME XVI. Fast Reactions
Edited by KENNETH KUSTIN

VOLUME XVII. Metabolism of Amino Acids and Amines (Parts A and B)
Edited by HERBERT TABOR AND CELIA WHITE TABOR

VOLUME XVIII. Vitamins and Coenzymes (Parts A, B, and C)
Edited by DONALD B. MCCORMICK AND LEMUEL D. WRIGHT

VOLUME XIX. Proteolytic Enzymes
Edited by GERTRUDE E. PERLMANN AND LASZLO LORAND

VOLUME XX. Nucleic Acids and Protein Synthesis (Part C)
Edited by KIVIE MOLDAVE AND LAWRENCE GROSSMAN

VOLUME XXI. Nucleic Acids (Part D)
Edited by LAWRENCE GROSSMAN AND KIVIE MOLDAVE

VOLUME XXII. Enzyme Purification and Related Techniques
Edited by WILLIAM B. JAKOBY

VOLUME XXIII. Photosynthesis (Part A)
Edited by ANTHONY SAN PIETRO

VOLUME XXIV. Photosynthesis and Nitrogen Fixation (Part B)
Edited by ANTHONY SAN PIETRO

VOLUME XXV. Enzyme Structure (Part B)
Edited by C. H. W. HIRS AND SERGE N. TIMASHEFF

VOLUME XXVI. Enzyme Structure (Part C)
Edited by C. H. W. HIRS AND SERGE N. TIMASHEFF

VOLUME XXVII. Enzyme Structure (Part D)
Edited by C. H. W. HIRS AND SERGE N. TIMASHEFF

VOLUME XXVIII. Complex Carbohydrates (Part B)
Edited by VICTOR GINSBURG

VOLUME XXIX. Nucleic Acids and Protein Synthesis (Part E)
Edited by LAWRENCE GROSSMAN AND KIVIE MOLDAVE

VOLUME XXX. Nucleic Acids and Protein Synthesis (Part F)
Edited by KIVIE MOLDAVE AND LAWRENCE GROSSMAN

VOLUME XXXI. Biomembranes (Part A)
Edited by SIDNEY FLEISCHER AND LESTER PACKER

VOLUME XXXII. Biomembranes (Part B)
Edited by SIDNEY FLEISCHER AND LESTER PACKER

VOLUME XXXIII. Cumulative Subject Index Volumes I–XXX
Edited by MARTHA G. DENNIS AND EDWARD A. DENNIS

VOLUME XXXIV. Affinity Techniques (Enzyme Purification: Part B)
Edited by WILLIAM B. JAKOBY AND MEIR WILCHEK

VOLUME XXXV. Lipids (Part B)
Edited by JOHN M. LOWENSTEIN

VOLUME XXXVI. Hormone Action (Part A: Steroid Hormones)
Edited by BERT W. O'MALLEY AND JOEL G. HARDMAN

VOLUME XXXVII. Hormone Action (Part B: Peptide Hormones)
Edited by BERT W. O'MALLEY AND JOEL G. HARDMAN

VOLUME XXXVIII. Hormone Action (Part C: Cyclic Nucleotides)
Edited by JOEL G. HARDMAN AND BERT W. O'MALLEY

VOLUME XXXIX. Hormone Action (Part D: Isolated Cells, Tissues, and Organ Systems)
Edited by JOEL G. HARDMAN AND BERT W. O'MALLEY

VOLUME XL. Hormone Action (Part E: Nuclear Structure and Function)
Edited by BERT W. O'MALLEY AND JOEL G. HARDMAN

VOLUME XLI. Carbohydrate Metabolism (Part B)
Edited by W. A. WOOD

VOLUME XLII. Carbohydrate Metabolism (Part C)
Edited by W. A. WOOD

VOLUME XLIII. Antibiotics
Edited by JOHN H. HASH

VOLUME XLIV. Immobilized Enzymes
Edited by KLAUS MOSBACH

VOLUME XLV. Proteolytic Enzymes (Part B)
Edited by LASZLO LORAND

VOLUME XLVI. Affinity Labeling
Edited by WILLIAM B. JAKOBY AND MEIR WILCHEK

VOLUME XLVII. Enzyme Structure (Part E)
Edited by C. H. W. HIRS AND SERGE N. TIMASHEFF

VOLUME XLVIII. Enzyme Structure (Part F)
Edited by C. H. W. HIRS AND SERGE N. TIMASHEFF

VOLUME XLIX. Enzyme Structure (Part G)
Edited by C. H. W. HIRS AND SERGE N. TIMASHEFF

VOLUME L. Complex Carbohydrates (Part C)
Edited by VICTOR GINSBURG

VOLUME LI. Purine and Pyrimidine Nucleotide Metabolism
Edited by PATRICIA A. HOFFEE AND MARY ELLEN JONES

VOLUME LII. Biomembranes (Part C: Biological Oxidations)
Edited by SIDNEY FLEISCHER AND LESTER PACKER

VOLUME LIII. Biomembranes (Part D: Biological Oxidations)
Edited by SIDNEY FLEISCHER AND LESTER PACKER

VOLUME LIV. Biomembranes (Part E: Biological Oxidations)
Edited by SIDNEY FLEISCHER AND LESTER PACKER

VOLUME 74. Immunochemical Techniques (Part C)
Edited by JOHN J. LANGONE AND HELEN VAN VUNAKIS

VOLUME 75. Cumulative Subject Index Volumes XXXI, XXXII, XXXIV–LX
Edited by EDWARD A. DENNIS AND MARTHA G. DENNIS

VOLUME 76. Hemoglobins
Edited by ERALDO ANTONINI, LUIGI ROSSI-BERNARDI, AND EMILIA CHIANCONE

VOLUME 77. Detoxication and Drug Metabolism
Edited by WILLIAM B. JAKOBY

VOLUME 78. Interferons (Part A)
Edited by SIDNEY PESTKA

VOLUME 79. Interferons (Part B)
Edited by SIDNEY PESTKA

VOLUME 80. Proteolytic Enzymes (Part C)
Edited by LASZLO LORAND

VOLUME 81. Biomembranes (Part H: Visual Pigments and Purple Membranes, I)
Edited by LESTER PACKER

VOLUME 82. Structural and Contractile Proteins (Part A: Extracellular Matrix)
Edited by LEON W. CUNNINGHAM AND DIXIE W. FREDERIKSEN

VOLUME 83. Complex Carbohydrates (Part D)
Edited by VICTOR GINSBURG

VOLUME 84. Immunochemical Techniques (Part D: Selected Immunoassays)
Edited by JOHN J. LANGONE AND HELEN VAN VUNAKIS

VOLUME 85. Structural and Contractile Proteins (Part B: The Contractile Apparatus and the Cytoskeleton)
Edited by DIXIE W. FREDERIKSEN AND LEON W. CUNNINGHAM

VOLUME 86. Prostaglandins and Arachidonate Metabolites
Edited by WILLIAM E. M. LANDS AND WILLIAM L. SMITH

VOLUME 87. Enzyme Kinetics and Mechanism (Part C: Intermediates, Stereochemistry, and Rate Studies)
Edited by DANIEL L. PURICH

VOLUME 88. Biomembranes (Part I: Visual Pigments and Purple Membranes, II)
Edited by LESTER PACKER

VOLUME 89. Carbohydrate Metabolism (Part D)
Edited by WILLIS A. WOOD

VOLUME 90. Carbohydrate Metabolism (Part E)
Edited by WILLIS A. WOOD

VOLUME 91. Enzyme Structure (Part I)
Edited by C. H. W. HIRS AND SERGE N. TIMASHEFF

VOLUME 92. Immunochemical Techniques (Part E: Monoclonal Antibodies and General Immunoassay Methods)
Edited by JOHN J. LANGONE AND HELEN VAN VUNAKIS

VOLUME 93. Immunochemical Techniques (Part F: Conventional Antibodies, Fc Receptors, and Cytotoxicity)
Edited by JOHN J. LANGONE AND HELEN VAN VUNAKIS

VOLUME 94. Polyamines
Edited by HERBERT TABOR AND CELIA WHITE TABOR

VOLUME 95. Cumulative Subject Index Volumes 61–74, 76–80
Edited by EDWARD A. DENNIS AND MARTHA G. DENNIS

VOLUME 96. Biomembranes [Part J: Membrane Biogenesis: Assembly and Targeting (General Methods; Eukaryotes)]
Edited by SIDNEY FLEISCHER AND BECCA FLEISCHER

VOLUME 97. Biomembranes [Part K: Membrane Biogenesis: Assembly and Targeting (Prokaryotes, Mitochondria, and Chloroplasts)]
Edited by SIDNEY FLEISCHER AND BECCA FLEISCHER

VOLUME 98. Biomembranes (Part L: Membrane Biogenesis: Processing and Recycling)
Edited by SIDNEY FLEISCHER AND BECCA FLEISCHER

VOLUME 99. Hormone Action (Part F: Protein Kinases)
Edited by JACKIE D. CORBIN AND JOEL G. HARDMAN

VOLUME 100. Recombinant DNA (Part B)
Edited by RAY WU, LAWRENCE GROSSMAN, AND KIVIE MOLDAVE

VOLUME 101. Recombinant DNA (Part C)
Edited by RAY WU, LAWRENCE GROSSMAN, AND KIVIE MOLDAVE

VOLUME 102. Hormone Action (Part G: Calmodulin and Calcium-Binding Proteins)
Edited by ANTHONY R. MEANS AND BERT W. O'MALLEY

VOLUME 103. Hormone Action (Part H: Neuroendocrine Peptides)
Edited by P. MICHAEL CONN

VOLUME 104. Enzyme Purification and Related Techniques (Part C)
Edited by WILLIAM B. JAKOBY

VOLUME 105. Oxygen Radicals in Biological Systems
Edited by LESTER PACKER

VOLUME 106. Posttranslational Modifications (Part A)
Edited by FINN WOLD AND KIVIE MOLDAVE

VOLUME 107. Posttranslational Modifications (Part B)
Edited by FINN WOLD AND KIVIE MOLDAVE

VOLUME 108. Immunochemical Techniques (Part G: Separation and Characterization of Lymphoid Cells)
Edited by GIOVANNI DI SABATO, JOHN J. LANGONE, AND HELEN VAN VUNAKIS

VOLUME 109. Hormone Action (Part I: Peptide Hormones)
Edited by LUTZ BIRNBAUMER AND BERT W. O'MALLEY

VOLUME 110. Steroids and Isoprenoids (Part A)
Edited by JOHN H. LAW AND HANS C. RILLING

VOLUME 111. Steroids and Isoprenoids (Part B)
Edited by JOHN H. LAW AND HANS C. RILLING

VOLUME 112. Drug and Enzyme Targeting (Part A)
Edited by KENNETH J. WIDDER AND RALPH GREEN

VOLUME 113. Glutamate, Glutamine, Glutathione, and Related Compounds
Edited by ALTON MEISTER

VOLUME 114. Diffraction Methods for Biological Macromolecules (Part A)
Edited by HAROLD W. WYCKOFF, C. H. W. HIRS, AND SERGE N. TIMASHEFF

VOLUME 115. Diffraction Methods for Biological Macromolecules (Part B)
Edited by HAROLD W. WYCKOFF, C. H. W. HIRS, AND SERGE N. TIMASHEFF

VOLUME 116. Immunochemical Techniques (Part H: Effectors and Mediators of Lymphoid Cell Functions)
Edited by GIOVANNI DI SABATO, JOHN J. LANGONE, AND HELEN VAN VUNAKIS

VOLUME 117. Enzyme Structure (Part J)
Edited by C. H. W. HIRS AND SERGE N. TIMASHEFF

VOLUME 118. Plant Molecular Biology
Edited by ARTHUR WEISSBACH AND HERBERT WEISSBACH

VOLUME 119. Interferons (Part C)
Edited by SIDNEY PESTKA

VOLUME 120. Cumulative Subject Index Volumes 81–94, 96–101

VOLUME 121. Immunochemical Techniques (Part I: Hybridoma Technology and Monoclonal Antibodies)
Edited by JOHN J. LANGONE AND HELEN VAN VUNAKIS

VOLUME 122. Vitamins and Coenzymes (Part G)
Edited by FRANK CHYTIL AND DONALD B. MCCORMICK

VOLUME 123. Vitamins and Coenzymes (Part H)
Edited by FRANK CHYTIL AND DONALD B. MCCORMICK

VOLUME 124. Hormone Action (Part J: Neuroendocrine Peptides)
Edited by P. MICHAEL CONN

VOLUME 125. Biomembranes (Part M: Transport in Bacteria, Mitochondria, and Chloroplasts: General Approaches and Transport Systems)
Edited by SIDNEY FLEISCHER AND BECCA FLEISCHER

VOLUME 126. Biomembranes (Part N: Transport in Bacteria, Mitochondria, and Chloroplasts: Protonmotive Force)
Edited by SIDNEY FLEISCHER AND BECCA FLEISCHER

VOLUME 127. Biomembranes (Part O: Protons and Water: Structure and Translocation)
Edited by LESTER PACKER

VOLUME 128. Plasma Lipoproteins (Part A: Preparation, Structure, and Molecular Biology)
Edited by JERE P. SEGREST AND JOHN J. ALBERS

VOLUME 129. Plasma Lipoproteins (Part B: Characterization, Cell Biology, and Metabolism)
Edited by JOHN J. ALBERS AND JERE P. SEGREST

VOLUME 130. Enzyme Structure (Part K)
Edited by C. H. W. HIRS AND SERGE N. TIMASHEFF

VOLUME 131. Enzyme Structure (Part L)
Edited by C. H. W. HIRS AND SERGE N. TIMASHEFF

VOLUME 132. Immunochemical Techniques (Part J: Phagocytosis and Cell-Mediated Cytotoxicity)
Edited by GIOVANNI DI SABATO AND JOHANNES EVERSE

VOLUME 133. Bioluminescence and Chemiluminescence (Part B)
Edited by MARLENE DELUCA AND WILLIAM D. MCELROY

VOLUME 134. Structural and Contractile Proteins (Part C: The Contractile Apparatus and the Cytoskeleton)
Edited by RICHARD B. VALLEE

VOLUME 135. Immobilized Enzymes and Cells (Part B)
Edited by KLAUS MOSBACH

VOLUME 136. Immobilized Enzymes and Cells (Part C)
Edited by KLAUS MOSBACH

VOLUME 137. Immobilized Enzymes and Cells (Part D)
Edited by KLAUS MOSBACH

VOLUME 138. Complex Carbohydrates (Part E)
Edited by VICTOR GINSBURG

VOLUME 139. Cellular Regulators (Part A: Calcium- and Calmodulin-Binding Proteins)
Edited by ANTHONY R. MEANS AND P. MICHAEL CONN

VOLUME 140. Cumulative Subject Index Volumes 102–119, 121–134

VOLUME 141. Cellular Regulators (Part B: Calcium and Lipids)
Edited by P. MICHAEL CONN AND ANTHONY R. MEANS

VOLUME 142. Metabolism of Aromatic Amino Acids and Amines
Edited by SEYMOUR KAUFMAN

VOLUME 143. Sulfur and Sulfur Amino Acids
Edited by WILLIAM B. JAKOBY AND OWEN GRIFFITH

VOLUME 144. Structural and Contractile Proteins (Part D: Extracellular Matrix)
Edited by LEON W. CUNNINGHAM

VOLUME 145. Structural and Contractile Proteins (Part E: Extracellular Matrix)
Edited by LEON W. CUNNINGHAM

VOLUME 146. Peptide Growth Factors (Part A)
Edited by DAVID BARNES AND DAVID A. SIRBASKU

VOLUME 147. Peptide Growth Factors (Part B)
Edited by DAVID BARNES AND DAVID A. SIRBASKU

VOLUME 148. Plant Cell Membranes
Edited by LESTER PACKER AND ROLAND DOUCE

VOLUME 149. Drug and Enzyme Targeting (Part B)
Edited by RALPH GREEN AND KENNETH J. WIDDER

VOLUME 150. Immunochemical Techniques (Part K: *In Vitro* Models of B and T Cell Functions and Lymphoid Cell Receptors)
Edited by GIOVANNI DI SABATO

VOLUME 151. Molecular Genetics of Mammalian Cells
Edited by MICHAEL M. GOTTESMAN

VOLUME 152. Guide to Molecular Cloning Techniques
Edited by SHELBY L. BERGER AND ALAN R. KIMMEL

VOLUME 153. Recombinant DNA (Part D)
Edited by RAY WU AND LAWRENCE GROSSMAN

VOLUME 154. Recombinant DNA (Part E)
Edited by RAY WU AND LAWRENCE GROSSMAN

VOLUME 155. Recombinant DNA (Part F)
Edited by RAY WU

VOLUME 156. Biomembranes (Part P: ATP-Driven Pumps and Related Transport: The Na,K-Pump)
Edited by SIDNEY FLEISCHER AND BECCA FLEISCHER

VOLUME 157. Biomembranes (Part Q: ATP-Driven Pumps and Related Transport: Calcium, Proton, and Potassium Pumps)
Edited by SIDNEY FLEISCHER AND BECCA FLEISCHER

VOLUME 158. Metalloproteins (Part A)
Edited by JAMES F. RIORDAN AND BERT L. VALLEE

VOLUME 159. Initiation and Termination of Cyclic Nucleotide Action
Edited by JACKIE D. CORBIN AND ROGER A. JOHNSON

VOLUME 160. Biomass (Part A: Cellulose and Hemicellulose)
Edited by WILLIS A. WOOD AND SCOTT T. KELLOGG

VOLUME 161. Biomass (Part B: Lignin, Pectin, and Chitin)
Edited by WILLIS A. WOOD AND SCOTT T. KELLOGG

VOLUME 162. Immunochemical Techniques (Part L: Chemotaxis and Inflammation)
Edited by GIOVANNI DI SABATO

VOLUME 163. Immunochemical Techniques (Part M: Chemotaxis and Inflammation)
Edited by GIOVANNI DI SABATO

VOLUME 164. Ribosomes
Edited by HARRY F. NOLLER, JR., AND KIVIE MOLDAVE

VOLUME 165. Microbial Toxins: Tools for Enzymology
Edited by SIDNEY HARSHMAN

VOLUME 166. Branched-Chain Amino Acids
Edited by ROBERT HARRIS AND JOHN R. SOKATCH

VOLUME 167. Cyanobacteria
Edited by LESTER PACKER AND ALEXANDER N. GLAZER

VOLUME 168. Hormone Action (Part K: Neuroendocrine Peptides)
Edited by P. MICHAEL CONN

VOLUME 169. Platelets: Receptors, Adhesion, Secretion (Part A)
Edited by JACEK HAWIGER

VOLUME 170. Nucleosomes
Edited by PAUL M. WASSARMAN AND ROGER D. KORNBERG

VOLUME 171. Biomembranes (Part R: Transport Theory: Cells and Model Membranes)
Edited by SIDNEY FLEISCHER AND BECCA FLEISCHER

VOLUME 172. Biomembranes (Part S: Transport: Membrane Isolation and Characterization)
Edited by SIDNEY FLEISCHER AND BECCA FLEISCHER

VOLUME 173. Biomembranes [Part T: Cellular and Subcellular Transport: Eukaryotic (Nonepithelial) Cells]
Edited by SIDNEY FLEISCHER AND BECCA FLEISCHER

VOLUME 174. Biomembranes [Part U: Cellular and Subcellular Transport: Eukaryotic (Nonepithelial) Cells]
Edited by SIDNEY FLEISCHER AND BECCA FLEISCHER

VOLUME 175. Cumulative Subject Index Volumes 135–139, 141–167

VOLUME 176. Nuclear Magnetic Resonance (Part A: Spectral Techniques and Dynamics)
Edited by NORMAN J. OPPENHEIMER AND THOMAS L. JAMES

VOLUME 177. Nuclear Magnetic Resonance (Part B: Structure and Mechanism)
Edited by NORMAN J. OPPENHEIMER AND THOMAS L. JAMES

VOLUME 178. Antibodies, Antigens, and Molecular Mimicry
Edited by JOHN J. LANGONE

VOLUME 179. Complex Carbohydrates (Part F)
Edited by VICTOR GINSBURG

VOLUME 180. RNA Processing (Part A: General Methods)
Edited by JAMES E. DAHLBERG AND JOHN N. ABELSON

VOLUME 181. RNA Processing (Part B: Specific Methods)
Edited by JAMES E. DAHLBERG AND JOHN N. ABELSON

VOLUME 182. Guide to Protein Purification
Edited by MURRAY P. DEUTSCHER

VOLUME 183. Molecular Evolution: Computer Analysis of Protein and Nucleic Acid Sequences
Edited by RUSSELL F. DOOLITTLE

VOLUME 184. Avidin–Biotin Technology
Edited by MEIR WILCHEK AND EDWARD A. BAYER

VOLUME 185. Gene Expression Technology
Edited by DAVID V. GOEDDEL

VOLUME 186. Oxygen Radicals in Biological Systems (Part B: Oxygen Radicals and Antioxidants)
Edited by LESTER PACKER AND ALEXANDER N. GLAZER

VOLUME 187. Arachidonate Related Lipid Mediators
Edited by ROBERT C. MURPHY AND FRANK A. FITZPATRICK

VOLUME 188. Hydrocarbons and Methylotrophy
Edited by MARY E. LIDSTROM

VOLUME 189. Retinoids (Part A: Molecular and Metabolic Aspects)
Edited by LESTER PACKER

VOLUME 190. Retinoids (Part B: Cell Differentiation and Clinical Applications)
Edited by LESTER PACKER

VOLUME 191. Biomembranes (Part V: Cellular and Subcellular Transport: Epithelial Cells)
Edited by SIDNEY FLEISCHER AND BECCA FLEISCHER

VOLUME 192. Biomembranes (Part W: Cellular and Subcellular Transport: Epithelial Cells)
Edited by SIDNEY FLEISCHER AND BECCA FLEISCHER

VOLUME 193. Mass Spectrometry
Edited by JAMES A. MCCLOSKEY

VOLUME 194. Guide to Yeast Genetics and Molecular Biology
Edited by CHRISTINE GUTHRIE AND GERALD R. FINK

VOLUME 195. Adenylyl Cyclase, G Proteins, and Guanylyl Cyclase
Edited by ROGER A. JOHNSON AND JACKIE D. CORBIN

VOLUME 196. Molecular Motors and the Cytoskeleton
Edited by RICHARD B. VALLEE

VOLUME 197. Phospholipases
Edited by EDWARD A. DENNIS

VOLUME 198. Peptide Growth Factors (Part C)
Edited by DAVID BARNES, J. P. MATHER, AND GORDON H. SATO

VOLUME 199. Cumulative Subject Index Volumes 168–174, 176–194 (in preparation)

VOLUME 200. Protein Phosphorylation (Part A: Protein Kinases: Assays, Purification, Antibodies, Functional Analysis, Cloning, and Expression)
Edited by TONY HUNTER AND BARTHOLOMEW M. SEFTON

VOLUME 201. Protein Phosphorylation (Part B: Analysis of Protein Phosphorylation, Protein Kinase Inhibitors, and Protein Phosphatases)
Edited by TONY HUNTER AND BARTHOLOMEW M. SEFTON

VOLUME 202. Molecular Design and Modeling: Concepts and Applications (Part A: Proteins, Peptides, and Enzymes)
Edited by JOHN J. LANGONE

VOLUME 203. Molecular Design and Modeling: Concepts and Applications (Part B: Antibodies and Antigens, Nucleic Acids, Polysaccharides, and Drugs)
Edited by JOHN J. LANGONE

VOLUME 204. Bacterial Genetic Systems
Edited by JEFFREY H. MILLER

VOLUME 205. Metallobiochemistry (Part B: Metallothionein and Related Molecules)
Edited by JAMES F. RIORDAN AND BERT L. VALLEE

VOLUME 206. Cytochrome P450
Edited by MICHAEL R. WATERMAN AND ERIC F. JOHNSON

VOLUME 207. Ion Channels
Edited by BERNARDO RUDY AND LINDA E. IVERSON

VOLUME 208. Protein–DNA Interactions
Edited by ROBERT T. SAUER

VOLUME 209. Phospholipid Biosynthesis
Edited by EDWARD A. DENNIS AND DENNIS E. VANCE

VOLUME 210. Numerical Computer Methods
Edited by LUDWIG BRAND AND MICHAEL L. JOHNSON

VOLUME 211. DNA Structures (Part A: Synthesis and Physical Analysis of DNA)
Edited by DAVID M. J. LILLEY AND JAMES E. DAHLBERG

VOLUME 212. DNA Structures (Part B: Chemical and Electrophoretic Analysis of DNA)
Edited by DAVID M. J. LILLEY AND JAMES E. DAHLBERG

VOLUME 213. Carotenoids (Part A: Chemistry, Separation, Quantitation, and Antioxidation)
Edited by LESTER PACKER

VOLUME 214. Carotenoids (Part B: Metabolism, Genetics, and Biosynthesis)
Edited by LESTER PACKER

VOLUME 215. Platelets: Receptors, Adhesion, Secretion (Part B)
Edited by JACEK J. HAWIGER

VOLUME 216. Recombinant DNA (Part G)
Edited by RAY WU

VOLUME 217. Recombinant DNA (Part H)
Edited by RAY WU

VOLUME 218. Recombinant DNA (Part I)
Edited by RAY WU

VOLUME 219. Reconstitution of Intracellular Transport
Edited by JAMES E. ROTHMAN

VOLUME 220. Membrane Fusion Techniques (Part A)
Edited by NEJAT DÜZGÜNEŞ

VOLUME 221. Membrane Fusion Techniques (Part B)
Edited by NEJAT DÜZGÜNEŞ

VOLUME 222. Proteolytic Enzymes in Coagulation, Fibrinolysis, and Complement Activation (Part A: Mammalian Blood Coagulation Factors and Inhibitors)
Edited by LASZLO LORAND AND KENNETH G. MANN

VOLUME 223. Proteolytic Enzymes in Coagulation, Fibrinolysis, and Complement Activation (Part B: Complement Activation, Fibrinolysis, and Nonmammalian Blood Coagulation Factors)
Edited by LASZLO LORAND AND KENNETH G. MANN

VOLUME 224. Molecular Evolution: Producing the Biochemical Data
Edited by ELIZABETH ANNE ZIMMER, THOMAS J. WHITE, REBECCA L. CANN, AND ALLAN C. WILSON

VOLUME 225. Guide to Techniques in Mouse Development
Edited by PAUL M. WASSARMAN AND MELVIN L. DEPAMPHILIS

VOLUME 226. Metallobiochemistry (Part C: Spectroscopic and Physical Methods for Probing Metal Ion Environments in Metalloenzymes and Metalloproteins)
Edited by JAMES F. RIORDAN AND BERT L. VALLEE

VOLUME 227. Metallobiochemistry (Part D: Physical and Spectroscopic Methods for Probing Metal Ion Environments in Metalloproteins)
Edited by JAMES F. RIORDAN AND BERT L. VALLEE

VOLUME 228. Aqueous Two-Phase Systems
Edited by HARRY WALTER AND GÖTE JOHANSSON

VOLUME 229. Cumulative Subject Index Volumes 195–198, 200–227 (in preparation)

VOLUME 230. Guide to Techniques in Glycobiology
Edited by WILLIAM J. LENNARZ AND GERALD W. HART

VOLUME 231. Hemoglobins (Part B: Biochemical and Analytical Methods)
Edited by JOHANNES EVERSE, KIM D. VANDEGRIFF AND ROBERT M. WINSLOW

VOLUME 232. Hemoglobins (Part C: Biophysical Methods)
Edited by JOHANNES EVERSE, KIM D. VANDEGRIFF AND ROBERT M. WINSLOW

VOLUME 233. Oxygen Radicals in Biological Systems (Part C)
Edited by LESTER PACKER

VOLUME 234. Oxygen Radicals in Biological Systems (Part D)
Edited by LESTER PACKER

VOLUME 235. Bacterial Pathogenesis (Part A: Identification and Regulation of Virulence Factors)
Edited by VIRGINIA L. CLARK AND PATRIK M. BAVOIL

VOLUME 236. Bacterial Pathogenesis (Part B: Integration of Pathogenic Bacteria with Host Cells)
Edited by VIRGINIA L. CLARK AND PATRIK M. BAVOIL

VOLUME 237. Heterotrimeric G Proteins
Edited by RAVI IYENGAR

VOLUME 238. Heterotrimeric G-Protein Effectors
Edited by RAVI IYENGAR

VOLUME 239. Nuclear Magnetic Resonance (Part C)
Edited by THOMAS L. JAMES AND NORMAN J. OPPENHEIMER

VOLUME 240. Numerical Computer Methods (Part B) (in preparation)
Edited by MICHAEL L. JOHNSON AND LUDWIG BRAND

VOLUME 241. Retroviral Proteases
Edited by LAWRENCE C. KUO AND JULES A. SHAFER

VOLUME 242. Neoglycoconjugates (Part A)
Edited by Y. C. LEE AND REIKO T. LEE

VOLUME 243. Inorganic Microbial Sulfur Metabolism (in preparation)
Edited by HARRY D. PECK, JR., AND JEAN LEGALL

VOLUME 244. Proteolytic Enzymes: Serine and Cysteine Peptidases (in preparation)
Edited by ALAN J. BARRETT

VOLUME 245. Extracellular Matrix Components (in preparation)
Edited by E. RUOSLAHTI AND E. ENGVALL

VOLUME 246. Biochemical Spectroscopy (in preparation)
Edited by KENNETH SAUER

VOLUME 247. Neoglycoconjugates (Part B: Biomedical Applications) (in preparation)
Edited by Y. C. LEE AND REIKO T. LEE

VOLUME 248. Proteolytic Enzymes: Aspartic and Metallo Peptidases (in preparation)
Edited by ALAN J. BARRETT

VOLUME 249. Enzyme Kinetics and Mechanisms (Part D) (in preparation)
Edited by DANIEL L. PURICH

VOLUME 250. Lipid Modifications of Proteins (in preparation)
Edited by PATRICK J. CASEY AND JANICE E. BUSS

VOLUME 251. Biothiols (in preparation)
Edited by LESTER PACKER

Section I

Neoglycoproteins

[1] Residualizing Glycoconjugates: Biologically Inert Tracers for Studies on Protein Endocytosis and Catabolism

By Suzanne R. Thorpe and John W. Baynes

Introduction

Neoglycoconjugates have been used to probe carbohydrate-binding proteins on cells and more recently to target proteins to particular cell types for diagnostic or therapeutic purposes. In this chapter the focus is on inert glycoconjugates which have been specifically engineered not to target the carrier protein to any particular tissue or cell type, but rather to serve as tracers to identify the tissue and cellular sites of protein uptake and degradation. The carbohydrate component of the glycoconjugate is used as a linker between protein and a reporter group detectable by spectroscopic or scintigraphic techniques. These tracers have been called residualizing labels (R-labels),[1] or trapped labels,[2] because the labels are designed to accumulate inside cells after uptake and catabolism of the carrier protein. An excellent presentation on the synthesis, attachment to protein, and applications of one of the earliest of these labels, tyramine-cellobiose (TC), as well as some more theoretical considerations on the use of trapped labels, has been presented in a previous volume in this series.[2] This chapter discusses aspects of the chemistry and applications of *N,N*-dilactitol-^{125}I-tyramine (^{125}I-DLT)[†] as a representative of another group of R-labels which provide an alternative means for protein labeling.

The R-labels were originally synthesized for use in experiments to identify the tissue sites of plasma protein catabolism *in vivo*. In general, circulating proteins, such as albumin, immunoglobulins, and lipoproteins, are delivered slowly (plasma half-lives at least 1 day) to tissues for catabolism, but once inside cells these proteins are degraded rapidly (intracellular half-lives less than 1 hr) to diffusible amino acids. If the protein is labeled conventionally, most typically with ^{125}I, the tracer will be lost from the site of catabolism at a rate too great to allow identification of the cell type(s) involved in the degradation of the protein. To circumvent this problem glycoconjugate labels have been developed which would not affect the biological properties of the protein, but which, because of their relatively large size, hydrophilicity, and resistance to degradation by cellu-

[1] S. R. Thorpe, J. W. Baynes, and Z. C. Chroneos, *FASEB J.* **7,** 399 (1993).
[2] R. C. Pittman and C. A. Taylor, Jr., this series, Vol. 129, p. 612.
[†] *I indicates that any isotope of iodine (^{123}I, ^{125}I, or ^{131}I) may be used.

Copyright © 1994 by Academic Press, Inc.
All rights of reproduction in any form reserved.

lar (lysosomal) glycosidases and hydrolases, would be retained inside cells following degradation of the carrier protein. The design of the R-labels has also incorporated a site for attachment of an easily detected reporter group, so that monitoring the accumulation of the reporter in tissues and cells allows identification of the site(s) of protein uptake and degradation. Figure 1 illustrates the structural features contributing to the cellular retention of ^{125}I-DLT-labeled amino acid or peptide degradation products.

The compound DLT was chosen as the first of the glycoconjugate labels developed in our laboratory, in part because it could be conjugated to protein via reductive amination, a mild (neutral pH, 37°) procedure, which minimizes chemical modification of the protein. Modification of the protein is restricted to the α-amino group or ε-amino groups of lysine residues. The DLT can be readily labeled with radioiodine, and applica-

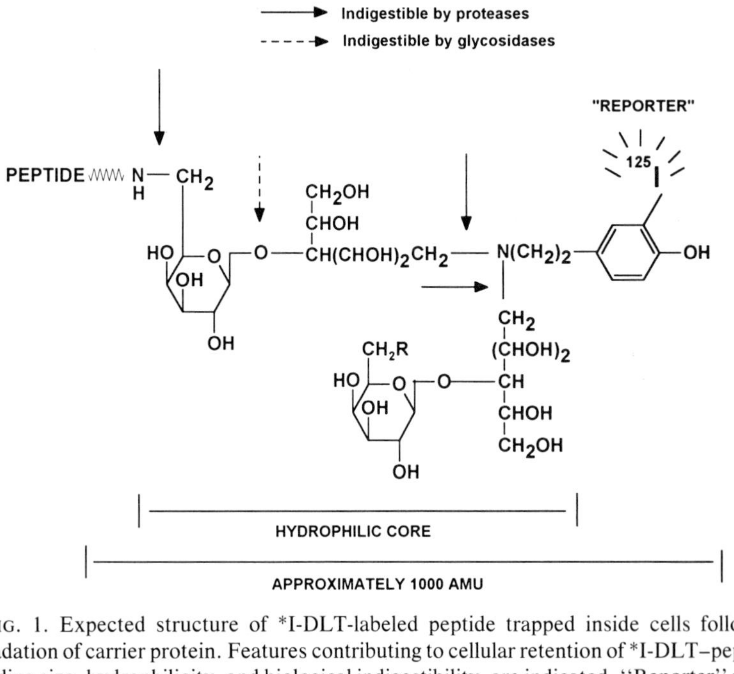

FIG. 1. Expected structure of *I-DLT-labeled peptide trapped inside cells following degradation of carrier protein. Features contributing to cellular retention of *I-DLT–peptide, including size, hydrophilicity, and biological indigestibility, are indicated. "Reporter" refers in the case of *I-DLT (in this example ^{125}I) to the radioactive tracer which allows detection of the R-labeled peptide. The C-6 carbon of one galactose residue in DLT is shown attached to the peptide. The second galactose residue is shown attached to an undefined R group, but protein cross-linking has not been observed.

tions involving the use of ^{125}I, ^{131}I, and ^{123}I have been reported.[1,3,4] In addition, DLT served as a model for the synthesis of other R-labels which can be detected by a variety of physical means, including *N,N*-dilactitol-*N'*-fluoresceinylethylenediamine (DLF),[5] which can be measured by fluorimetry and fluorescence microscopy[6] or fluorescence-activated cell sorting[7]; *N,N*-dilactitol-3,5-bis(trifluoromethyl)benzylamine (DLBA),[8] which because of its high content of ^{19}F can be detected by nuclear magnetic resonance (NMR) spectroscopy; and *N*-lactitol-*S*-([^{18}F]fluorophenacyl)-cysteamine, which can be detected by positron emission tomography (PET).[9]

The strategy in the design of all of these glycoconjugate labels was to take advantage of the selectivity of galactose oxidase in oxidizing the C-6 hydroxymethyl group of galactose to an aldehyde residue, providing a site for attachment of the labels to protein by reductive amination using $NaBH_3CN$. The syntheses also included incorporation of an amine (tyramine in DLT) which could be derivatized with a readily detectable tracer. Figure 2 illustrates schematically the sequence of reactions used for conjugation of *I-DLT to protein. Although some aspects of the synthesis and conjugation of each of the labels differ from the procedure used for DLT, the following discussion focuses on DLT as a prototype for R-labels conjugated to protein by reductive amination.

Synthesis of *N,N*-Dilactitol-tyramine

Lactose and tyramine in a 10 : 1 molar ratio (typically 100–130 m*M* tyramine) are dissolved in 0.2 *M* potassium borate buffer, pH 9.0, and heated at 65° in a screw-capped test tube for 30 min to allow Schiff

[3] A. Daugherty, S. R. Thorpe, L. G. Lange, B. E. Sobel, and G. Schonfeld, *J. Biol. Chem.* **260,** 14564 (1985).

[4] J. M. Ord, J. Hasapes, A. Daugherty, S. R. Thorpe, S. R. Bergmann, and B. E. Sobel, *Circulation* **85,** 288 (1992).

[5] J. W. Baynes, J. L. Maxwell, K. M. Rahman, and S. R. Thorpe, *Anal. Biochem.* **170,** 382 (1988).

[6] J. L. Maxwell, L. Terracio, T. K. Borg, J. W. Baynes, and S. R. Thorpe, *Biochem. J.* **267,** 155 (1990).

[7] J. L. Maxwell, J. W. Baynes, and S. R. Thorpe, *In* "The Pharmacology and Toxicology of Proteins" (J. S. Holcenberg and J. L. Winkelhake, eds.), p. 59. Alan R. Liss, New York, 1987.

[8] A. Daugherty, N. N. Becker, L. A. Scherrer, B. E. Sobel, J. J. H. Ackerman, J. W. Baynes, and S. R. Thorpe, *Biochem. J.* **264,** 829 (1989).

[9] A. Daugherty, M. R. Kilbourn, C. S. Dence, B. E. Sobel, and S. R. Thorpe, *Nucl. Med. Biol.* **19,** 411 (1992).

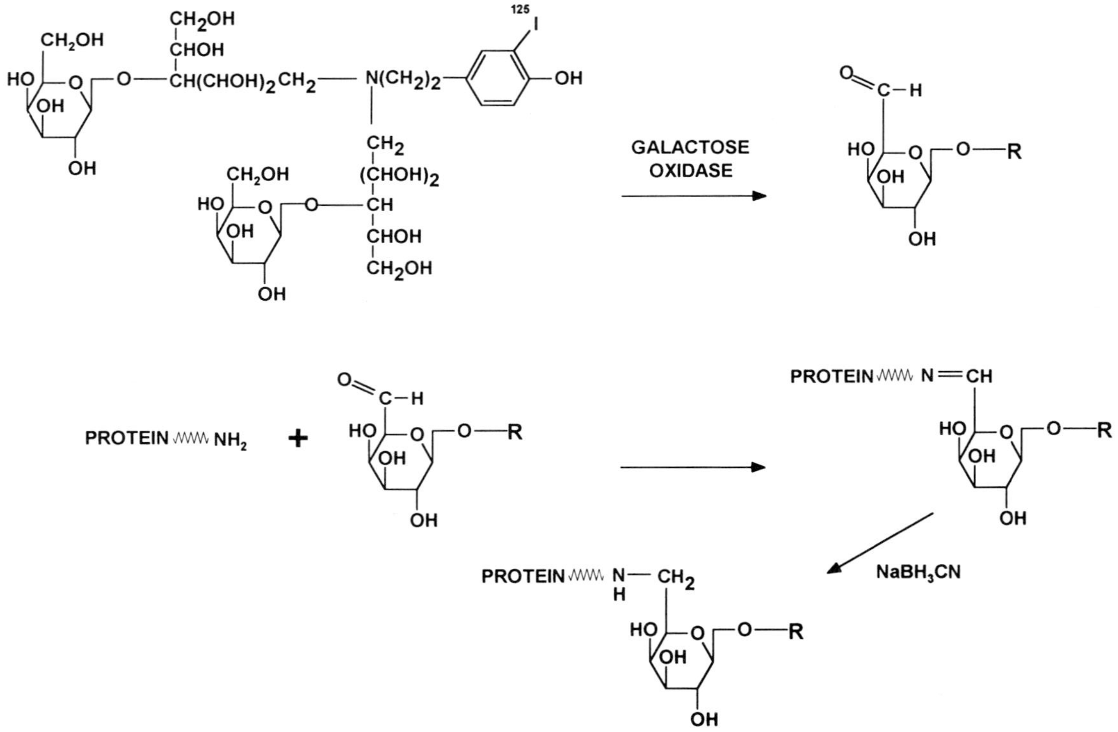

FIG. 2. Scheme showing steps involved in conjugation of *I-DLT to protein. After iodination of DLT, the *I-DLT is treated with galactose oxidase to generate aldehyde residues. The *I-DLT aldehyde forms a Schiff base with primary amino groups on protein, primarily the ε-amino group of lysine residues, and is then covalently attached to protein by reduction with $NaBH_3CN$.

base formation.[10] The solution is cooled to room temperature, and solid $NaBH_3CN$ (4-fold molar excess over tyramine) is then added; since HCN is a side product of the reaction, the synthesis should be conducted in a fume hood. The incubation mixture is heated overnight at 65°. These reaction conditions produce a mixture of DLT and monolactitol-tyramine (MLT) in a ratio of about 9 : 1.

Isolation of *N,N*-Dilactitol-tyramine

The DLT can be isolated either by conventional cation-exchange chromatography[10] or reversed-phase high-performance liquid chromatography (RP-HPLC).[5] Purification by conventional cation-exchange chromatography allows the preparation of large quantities of material in a single run and has been particularly suited for isolation of DLBA used for NMR studies.[11] Alternatively, isolation of DLT and other glycoconjugates, including TC, DLF, and DLBA, by either analytical or preparative RP-HPLC[5,8] is convenient for preparing smaller amounts of the labels suitable for numerous experiments.

Cation-Exchange Chromatography. A 2.5 × 8 cm Dowex 50-X2 column equilibrated in 50 m*M* ammonium acetate, pH 4.6, is used for purification of DLT from a 5-ml incubation mixture (lactose : $NaBH_3CN$: tyramine, 10 : 4 : 1, 130 m*M* tyramine). The reaction mixture is diluted to 50 ml with deionized water and adjusted to pH 4.6 using glacial acetic acid. After loading the sample onto the ion-exchange column, the column is eluted with 50 m*M* ammonium acetate, pH 4.6, until the absorbance at 280 nm is less than 0.05 and the effluent is negative for carbohydrate, based on the anthrone assay.[12] This washing step removes unreacted sugar and some salts. The DLT and MLT are eluted from the column using 1 *M* ammonium acetate, pH 7. Elution of products is monitored by measuring A_{280}, and the concentration is determined using an extinction coefficient of 1360 M^{-1} for tyramine. The identity of the products is established by measuring the carbohydrate content of the fractions using the anthrone assay,[12] with galactose as standard, and determining the ratio of carbohydrate to tyramine. The DLT elutes first from the column, whereas MLT elutes later and tyramine last. The DLT can be pooled, then stored as a concentrate at −20° for several years. Residual ammonium can be removed by repeated evaporation from a basic solution prepared by the addition of 1–2 drops of saturated Na_2CO_3.

[10] J. L. Strobel, J. W. Baynes, and S. R. Thorpe, *Arch. Biochem. Biophys.* **240,** 635 (1985).

[11] L. A. Meeh, J. J. H. Ackerman, S. R. Thorpe, and A. Daugherty, *Biochem. J.* **286,** 785 (1992).

[12] R. G. Spiro, this series, Vol. 8, p. 3.

Reversed-Phase Chromatography. An aliquot of a DLT reaction mixture (0.5–1 μmol tyramine) is applied to a 4.6 mm × 15 cm Varian SPC 18-5 column (Varian Instruments, Palo Alto, CA) in deionized water pumped at 1 ml/min. The DLT is separated from MLT using gradient elution with buffer A (deionized water) and buffer B (deionized water containing 0.1% (v/v) heptafluorobutyric acid, 50% acetonitrile) and the following gradient: 0 min, 0% B; 2.5 min, 5% B; 22.5 min, 15% B; 30 min, 100% B. Under these conditions DLT elutes at 11.8 min and MLT about 3 min later. The column effluent is monitored at 280 nm and then collected in a fraction collector for recovery of DLT. Fractions are read for A_{280} and peaks pooled for carbohydrate analysis. It is important to ensure complete removal of acetonitrile by rotary evaporation under reduced pressure prior to carbohydrate analysis, because CH_3CN causes substantial interference in the anthrone assay. Rechromatography of the material collected from the DLT peak should yield a single peak with carbohydrate to tyramine in a 2 : 1 ratio. From an injection of an aliquot of an incubation mixture containing 1 μmol tyramine, approximately 600 nmol DLT was recovered. As 10 nmol DLT is typically used for preparation of *I-DLT-labeled proteins (see below), isolation of DLT by RP-HPLC provides adequate amounts of material for many experiments.

Labeling of *N,N*-Dilactitol-tyramine and Coupling to Protein

Reagents

Iodogen (1,3,4,6-tetrachloro-3α,6α-diphenylglycouril) (Pierce Chemical Co., Rockford, IL) coated tubes: Prepare a solution of Iodogen in $CHCl_3$ (1 mg/ml) and then dispense 20-μl aliquots into 0.5-ml Eppendorf centrifuge tubes. The $CHCl_3$ is removed under a gentle stream of N_2, so that there is a film of Iodogen coating the sides of the tube, equivalent to a volume of 20–40 μl. Letting the $CHCl_3$ evaporate, rather than blowing it off with N_2, produces small beads of reagent, which results in less efficient iodination. Iodogen-coated tubes can be stored at room temperature for at least 6 months.

DLT: A 1 m*M* solution of DLT in 0.5 *M* potassium phosphate buffer, pH 7.6, can be conveniently stored in 50-μl aliquots at −20°. Aliquots can be frozen and thawed several times with no deleterious effects.

$Na^{125}I$: Amersham (Arlington Heights, IL), New England Nuclear (Boston, MA), and ICN (Costa Mesa, CA) all supply concentrated forms (100 mCi/ml) of ^{125}I in dilute NaOH.

Galactose oxidase (Sigma Chemical Co., St. Louis, MO; Cat. No. G3385): The enzyme is supplied as a lyophilized powder which should be diluted to 2 units/μl using 0.1 *M* potassium phosphate, pH 7.0, at 4°. Dispense aliquots of 4–10 units (2–5 μl) into 0.5-ml Eppendorf centrifuge tubes and store at −20°. Because galactose oxidase is reported to lose about 10% of activity with each freeze–thaw, the enzyme solution is stored in aliquots so that any given tube is subjected to a minimum number of freeze–thaw cycles. We have not observed any decrease in the overall efficiency of protein labeling using enzyme which has been through two cycles of freezing and thawing.

$NaBH_3CH$: A 2 *M* $NaBH_3CN$ solution in 0.5 *M* potassium phosphate buffer, pH 7.7, can be used for up to 1 week if stored at 4°.

Method. The procedure takes 2–3 hr but can be interrupted at several points as indicated below.

1. Add 10 μl of 0.5 *M* potassium phosphate buffer, pH 7.6, to an Iodogen-coated tube, followed by addition of 10 nmol (10 μl) DLT.
2. Immediately add $Na^{125}I$; cover and let stand covered in a fume hood at room temperature. Up to 3 mCi of $Na^{125}I$ can be used depending on the desired final specific activity of the proteins.
3. After 15–20 min transfer (taking care not to scrape the sides of the tube) the entire iodination mixture to a fresh 0.5-ml Eppendorf tube containing 4 units (2 μl) galactose oxidase. If desired the ^{125}I-DLT can be transferred to a fresh Eppendorf tube and stored at −20° before proceeding to the enzyme step, but the reagent should not be left in contact with the Iodogen. Note that there is about 10% free ^{125}I in the reaction mixture at this point, so that all transfers should be carried out in a fume hood and the tube kept covered.
4. Once the ^{125}I-DLT is added to galactose oxidase, the mixture is incubated at 37° for 45 min. If it is not convenient to proceed further, the tube containing the enzyme-treated ^{125}I-DLT can be frozen at −20° until needed. In our experience the aldehyde stored at −20° for several days still couples efficiently to protein.
5. Protein, 0.2–1 mg in 50–200 μl phosphate buffer, pH 7.5–7.7, is added to the enzyme-treated ^{125}I-DLT. The entire mixture is made approximately 40 m*M* in $NaBH_3CN$ by addition of a small aliquot of 2 *M* $NaBH_3CN$, then incubated at 37° for 1–2 hr. The coupling reaction is generally 80% complete in 1 hr, and for proteins which should be kept at 37° for a minimum time, 1 hr is usually sufficient to achieve an adequate specific activity. The extent of protein labeling can be estimated by trichlo-

roacetic acid (TCA) precipitation of an aliquot of the incubation mixture at different time points, to determine the time required for maximal labeling. The smaller the reaction volume, the more efficient the coupling reaction. Albumin is a convenient test protein for trying out the labeling procedure, since 50–70% of the ^{125}I-DLT is typically coupled to albumin. Coupling efficiencies range from 10% for some lipoproteins[3] up to 70% for albumin.

6. Unconjugated label can be separated from ^{125}I-DLT-labeled protein by any of a number of conventional techniques, including centrifugal chromatography[13] on Sephadex G-25, conventional gel-permeation chromatography, or extensive dialysis. After centrifugal chromatography the material voided from the column is typically at least 97% TCA-precipitable, and generally 90–95% of applied protein is recovered.

A sample protocol for labeling 1 mg of albumin to achieve a final specific activity of approximately 1.5×10^6 counts/min (cpm)/μg is as follows:

1. Place 10 μl of 0.5 *M* potassium phosphate buffer, pH 7.6, in an Iodogen-coated tube followed by 10 μl (10 nmol) DLT.
2. Add 1 mCi (10 μl) Na^{125}I and let the covered tube stand at room temperature for 15–20 min.
3. Transfer the reaction mixture (using an adjustable pipettor with plastic tip) to a 0.5 ml Eppendorf tube containing 4 units (2 μl) galactose oxidase and place at 37° for 45 min.
4. Add 1 mg albumin in 100 μl phosphate-buffered saline (PBS), pH 7.4, followed by addition of 3 μl of 2 *M* $NaBH_3CN$ (final concentration ~44 m*M* $NaBH_3CN$), and place at 37° for 1 hr.
5. Apply the sample to a 1 ml Sephadex G-25 column and centrifuge at 150 *g* for 60 sec at room temperature.
6. Add a 1-μl aliquot of the Sephadex G-25 eluate to 1 mg carrier albumin and precipitate with a final concentration of 10% TCA. Measure the radioactivity in supernatant and pellet to estimate the percent protein-bound radioactivity.

Application of Radioiodinated *N,N*-Dilactitol-tyramine-Labeled Protein for Studies on Protein Uptake and Catabolism *in Vivo*

General Considerations. Following catabolism of *I-DLT-labeled protein, amino acids or small peptides tagged with *I-DLT remain entrapped within cellular lysosomes. Thus, in principle once a *I-DLT-labeled protein has been cleared from the circulation, measuring the radioactivity remaining in tissues should give an estimate of the amount of protein

[13] H. S. Penefsky, this series, Vol. 56, p. 527.

degraded in that tissue. However, plasma proteins leave the circulation by equilibration into intra- and extravascular compartments, as well as because of catabolic processes. Thus, tissue radioactivity includes intact protein in vascular and extravascular compartments outside the cell, as well as intracellular degradation products. Intravascular radioactivity can be largely eliminated by thorough perfusion of the animal prior to removal of tissues. However, following perfusion, intact *I-DLT-labeled protein will still be present in extravascular spaces. Precipitation of tissue homogenates or alkaline extracts with TCA is used to differentiate intracellular from extravascular radioactivity. The assumption in this procedure is that radioactivity in a catabolized form will remain soluble during the acid precipitation, whereas that still attached to protein will be precipitated.

An alternative to acid precipitation of tissues is to carry out double-label experiments in which an aliquot of the protein is labeled with ^{125}I-DLT and a separate aliquot is directly labeled with ^{131}I.[2] The assumption in this approach is that the degradation products from the ^{131}I-labeled protein will be rapidly lost from cells and excreted, so that any ^{131}I radioactivity remaining in tissues will represent radioactivity still attached to intact protein in extravascular spaces. In contrast, radioactivity from ^{125}I-DLT–protein will represent both intracellular and extravascular radioactivity. The contribution of ^{125}I-DLT radioactivity from extravascular protein can then be calculated based on the ^{131}I content of the tissue. This assumption works well in tissues such as liver and kidney which have relatively small extracellular volumes and in which catabolic activity is relatively high, but it is compromised in tissues such as muscle and skin which have large extravascular volumes and relatively low catabolic rates. Released *I or *I-tyrosine appears to be retained in the extracellular matrix of muscle and skin so that in these tissues acid precipitation is still necessary to correct for intact protein.[2,3]

One further consideration in the use of R-labels is that although R-labels are retained inside cells with relatively high efficiency compared to conventional labels, they do leak slowly from cells. In the case of *I-DLT-labeled catabolites, the half-life for retention inside cells is about 2 days. Figure 3 illustrates the relative whole-body recoveries in the rat of radioactivity from asialofetuin (ASF) and rat serum albumin (RSA) labeled either conventionally or with *I-DLT. The respective circulating half-lives of the two proteins are less than 5 min for ASF and around 2 days for RSA. In the case of *I-ASF, which is rapidly and specifically cleared by liver,[14] less than 10% of the injected radioactivity is recovered in the body after

[14] K. G. Rice and Y. C. Lee, *Adv. Enzymol. Relat. Areas Mol. Biol.* **66,** 41 (1993).

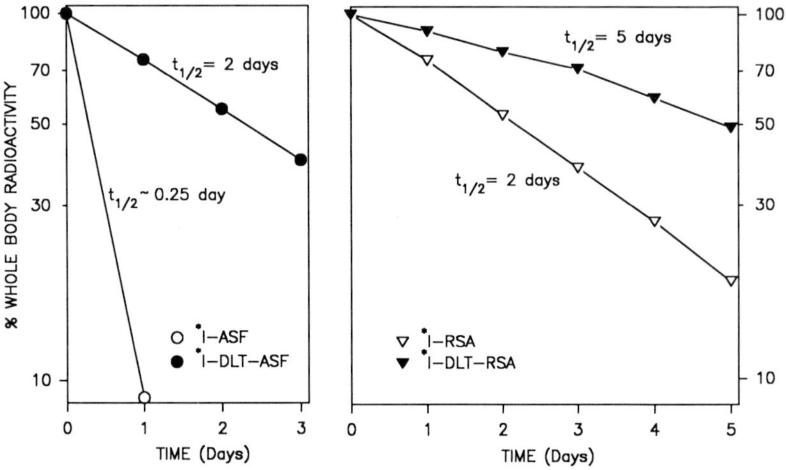

FIG. 3. Comparison of the whole-body recovery of radioactivity from *I-labeled and *I-DLT-labeled proteins. The graphs show the recovery of radioactivity for ASF, which is rapidly ($t_{1/2} < 5$ min) removed from the circulation, and RSA, which is cleared more slowly ($t_{1/2} \approx 2$ days). In both cases there is a significant increase in the half-life for retention of radioactivity in the body for the *I-DLT-labeled compared to the *I-labeled protein. The data are adapted from studies in the authors' laboratory.[7,15]

1 day, whereas about 50% of *I-DLT radioactivity is still in the body after 2 days. For *I-RSA the half-life of whole-body radioactivity for the conventionally iodinated protein is about 2 days, similar to its plasma half-life,[15] emphasizing the rapid excretion of degradation products from conventionally labeled protein following uptake from the circulation. In contrast, radioactivity from *I-DLT–RSA was recovered with a half-life of around 5 days, because of retention of degradation products inside cells.

In selecting the time at which to sacrifice the animal, the kinetics both of clearance of the protein from the circulation and of leakage of *I-DLT-labeled catabolites from cells have to be taken into account. Studies conducted at the half-life and twice the half-life of the protein provide data at times when 50 and 75%, respectively, of administered protein have been catabolized. The relative activity of various tissues in the catabolism of the protein should be comparable at these two times, but at the later times the absolute amount of radioactivity in tissues will be decreased, thus decreasing sensitivity. In the case of RSA the average recovery of

[15] J. L. Strobel, S. G. Cady, T. K. Borg, L. Terracio, J. W. Baynes, and S. R. Thorpe , *J. Biol. Chem.* **261,** 7989 (1986).

radioactivity in the body at 2 and 4 days was approximately 80 and 60%, respectively.[15] Experimentally the actual amount of catabolites lost from the body can be determined either by collecting excreta using a metabolic cage or by serial measurements of whole-body radioactivity. The percentage of dose recovered in feces is sometimes added to that recovered in liver, to estimate the total amount of protein degraded in liver. This approach assumes that degradation products arising from extrahepatic tissue are not excreted through the bile, an assumption which has been questioned.[10]

Procedure for Obtaining Tissues and Measuring Total and Degraded Protein. Prior to sacrifice, the anesthetized animal is weighed and the fur on an area approximately 4 × 4 cm on the ventral side is removed by shaving with a razor. After sacrificing, animals are perfused through the heart or portal vein with PBS at 4° until the perfusate is cleared of blood, about 60 ml for a 200-g rat or 250 ml for a 2-kg rabbit. Organs are dissected out as quickly as possible, rinsed in iced PBS, blotted, and placed in a convenient vial on ice. The shaved area of skin is separated from the underlying muscle, and samples of both are each placed in a vial on ice. Tissues should be kept cold as much as possible to avoid continued protein degradation because of autolysis.

Processing of Discrete Tissues. Radioactivity in each kidney, the whole spleen and heart, and weighed portions (~1-g pieces) of liver or lung is measured. Total radioactivity in the whole organ is then calculated from total organ weight and the measured counts per minute per gram. The weighed tissue aliquots are placed in a Potter–Elvehjem glass homogenizer vessel and 3 volumes of iced distilled water added per gram of tissue. The tissue is then homogenized, keeping the mortar in an ice bath. An aliquot of the homogenate, typically 250 μl, is placed in a 13 × 100 mm test tube and precipitated with an equal volume of 40% TCA. After centrifugation, radioactivity is measured in the supernatant and pellet. Radioactivity in the supernatant is expressed as a percentage of the total radioactivity in the supernatant and pellet to estimate the percentage of degraded protein in the sample.

Processing of Dispersed Tissues. Radioactivity in several weighted portions (~0.3 g) of skin or muscle is measured. Total radioactivity is then calculated from the average radioactivity per gram times total tissue weight estimated from published values, based on animal body weight.[16–18]

[16] H. H. Donaldson, "The Rat," 2nd Ed. Wistar Institute, Philadelphia, Pennsylvania, 1924.

[17] W. O. Caster, J. Poncelet, A. B. Simon, and A. D. Armstrong, *Proc. Soc. Exp. Biol. Med.* **91,** 122 (1956).

[18] H. B. Latimer and P. B. Sawin, *Anat. Rec.* **128,** 457 (1957).

The individual pieces are placed in separate 20-ml glass vials, 2 ml of 0.1 *N* NaOH is added, and the tissue is minced thoroughly with a pair of scissors. The vials are shaken at room temperature for 2 hr followed by acid precipitation of an aliquot (300 μl) of the base extract with an equal volume of 40% TCA, as described above.

Gastrointestinal Tract. The stomach contents and intestinal contents are separated from the organ, and radioactivity in the tissue and contents is measured separately.

Interpretation of Results. Generally, two types of information are obtained from analysis of the data. First, an estimate of the overall distribution of the protein plus degradation products in the body can be made from the total recovery of radioactivity in individual tissues, expressed as a percentage of the injected dose. Second, the relative contribution of each tissue to protein degradation can be obtained by multiplying the percentage of the injected dose recovered in the tissue by the fraction of radioactivitiy which is in acid-soluble form:

$$\frac{\%\ \text{injected dose}}{\text{tissue}} \times \%\ \text{acid-soluble radioactivity} = \frac{\%\ \text{injected dose catabolized}}{\text{tissue}}$$

Further, dividing the percentage of the injected dose catabolized in a tissue by the tissue weight gives an estimate of the specific catabolic activity of that tissue:

$$\frac{\%\ \text{injected dose catabolized}}{\text{tissue weight}} = \frac{\text{specific catabolic activity}}{\text{g tissue}}$$

This latter calculation allows comparisons of relative catabolic activities among tissues on a weight basis.

Applications of Radioiodinated *N,N*-Dilactitol-tyramine for Studies on Specific Proteins

Table I lists some of the proteins which have been labeled with DLT and the animal species in which they have been studied.[19–21] The diversity of proteins and range of animal sizes emphasize that sufficient specific

[19] D. Potter, Z. C. Chroneos, J. W. Baynes, P. R. Sinclair, H. Gorman, H. H. Liem, U. Muller-Eberhard, and S. R. Thorpe, *Arch. Biochem. Biophys.* **300,** 98 (1993).

[20] S. Demignot, M. V. Pimm, S. R. Thorpe, and R. W. Baldwin, *Cancer Immunol. Immunother.* **33,** 359 (1991).

[21] J. Bichler, J. W. Baynes, and S. R. Thorpe, *Biochem. J.* **296,** 771 (1993).

TABLE I
PROTEINS LABELED WITH RADIOIODINATED *N,N*-DILACTITOL-TYRAMINE AND SPECIES USED FOR *in Vivo* STUDIES

Protein	Animal	Refs.
Plasma proteins		
Albumin	Rat	15
Immunoglobulins	Mouse	*a*
IgA	Monkey	*b*
IgG	Mouse	*a*, 20
IgM	Rat	1
Low-density lipoproteins	Rabbit	3
Very low-density lipoproteins	Rabbit	3
Hemopexin	Rat	19
Thrombin	Rat	21
Retinol-binding protein	Rat	*c*
Other proteins		
Asialofetuin	Rat	10
Hirudin	Rat	21
Tissue plasminogen activator	Rabbit, dog	4
Gelatin	Rat	*d*

[a] Z. Moldoveanu, J. M. Epps, S. R. Thorpe, and J. Mestecky, *J. Immunol.* **140,** 208 (1988).

[b] Z. Moldoveanu, I. Moro, J. Radl, S. R. Thorpe, K. Komiyama, and J. Mestecky, *Scand. J. Immunol.* **32,** 577 (1990).

[c] K. E. Creek and R. Bullard-Dillard, personal communication (1993).

[d] F. A. Blumenstock, D. L. Celle, A. Herrmannsdoerfer, C. Giunta, F. L. Minnea, E. Cho, and T. M. Shaba, *J. Leukocyte Biol.* **54,** 56 (1993).

activity can be achieved so that administration of *I-DLT-labeled protein does not alter the mass of endogenous protein, and also allows detection in animals as large as monkeys and dogs. As with any chemical modification of protein, it is important to document that *I-DLT labeling does not alter the chemical and biological properties of the protein. When *I-DLT-labeled proteins have been analyzed by gel electrophoresis under reducing conditions, there are no reported instances of protein cross-linking. In addition, no *I-DLT radioactivity is recovered in the lipid moiety of various lipoproteins.[3] For plasma and other proteins, the most commonly tested biological property has been the circulating half-life, with a comparison made between the conventionally radioiodinated protein and the *I-DLT-labeled material. In general, for proteins labeled by either method, the plasma half-lives for *I-versus *I-DLT-labeled proteins, estimated from the linear phase of the clearance curve, are essentially identical. In some instances there have been slight differences in the equilibration phase, with more of the conventionally labeled material lost from the plasma compartment during the initial period of protein clearance.[19] It is possible that the oxidative conditions used for direct iodination result in small amounts of denatured protein, leading to their rapid removal from the circulation.

Where possible, biological activities of the native and *I-DLT-labeled protein should also be compared. For instance, *I-DLT labeling did not alter either the heme binding capacity of hemopexin[19] or the antigen specificity of immunoglobulin G (IgG).[20] Similarly, labeling with *I-DLT did not affect the saturable binding of low-density lipoprotein (LDL) or very low-density lipoprotein (VLDL) to fibroblasts[3] or of tissue plasminogen activator (tPA)[4] to preformed clots in *in vitro* assays. In these instances it was particularly important to evaluate the behavior of the modified proteins because lysine residues, the likely amino acid modified by DLT, are involved in the binding of lipoproteins to cellular receptors involved in their catabolism and in the binding of tPA to fibrin. An interesting example of where DLT modification did interfere with normal protein–protein interaction was the case of the 1 : 1 complex normally formed between thrombin and the leech anticoagulant protein hirudin.[21] In this case, only limited amounts of *I-DLT-labeled hirudin were bound to thrombin, presumably because lysine residues involved in complex formation were modified by the glycoconjugate.

Use of Radioiodinated *N,N*-Dilactitol-tyramine-Labeled Protein for Studies on Protein Uptake and Catabolism *in Vitro*

Although this chapter focuses on the use of R-labels for *in vivo* studies, it should be mentioned that the labels are equally useful for following the kinetics of protein uptake and degradation in cell culture. Although cell culture incubations are generally short-term (1–24 hr), proteins can be degraded intracellularly within 1 hr, so that typically radioactivity from proteins labeled with conventional tracers is recovered in acid-soluble form in the cell culture medium within 30 min and gradually increases with continued protein uptake. Because *I-DLT-labeled degradation products are efficiently retained inside lysosomes, there is minimal leakage of catabolites from *I-DLT-labeled proteins into the culture medium. Thus, in studies on the processing of *I-DLT–VLDL and *I-DLT–LDL in fibroblasts and macrophages[3,22] or *I-DLT-labeled high-density lipoprotein

[22] A. Daugherty and D. L. Rateri, *J. Biol. Chem.* **266,** 17269 (1991).

(HDL)[23] in granulosa cells, only the cells themselves were analyzed for total and TCA-soluble radioactivity, since negligible radioactivity was recovered in the medium. The use of the R-label requires minimal manipulations for assessing protein uptake and degradation, compared to the use of conventional labels which requires processing of both cells and media. The concentration of labeled degradation products inside cells also provides greater sensitivity for measuring protein uptake and catabolism.

Acknowledgment

Research in the authors' laboratory was supported by National Institutes of Health Research Grant DK-25373.

[23] S. Azhar, L. Tsai, and E. Reaven, *Biochim. Biophys. Acta* **1047,** 148 (1990).

[2] Ganglioside-Based Neoglycoproteins

By JAMES A. MAHONEY and RONALD L. SCHNAAR

Introduction

Gangliosides are a varied and widely distributed family of cell membrane glycosphingolipids.[1,2] They contain three structural components: a long-chain amino alcohol (e.g., sphingosine), a fatty acid amide, and an oligosaccharide chain containing at least one sialic acid residue. Increasing evidence demonstrates that gangliosides act both as recognition signals and as regulators of membrane protein function.[3,4] Some of the physiological effects of gangliosides may occur via their binding to complementary receptors on the same or apposing cell membranes. Identification and characterization of ganglioside-binding proteins are important steps in describing the mechanisms by which gangliosides regulate cell functions.

The amphipathic nature of gangliosides results in their hydrophobic binding to and spontaneous insertion into cell membranes. Experimental membrane insertion of exogenously added gangliosides is useful for invest-

[1] R. K. Yu and M. Saito, *in* "Neurobiology of Glycoconjugates" (R. U. Margolis and R. K. Margolis, eds.), p. 1. Plenum, New York, 1989.
[2] C. L. M. Stults, C. C. Sweeley, and B. A. Macher, this series, Vol. 179, p. 167.
[3] S. Hakomori, *J. Biol. Chem.* **265,** 18713 (1990).
[4] R. L. Schnaar, *Glycobiology* **1,** 477 (1991).

Copyright © 1994 by Academic Press, Inc.
All rights of reproduction in any form reserved.

igating some ganglioside functions.[5,6] However, nonspecific hydrophobic binding and membrane insertion limit the usefulness of radiolabeled gangliosides as high-affinity probes for carbohydrate-specific ganglioside receptors on cell membranes. To circumvent this limitation we synthesized neoganglioproteins consisting of gangliosides covalently linked, via the lipid moiety, to bovine serum albumin.[7] Neoganglioproteins prepared as described below[8] retain nearly all of their original structure, including most of the ceramide moiety and the entire oligosaccharide chain. Although they retain the lipid structure, the covalently bound gangliosides are precluded from inserting into cell membranes, improving their utility for detection of complementary binding proteins. Neoganglioproteins have the additional advantage of presenting the ganglioside determinants in a polyvalent array, which may be important for high-affinity binding to complementary receptors.[9,10] Ganglioside–protein conjugates are readily radiolabeled to high specific activity by standard protein radioiodination techniques, allowing their use as radioligands to detect complementary ganglioside-bindging activities.[7,11]

As an alternative to neoganglioproteins, the oligosaccharides from most gangliosides (and other glycosphingolipids) can be enzymatically released using ceramide glycanase[12] and attached to carrier proteins via reductive amination.[13] Such oligosaccharide-based neoglycoconjugates are easier to synthesize than the neoganglioproteins described below. However, they lack the reducing terminal glucose ring and ceramide moieties which are required for binding to some ganglioside receptors. Therefore, use of the more complex neoganglioproteins as radioligands for initial discovery and characterization of previously undefined ganglioside-binding proteins is judicious. Oligosaccharide-based neoglycoconjugates can then be confidently evaluated as alternatives to the neoganglioproteins.

[5] J. Moss, P. H. Fishman, V. C. Manganiello, M. Vaughan, and R. O. Brady, *Proc. Natl. Acad. Sci. U.S.A.* **73,** 1034 (1976).

[6] T. Pacuszka, R. M. Bradley, and P. H. Fishman, *Biochemistry* **30,** 2563 (1991).

[7] M. Tiemeyer, Y. Yasuda, and R. L. Schnaar, *J. Biol. Chem.* **264,** 1671 (1989).

[8] J. A. Mahoney and R. L. Schnaar, *in* "Neoglycoconjugates: Preparation and Applications" (Y. C. Lee and R. T. Lee, eds.), p. 445. Academic Press, San Diego, 1994.

[9] Y. C. Lee, *FASEB J.* **6,** 3193 (1992).

[10] M. Tiemeyer and R. L. Schnaar, *in* "Trophic Factors and the Nervous System" (L. A. Horrocks, N. H. Neff, A. J. Yates, and M. Hadjiconstantinou, eds.), p. 119. Raven, New York, 1990.

[11] M. Tiemeyer, P. Swank-Hill, and R. L. Schnaar, *J. Biol. Chem.* **265,** 11990 (1990).

[12] B. Zhou, S.-C. Li, R. A. Laine, R. T. C. Huang, and Y. T. Li, *J. Biol. Chem.* **264,** 12272 (1989).

[13] L. K. Needham and R. L. Schnaar, *J. Cell Biol.* **121,** 397 (1993).

Radiolabeled neoganglioproteins have been used to detect and characterize ganglioside-specific binding activities on cell membranes and on proteins solubilized from those membranes using filter binding assays.[7,11] They also have been used as histochemical markers by overlaying tissue sections with radioligand, washing, and subjecting the sections to autoradiography.[11] Accordingly, neoganglioproteins have utilities similar to other radioligands and other neoglycoconjugates.

Synthesis of Neoganglioproteins

Principle

Neoganglioproteins are prepared as outlined in Scheme 1.[8] The fatty acid amide is hydrolyzed from the intact ganglioside, converting it to its "lyso" form which has a unique primary amine at the 2-position of sphingosine.[14,15] The lysoganglioside is treated with a bifunctional cross-linking reagent, succinimidyl 4-(*N*-maleimidomethyl)cyclohexane 1-carboxylate (SMCC),[16] which forms an amide bond to the 2-position of sphingosine and results in a sulfhydryl-reactive maleimidyl moiety attached, through a linker arm, to the original position of the fatty acid amide on the ceramide portion of the ganglioside. The carrier protein bovine serum albumin (BSA) is treated with a reagent, *N*- succinimidyl *S*-acetylthioacetate (SATA),[17] which converts lysine ε-amino groups to acetylated sulfhydryls. Subsequent treatment with hydroxylamine reveals the desired free sulfhydryls. Treatment of sulfhydryl-derivatized BSA with maleimidyl-derivatized ganglioside results in a stable thioether linkage between the ganglioside and the protein. The final product is chromatographically purified and characterized by protein and carbohydrate analysis prior to radiolabeling and use as a ligand.

Procedures

Lysoganglioside. Lysoganglioside **(I)** is prepared based on the method of Neuenhofer *et al.*[14] For example, bovine brain G_{T1b} (10 μmol, Sigma Chemical Co., St. Louis, MO) is dissolved in 5 ml of 1 *M* KOH in methanol, sealed under argon in a glass tube, and heated at 100° for 24 hr with stirring. After neutralizing with acetic acid, the solvent is evaporated

[14] S. Neuenhofer, G. Schwarzmann, H. Egge, and K. Sandhoff, *Biochemistry* **24,** 525 (1985).
[15] G. Schwarzmann and K. Sandhoff, this series, Vol. 138, p. 319.
[16] S. Yoshitake, Y. Yamada, E. Ishikawa, and R. Masseyeff, *Eur. J. Biochem.* **101,** 395 (1979).
[17] R. J. Duncan, P. D. Weston, and R. Wrigglesworth, *Anal. Biochem.* **132,** 68 (1983).

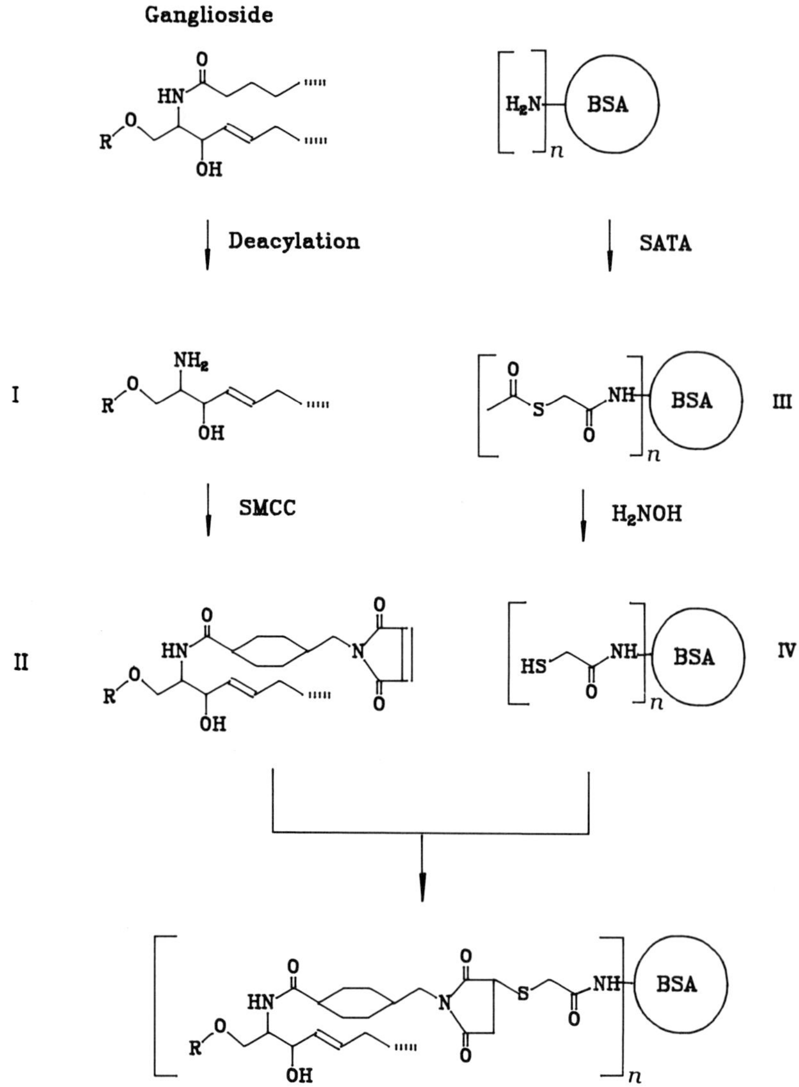

SCHEME 1. Neoganglioprotein synthesis: Coupling of lysogangliosides to BSA.[8] A generalized ganglioside is represented at top, with the oligosaccharide chain designated *R*. Strong alkaline hydrolysis followed by selective re-N-acetylation of ganglioside sialic acids results in production of lysoganglioside **(I)**. Subsequent incubation with the heterobifunctional cross-linking reagent succinimidyl 4-(*N*-maleimidomethyl)cyclohexane 1-carboxylate (SMCC) generates a maleimide-activated intermediate **(II)** which is purified chromatographically. The BSA is first reacted with *N*-succinimidyl *S*-acetylthioacetate (SATA) to generate a protein polyvalently derivatized (at lysine residues) with acetyl-blocked thiol groups **(III)**. Subsequent treatment with hydroxylamine generates free sulfhydryls on the BSA **(IV)**. The sulfhydryl-containing BSA is added to the maleimide-activated glycosphingolipid, resulting in formation of stable thioether bonds linking the derivatized glycosphingolipid to the protein. Details are presented in the text.

under a stream of nitrogen and the resulting residue resuspended in 10 ml of water and dialyzed for 90 min against 4 liters of water using Spectra/Por 6 dialysis tubing, 1000 molecular weight cutoff (Spectrum Medical Industries, Inc., Los Angeles, CA). The resulting product, which has primary amino groups both on the sphingosine and the sialic acids, is further treated to reacetylate the sialic acids selectively. The dialyzed solution is evaporated, suspended in 5 ml anhydrous ethyl ether, and sonicated. After addition of 5 ml of 0.5 *M* sodium bicarbonate, the mixture is further sonicated, then frozen in a 2-propanol–dry ice bath. 9-Fluorenylmethylchloroformate (Fmoc, 20 μmol in 0.5 ml hexane, Aldrich Chemical Co., Milwaukee, WI) is added to the ether layer, and the resulting suspension is mixed gently at room temperature until the aqueous layer thaws, then mixed further until emulsified, and finally stirred in the cold for 16 hr. This results in selective and reversible Fmoc blocking of the spinogosine amine.

The phases are separated by centrifugation, and the aqueous phase is collected and warmed to ambient temperature. Aliquots (20 μl) of acetic anhydride are added until the pH drops to 5. The resulting solution is stirred for 3 hr, during which 5 additional 20-μl aliquots of acetic anhydride are added. This results in quantitative reacetylation of the sialic acid amines. On silica gel high-performance thin-layer chromatography (HPTLC) analysis in chloroform–methanol–water (10 : 10 : 3, v/v/v) the product is positive for sialic acids (via resorcinol stain[18]), negative for free amino groups (via ninhydrin stain),[19] and fluorescent, indicating the successful production of an Fmoc-blocked lysoganglioside. The reaction solvents are evaporated, and the residue is resuspended in water, dialyzed (as above) for 90 min, reevaporated, resuspended in 5 ml of concentrated ammonium hydroxide, and stirred at ambient temperature for 16 hr to remove the Fmoc group. The resulting solution, containing deblocked lysoganglioside and released Fmoc, is evaporated, resuspended in 3 ml of water, stirred for 1 hr at ambient temperature, and reevaporated. The resulting residue is dissolved in 2 ml of chloroform–methanol–water (6 : 4 : 1, v/v/v) and applied to an Iatrobeads (Iatron Corp., Tokyo, Japan) silica bead column (20 ml column volume) previously equilibrated in the same solvent. After washing in the same solvent, lyso-G_{T1b} is eluted with chloroform–methanol–water (10 : 10 : 3, v/v/v). Fractions containing product are identified by HPTLC in chloroform–methanol–water (6 : 4 : 1, v/v/v) with ninhydrin and resorcinol detection (approximate R_f values: G_{T1b}, 0.43; lyso-G_{T1b}, 0.22).

[18] L. Svennerholm, *Biochim. Biophys. Acta* **24,** 604 (1957).

[19] P. H. Weigel, M. Naoi, S. Roseman, and Y. C. Lee, *Carbohydr. Res.* **70,** 83 (1979).

Maleimidyl-Derivatized Ganglioside. Lysoganglioside is treated with SMCC (Pierce Chemical Co., Rockford, IL) to incorporate a reactive maleimidyl group on a linker arm at the position of the original fatty acid amide, producing maleimidyl-derivatized ganglioside **(II)**. For example, 1 μmol of lyso-G_{T1b} is dissolved in 1.5 ml of anhydrous dimethylformamide (DMF) containing 0.1 *M* redistilled triethylamine. A 0.5-ml aliquot of DMF containing 10 μmol of SMCC is added and the solution incubated at 45° under nitrogen in a 13 × 100 mm screw capped tube. After about 24 hr the reaction is complete as judged by thin-layer chromatography (TLC) in solvent A [chloroform–methanol–0.1 *M* aqueous acetic acid (6 : 4 : 1, v/v/v)], with ganglioside detected using a resorcinol stain (approximate R_f values: lyso-G_{T1b}, 0.13; SMCC-derivatized G_{T1b}, 0.19). The reaction is evaporated to dryness under nitrogen or in a Speed-Vac concentrator (Savant Instruments, Farmingdale, NY), the resulting residue is redissolved in 0.8 ml of solvent B [chloroform–methanol–water (6 : 4 : 1, v/v/v)] and loaded onto a 0.8 × 6 cm Iatrobead column preequilibrated in solvent B, and the column is washed with 6 column volumes of solvent B. Product is then eluted with chloroform–methanol–water (4 : 8 : 3, v/v/v). Fractions containing purified maleimidyl-derivatized lyso-G_{T1b} are identified by TLC as above, pooled, evaporated to dryness, and kept at −20° under nitrogen until freshly deblocked sulfhydryl-derivatized BSA is added (see below).

Thioester-Derivatized Bovine Serum Albumin. Excess free sulfhydryl groups are introduced into BSA in a protected acetyl thioester form **(III)** using a commercial reagent, SATA (Pierce). This allows high levels of derivatization without extensive spontaneous disulfide cross-linking of the derivatized protein.[17] Six microliters of dimethyl sulfoxide (DMSO) containing 1.1 mg (4.8 μmol) of SATA is added to a stirring, ice-cold solution containing 10 mg (0.15 μmol) of BSA (fatty acid and globulin free, Cat. No. A0281, Sigma) in 0.5 ml of 50 mM *N*-2-hydroxyethylpiperazine-*N*′-2-ethanesulfonic acid buffer, pH 8. After incubation for 4 hr at 4°, the reaction is loaded onto a 0.8 × 20 cm Sephadex G-25 column preequilibrated and eluted with 10 m*M* sodium phosphate buffer, pH 7. Protein-containing fractions are identified by UV absorbance at 280 nm, pooled, and stored frozen in aliquots.

Ganglioside–Protein Coupling. Just prior to mixing with the maleimidyl-derivatized lysoganglioside (see above), an aliquot of the SATA-reacted BSA **(III)** is deblocked to reveal free sulfhydryls **(IV)**. An aliquot of SATA–BSA product (0.34 ml, 4 mg BSA) is mixed with 34 μl of deacetylation solution (0.5 *M* hydroxylamine hydrochloride, 40 m*M* sodium phosphate buffer, pH 7.5, and 25 m*M* EDTA). After 2 hr at ambient temperature, a 7.5-μl aliquot of 100 m*M* aqueous dithiothreitol (DTT) is added. After 15 min, the reaction solution is applied to a 0.8 × 20 cm

Sephadex G-25 column preequilibrated and eluted with 10 mM sodium phosphate, pH 7. Protein-containing fractions are pooled, the protein concentration is determined, and the derivatized BSA (50 nmol, 3.3 mg) is added to the dried maleimidyl-derivatized lysoganglioside (~1 μmol, see above). The reaction is sonicated in a bath sonicator to ensure dispersal of the glycolipid derivative. After 12–24 hr at 4°, TLC in solvent A reveals that most of the resorcinol-positive material does not migrate away from the origin, indicating successful conjugation of ganglioside to BSA. Purification and analysis (see below) typically demonstrate 8–13 G_{T1b} molecules bound per BSA molecule.

Neoglycoprotein Purification. Ganglioside-conjugated BSA derivatives are purified by sequential ion-exchange and size-exclusion high-performance liquid chromatography (HPLC).[7] Aliquots of the above reaction (up to 1.5 ml containing up to 0.5 mg protein) are injected onto a 0.75 × 7.5 cm TSK-Gel DEAE-5PW HPLC column (TosoHaas, Montgomeryville, PA) preequilibrated in buffer A (10 mM Tris buffer, pH 7.6). The column is eluted with a gradient from 0 to 55% buffer B (1 M NaCl in 10 mM Tris buffer, pH 7.6) over 10 min, then kept at 55% buffer B for 15 min. Protein is detected by UV absorbance at 280 nm. Under these conditions neoganglioproteins elute at approximately 12 min, fully resolved from underivatized BSA which elutes around 10 min [any unconjugated ganglioside is retained on the column until eluted with methanol–buffer B (1 : 1, v/v)]. Fractions are collected, and those containing protein are pooled, lyophilized, and resuspended in 0.5 ml of water.

Any excess sulfhydryl groups on the purified conjugate are blocked by adding 20 μl of 30 mM dithiothreitol in 10 mM sodium phosphate buffer, pH 7, incubating at 37° for 1 hr, then adding 20 μl of 150 mM iodoacetamide in the same buffer and incubating at 37° for 1 hr. Finally, 20 μl of 750 mM dithiothreitol is added. After 1 hr at 37° the same is applied to a Sephadex G-25 column and eluted as described above. The final product is stored frozen in aliquots until use.

Characterization and Radiolabeling of Neoganglioproteins

Neoganglioprotein characterization involves determining the carbohydrate composition and structure, and confirming the covalent nature of the carbohydrate–protein bond. Protein values are determined by any suitable protein microassay. Sialic acid content is determined by mild acid hydrolysis followed by quantitation of released *N*-acetylneuraminic acid using a Dionex (Sunnyvale, CA) analytical HPLC.[20] Aliquots (50 μl) of

[20] Y. C. Lee, *Anal. Biochem.* **189,** 151 (1990).

standards or experimental samples containing 0.1–1 nmol of sialic acid are placed in 0.5-ml microcentrifuge tubes with 50 μl of hydrolysis solution containing 0.2 *M* HCl and 0.5 *M* NaCl. Tubes are incubated at 80° for 4 hr, then cooled, and 10-μl aliquots are directly injected into a Dionex HPLC system using an HPIC-AS6 column running isocratically in 100 m*M* sodium acetate in 113 m*M* sodium hydroxide, with quantitation via pulsed amperometric detection. Similarly, neutral and amino sugar content is determined after hydrolysis of an aliquot in 2 *N* trifluoroacetic acid for 5 hr at 100°. The hydrolyzate is evaporated to dryness to remove trifluoroacetic acid, the residue is resuspended in water, and aliquots containing 50–250 pmol sugar are injected onto the Dionex system under conditions described previously.[21] The ratio of monosaccharides released reflects the linked oligosaccharide structure, and the ratio of saccharides to protein reveal the valence of the conjugate.[7] Neoganglioprotein nomenclature is modeled after the nomenclature for neoglycoproteins.[22] For instance, the G_{T1b}-derivatized BSA described here is denoted $(G_{T1b})_n$BSA, where the type of ganglioside covalently derivatized to the protein is in parentheses and the average number of ganglioside moieties attached to each protein molecule is indicated by the subscript *n*.

The covalent nature of the glycolipid–protein linkage is tested by sodium dodecyl sulfate–polyacrylamide gel electrophoresis (SDS–PAGE).[7,23] Neoganglioproteins migrate less rapidly than the parent BSA and, unlike the parent, can be detected using saccharide-specific reagents. For example, G_{M1}-conjugated BSA subjected to SDS–PAGE and blotted to nitrocellulose binds ^{125}I-labeled cholera toxin B subunit. In contrast, G_{T1b}-conjugated BSA does not bind the toxin unless the blot is first incubated in a solution of *Vibrio cholera* neuraminidase to convert the G_{T1b} structure to G_{M1}.[24] When available, saccharide-specific antibodies can be used to confirm the covalent linkage of a ganglioside to the carrier via typical Western blotting techniques.

Radioiodination of neoganglioproteins is performed using a commercially available immobilized oxidizing agent. A small sample of neoganglioprotein [e.g., $(G_{T1b})_{13}$BSA, 50 μg] is incubated for 30 min with 1 mCi Na^{125}I and 2 Iodobeads (Pierce) in 0.5 ml of 0.1 *M* sodium phosphate buffer, pH 7. The reaction solution is then loaded on a 0.6 × 18 cm Sephadex G-25 column preequilibrated in the same buffer (and prerun with BSA to block nonspecific adsorption), and void volume fractions containing radiolabel

[21] M. R. Hardy, R. R. Townsend, and Y. C. Lee, *Anal. Biochem.* **170,** 54 (1988).

[22] T. B. Kuhlenschmidt and Y. C. Lee, *Biochemistry* **23,** 3569 (1984).

[23] U. K. Laemmli, *Nature (London)* **227,** 680 (1970).

[24] R. Schauer, W. V. Rudiger, M. Sander, A. P. Corfield, and H. Wiegandt, *Adv. Exp. Med. Biol.* **125,** 283 (1980).

are pooled. The resulting radioligand comigrates with unlabeled neoganglioprotein[7] and is typically radiolabeled to very high specific activity (up to 1 mCi/nmol). After determination of protein and radiolabel recovery, excess BSA is added (5 mg/ml) to enhance stability during storage at −20°.

Applications of Neogangliproteins

Neoganglioproteins have been used much like neoglycoproteins for detection and characterization of complementary binding proteins from a variety of biological sources. For example, incubation of rat brain membranes with ^{125}I-labeled $(G_{T1b})_{13}BSA$ followed by collection of bound radiolabel on glass fiber filters resulted in high-affinity, specific, and saturable binding.[7,11] Addition of unlabeled gangliosides as inhibitors revealed specificity for the "1b" ganglioside structure. A similar assay has been used for detection of ganglioside binding activity after detergent solubilization from these membranes. Soluble receptor–ligand complex is selectively precipitated by addition of an equal volume of 15% aqueous polyethylene glycol (PEG 8000, Sigma), followed by collection of bound radioligand on glass fiber filters. A soluble ganglioside binding activity has been detected in rat muscle extracts using a similar PEG precipitation assay.

The high affinity and low dissociation rate of ^{125}I-labeled $(G_{T1b})BSA$ bound to its brain membrane receptor have allowed the use of this neoganglioprotein for histochemical staining of brain sections[11] using autoradiographic methods modeled after previous studies.[25] Lightly fixed tissue sections on microscope slides are overlaid with buffer containing ^{125}I-labeled $(G_{T1b})BSA$. After allowing binding to occur at 4°, unbound radioligand is removed by washing in ice-cold buffer, and the slides are air-dried and exposed to preflashed X-ray film.[26] Companion sections incubated identically, but in the presence of unlabeled G_{T1b} or G_{M1}, reveal the same "1b" ganglioside specificity of neoganglioprotein binding determined using isolated membranes.

The finding that different types of well-established radioligand receptor binding assays are readily adapted for use with neoganglioproteins is encouragement for their broader application.

Comments and Conclusions

Notes on Procedures. Overall yield of lysoganglioside from the three synthetic steps (hydrolysis, Fmoc blocking/reacetylation, deblocking) is

[25] W. S. Young, III and M. J. Kuhar, *Brain Res.* **179,** 255 (1979).

[26] R. A. Laskey and A. D. Mills, *FEBS Lett.* **82,** 314 (1977).

low, typically 30–35% based on the starting ganglioside.[15] In contrast, conversion to the maleimidyl derivative is efficient, although modest losses are experienced during purification. Sulfhydryl-derivatized BSA is readily and reproducibly synthesized in relatively large quantities. The degree of sulfhydryl derivatization is routinely determined by deblocking a test aliquot and determining the free sulfhydryl content using Ellman's reagent.[27] In the example above, the degree of sulfhydryl derivatization was 13 per BSA molecule, and subsequent attachment of maleimidyl-derivatized ganglioside was quantitative. Although the degree of sulfhydryl derivatization can be increased by increasing the SATA concentration or its time of incubation with BSA, there is no improvement in the subsequent amount of ganglioside attached. This may be due to steric factors. Ganglioside G_{T1b} is relatively large and highly anionic, and a high density of bound G_{T1b} molecules may restrict access by additional unbound G_{T1b} molecules. Alternatively, only a portion of the 60 lysines on BSA may be sterically positioned (after sulfhydryl derivatization) to react with maleimidyl-derivatized G_{T1b}.

The maleimidyl groups used in the coupling scheme are reported to be stable for several hours in aqueous solution. However, storage of the maleimidyl-derivatized ganglioside is not recommended, since maleimidyl groups will eventually deteriorate. The acetyl thioesters on SATA-derivatized BSA are more stable, but they will become deblocked after months of storage, even at −20°, resulting in disulfide cross-linking between BSA molecules. Therefore, the use of fresh preparations of maleimidyl-derivatized ganglioside and SATA-derivatized BSA is prudent. Lysogangliosides, in contrast, are stable indefinitely when sealed and stored at −20° either dry or in chloroform–methanol–water mixtures.

In comparison to neoglycoproteins, neoganglioproteins adsorb more readily to certain surfaces if they are exposed at low protein concentration. Therefore, it has been useful to store neoganglioproteins in polypropylene tubes (such as standard microcentrifuge tubes) and to prerun a bolus of underivatized BSA over Sephadex G-25 columns prior to purification of radioiodinated neoganglioproteins (see above). The column is then washed extensively with buffer to elute underivatized BSA prior to application of the radioiodinated sample.

Limitations. Because lysoganglioside production requires incubation in alkali under harsh conditions, gangliosides (or other glycosphingolipids) having alkali-labile groups cannot be converted to neoganglioproteins using the current techniques. Although most gangliosides are highly alkali stable, glycosphingolipids having *O*-acetyl groups (e.g., on sialic acids)

[27] G. L. Ellman, *Arch. Biochem. Biophys.* **82,** 70 (1959).

or *O*-sulfate groups are irreversibly altered during the first step of lysoglycolipid preparation. Alternative procedures, such as enzymatic removal of the oligosaccharide and its attachment to BSA via reductive amination,[13] should be considered for these glycoconjugates.

Conclusions. The high specific activity, high affinity, and low background binding characteristic of polyvalent BSA neoglycoconjugates, as exemplified by their utility in the study of other vertebrate lectins,[9] encouraged the search for ganglioside-specific binding proteins using neoganglioproteins. The results to date have been encouraging, with indications of ganglioside-specific binding activities in vertebrate brain and muscle. The application of neoganglioproteins to other tissue systems may reveal additional cell surface lectins important in cell–cell recognition or modulation of membrane function. As a greater variety of neoganglioproteins and related neoglycoconjugates become available and are applied to additional systems where glycosphingolipids are posited to play physiological and/or pathological roles, important new lectins may be revealed. Neoganglioproteins are also applicable to purification and molecular characterization of lectins. As these studies progress, additional biochemical, immunological, and molecular tools needed to probe the physiological roles of gangliosides and their receptors may emerge.

Acknowledgment

This work was supported by Grant HD14010 from the National Institutes of Health.

[3] Leprosy-Specific Neoglycoconjugates: Synthesis and Application to Serodiagnosis of Leprosy

By PATRICK J. BRENNAN, DELPHI CHATTERJEE, TSUYOSHI FUJIWARA, and SANG-NAE CHO

Phenolic Glycolipid I of *Mycobacterium leprae*

The discovery of a glycolipid surrounding *Mycobacterium leprae* bacilli in large amounts in infected tissues[1] proved to be of considerable importance in understanding aspects of the immunopathogenesis of leprosy and providing tools for serodiagnosis of leprosy.[2] The glycolipid population, which is dominated by one component, called phenolic glyco-

[1] S. W. Hunter and P. J. Brennan, *J. Bacteriol.* **147,** 728 (1981).
[2] H. Gaylord and P. J. Brennan, *Annu. Rev. Microbiol.* **41,** 49 (1987).

Copyright © 1994 by Academic Press, Inc.
All rights of reproduction in any form reserved.

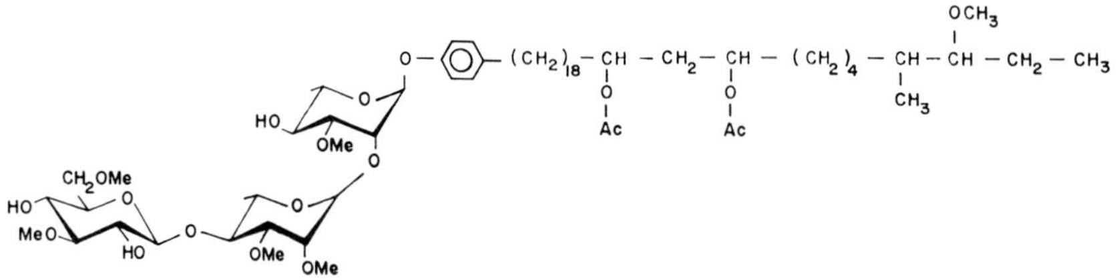

FIG. 1. Structure of phenolic glycolipid I from *Mycobacterium leprae*. Me, CH_3; Ac (acyl), mixture of C_{32}, C_{34}, and C_{36} mycocerosic acids.

lipid I (PGL-I), consists of glycosides of phenolphthiocerol dimycocerosate. This class of glycolipid is widely distributed within members of the genus *Mycobacterium*.[3,4] The aglycan structures of all of them are quite similar. However, the oligosaccharide units are generally characteristic of a single species and are largely responsible for the specificity of the antibody response in infections. The carbohydrate composition of the PGL-I from *M. leprae* consists of 3,6-di-*O*-methyl-β-D-glucopyranose, 2,3-di-*O*-methyl-α-L-rhamnopyranose, and 3-*O*-methyl-α-L-rhamnopyranose, and these are combined as the trisaccharide unit, *O*-(3,6-di-*O*-methyl-α-D-glucopyranosyl)-(1 → 4)-*O*-(2,3-di-*O*-methyl-α-L-rhamnopyranosyl)-(1 → 2)-*O*-3-*O*-methyl-α-L-rhamnopyranosyl, glycosidically attached to the phenol of the phenolphthiocerol dimycocerosate[5] (Fig. 1). The acyl functions of PGL-I are all tetramethyl-substituted, branched C_{30}, C_{32}, and C_{34} mycocerosic acids, and reversed-phase high-performance liquid chromatography (HPLC) of highly purified preparations of PGL-I combined with plasma desorption–mass spectrometry showed considerable heterogeneity owing to the existence of the C_{30}/C_{30}, the C_{32}/C_{32}, and the C_{34}/C_{34} forms of the dimycocerosate.[6] There are two less abundant variants of PGL-I in *M. leprae*, known as PGL-II and PGL-III, which show minor variations in the methylation pattern of the glycosyl residues. These proved to be of importance in defining the molecular basis of the antigenicity of PGL-I and leading to the first generation of leprosy-specific neoglycoconjugates.

Seroreactivity of Phenolic Glycolipid I

In early experiments it was demonstrated that antibodies in pooled serum from lepromatous leprosy patients reacted readily with the native

[3] G. Puzo, *Rev. Microbiol.* **17,** 305 (1990).
[4] P. J. Brennan, *in* "Microbial Lipids" (C. Ratledge and S. G. Wilkinson, eds.), p. 203. Academic Press, London, 1988.
[5] S. W. Hunter, T. Fujiwara, and P. J. Brennan, *J. Biol. Chem.* **257,** 15072 (1982).
[6] I. Jardine, G. Scanlan, M. McNeil, and P. J. Brennan, *Anal. Chem.* **61,** 416 (1989).

and the deacylated form of PGL-I, but not with the nonglycosylated product nor with mycocerosic acid or the phenolphthiocerol core.[7] Also, when the terminal 3,6-di-*O*-methylglucopyranose unit was removed by partial acid hydrolysis and the partially deglycosylated product was purified and tested, this also had lost most of its serological activity.[8]

It was also demonstrated that a minor relative of PGL-I, PGL-III, which is devoid of the methyl ether group at O-6 of the 3,6-di-*O*-methyl-β-D-glucopyranose residue, did not bind to anti-PGL-I antibodies in serum from lepromatous leprosy patients.[8] Accordingly, the terminal intact 3,6-di-*O*-methyl-β-D-glucopyranose residue was implicated as the primary epitope. Fujiwara *et al.*[8–10] synthesized the di- and trisaccharides *O*-(3,6-di-*O*-methyl-β-D-glucopyranosyl)-(1 → 4)-2,3-di-*O*-methyl-L-rhamnopyranose and *O*-(3,6-di-*O*-methyl-β-D-glucopyranosyl)-(1 → 4)-2,3-di-*O*-methylα-L-rhamnopyranosyl)-(1→2)-*O*-3-*O*-methyl-L-rhamnopyranose together with less fully methylated analogs and also products with the incorrect anomers. The structurally complete haptens, both the terminal disaccharide and the complete trisaccharide, substantially inhibited binding between anti-PGL-I antibodies and the native PGL-I antigen in a competitive inhibition enzyme-linked immunosorbent assay (ELISA). Gigg *et al.*[11] reached essentially the same conclusion through synthesis of the terminal disaccharide and its propyl α-glycoside; the latter was the first oligosaccharide hapten to contain both sugar residues in the correct, defined anomeric configurations.

Synthesis of Neoglycoconjugates Based on Phenolic Glycolipid I

Fujiwara *et al.*[8] prepared the first generation of neoglycoconjugates (NGCs) capable of specific recognition of anti-PGL-I antibodies in sera from patients with lepromatous leprosy. Conjugation of the synthetic terminal disaccharide to the ε-amino groups of the lysine residues in bovine serum albumin (BSA) by reductive animation with sodium borohydride yielded a NGC which emulated the natural PGL-I in an ELISA against sera from lepromatous leprosy patients. This type of synthesis, however, was wasteful, in that the reducing residue of the disaccharide (which

[7] S.-N. Cho, D. L. Yanagihara, S. W. Hunter, R. H. Gelber, and P. J. Brennan, *Infect. Immun.* **41,** 1077 (1983).

[8] T. Fujiwara, S. W. Hunter, S.-N. Cho, G. O. Aspinall, and P. J. Brennan, *Infect. Immun.* **43,** 245 (1984).

[9] T. Fujiwara, S. W. Hunter, and P. J. Brennan, *Carbohydr. Res.* **148,** 287 (1986).

[10] T. Fujiwara, G. O. Aspinall, S. W. Hunter, and P. J. Brennan, *Carbohydr. Res.* **163,** 41 (1987).

[11] R. Gigg, S. Payne, and R. Conant, *J. Carbohydr. Chem.* **2,** 207 (1983).

does contribute to the activity inherent in the disaccharide) is altered on conjugation into an acyclic 1-amino-1-deoxyalditol spacer arm.

Evaluation of the subsequent strategy for the synthesis of leprosy-specific NGCs and also the matter of improved synthesis of the oligosaccharides themselves and their glycosides have been reviewed by Aspinall *et al.*[12] In this chapter, we concentrate on the NGCs synthesized in the laboratories of Brennan and co-workers,[12] Fujiwara *et al.*,[8] and Gigg *et al.*,[11] since these products are readily available and are in wide use.

All of the syntheses, whether involving the entire trisaccharide unit or the outermost disaccharide, were accomplished by stepwise glycosylation from the innermost residue which contained a glycosidically attached linker arm in configurationally correct form. Two basic strategies were used for purposes of conjugation to protein, usually bovine serum albumin, through covalent linkage to the ε-amino groups of lysine residues. In one approach, the innermost glycosyl residue was assembled as a glycoside bearing an alkoxycarbonyl group for conversion via acylhydrazide to an acyl azide for coupling to lysine residues. This approach was based on the linker arm method of Lemieux *et al.*[13,14] Chatterjee *et al.*[15–17] employed this procedure with the 8-methoxycarbonyloctyl glycoside as a lipophilic spacer to construct a variety of PGL-I-based NGCs containing the terminal monosaccharide, disaccharide, and full trisaccharide. The latter two products are in wide usage and are readily available from the authors; they have become known as natural disaccharide-octyl-BSA (ND-O-BSA) and natural trisaccharide-octyl-BSA (NT-O-BSA) (Fig. 2).

The same principle was adopted by Fujiwara *et al.*,[18,19] who prepared a similar array of glycosides with a methyl-3-(*p*-hydroxyphenyl)propionate linker arm. These products are now known as natural disaccharide- and natural trisaccharide-propionyl-BSA (ND-P-BSA and NT-P-BSA) (Fig. 2) and have been incorporated into an agglutination kit.

[12] G. P. Aspinall, D. Chatterjee, and P. J. Brennan, *Adv. Carbohydr. Chem. Biochem.* in press.

[13] R. U. Lemieux, D. R. Bundle, and D. A. Baker, *J. Am. Chem. Soc.* **97,** 4076 (1975).

[14] R. U. Lemieux, D. A. Baker, and D. R. Bundle, *Can. J. Biochem.* **55,** 507 (1977).

[15] D. Chatterjee, J. T. Douglas, S.-N. Cho, T. H. Rea, R. H. Gelber, G. O. Aspinall, and P. J. Brennan, *Glycoconjugate J.* **2,** 187 (1985).

[16] D. Chatterjee, S.-N. Cho, P. J. Brennan, and G. O. Aspinall, *Carbohydr. Res.* **156,** 39 (1986).

[17] D. Chatterjee, S.-N. Cho, C. Stewart, J. T. Douglas, T. Fujiwara, and P. J. Brennan, *Carbohydr. Res.* **183,** 241 (1988).

[18] T. Fujiwara, S. Izumi, and P. J. Brennan, *Agric. Biol. Chem.* **49,** 2301 (1985).

[19] T. Fujiwara and S. Izumi, *Agric. Biol. Chem.* **51,** 2539 (1987).

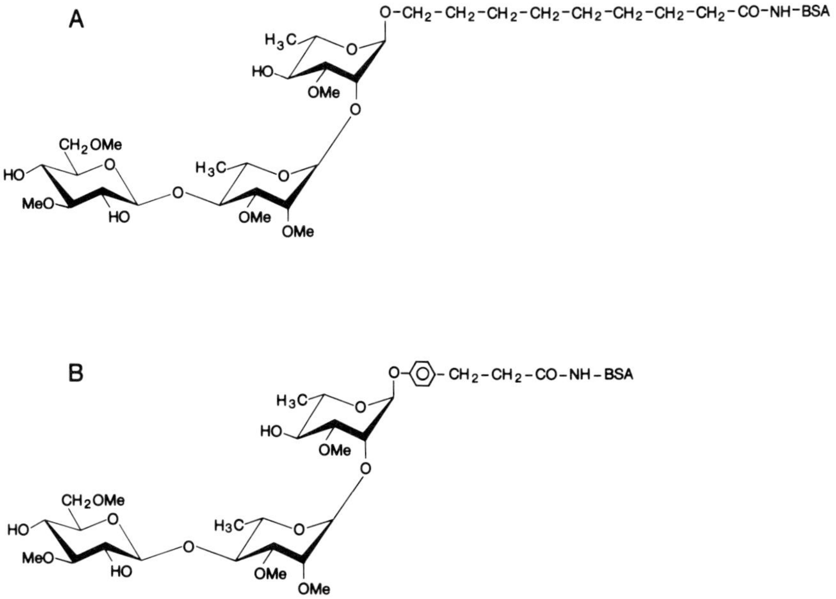

FIG. 2. Structures of two of the more widely used trisaccharide-containing neoglycoconjugates based on phenolic glycolipid I. (A) *O*-(3,6-Di-*O*-methyl-β-D-glucopyranosyl)-(1 → 4)-*O*-(2,3-di-*O*-methyl-α-L-rhamnopyranosyl)-(1 → 2)-*O*-(3-*O*-methyl-α-L-rhamnopyranosyl)-(1 → 9)-oxynonanoyl-BSA. (B) *O*-(3,6-Di-*O*-methyl-β-D-glucopyranosyl)-(1→4)-*O*-(2,3-Di-*O*-methyl-α-L-rhamnopyranosyl)-(1 → 2)-*O*-(3-*O*-methyl-α-L-rhamnopyranosyl)-(1 → 4′)-oxy-(3-phenylpropionyl)-BSA.

In the second approach, Gigg *et al.*[20] prepared allyl glycosides for conversion to formylmethyl glycosides for reductive amination to ε-amino groups of lysine residues in the carrier protein, invariably BSA. Synthesis of the protected allyl glycoside of the disaccharide (the disaccharide only was produced) was followed by epoxidation and basic epoxide opening and removal of benzyl ether substituents by hydrogenolysis. The resulting 1-glyceryl glycoside was then converted to the formylmethyl glycoside by oxidation with sodium metaperiodate.[20] The product, which is available through the IMMYC (Immunology of Mycobacterial Diseases) Program at the World Health Organization (WHO) (Geneva, Switzerland), is known as natural disaccharide-allyl-BSA (ND-A-BSA).

Aspinall *et al.*[12] have extensively reviewed the literature on alternative procedures for synthesis of the glycosides and their conjugation to pro-

[20] J. Gigg, R. Gigg, S. Payne, and R. Conant, *J. Chem. Soc. Perkins Trans. 1*, 1165 (1987).

teins. However, the products of these procedures have not been widely implemented, and the protocols are of less practical benefit.

Synthesis of More Common Leprosy-Specific Neoglycoconjugates

Chatterjee *et al.*[15–17] have described in detail the synthesis of the 8-(methoxycarbonyl)octyl *O*-(3,6-di-*O*-methyl-β-D-glucopyranosyl)-(1 → 4)-2,3-di-*O*-methyl-α-L-rhamnopyranoside, its subsequent conversion to the crystalline hydrazide, and its coupling to BSA, via intermediate acyl azide formation, to produce the corresponding NGC, *O*-(3,6-di-*O*-methyl-β-D-glucopyranosyl)-(1 → 4)-*O*-(2,3-di-*O*-methyl-α-L-rhamnopyranosyl)-(1 → 9)-oxynonanoyl-BSA (ND-O-BSA), containing about 34 mol of hapten per mole of BSA. Chatterjee *et al.* also described the synthesis of the *O*-(3,6-di-*O*-methyl-β-D-glucopyranosyl)-(1 → 4)-*O*-(2,3-di-*O*-methyl-α-L-rhamnopyranosyl)-(1 → 2)-*O*-(3-*O*-methyl-α-L-rhamnopyranosyl)-(1 → 9)-oxynonanoyl-BSA (NT-O-BSA). Likewise, Fujiwara *et al.*[18,19] have described in detail the synthesis of the corresponding 3-phenylpropionyl-linked NGCs, namely, *O*-(3,6-di-*O*-methyl-β-D-glucopyranosyl)-(1 → 4)-*O*-(2,3-di-*O*-methyl-α-L-rhamnopyranosyl)-(1 → 4′)-oxy-(3-phenylpropionyl)-BSA (ND-P-BSA) and *O*-(3,6-di-*O*-methyl-β-D-glucopyranosyl)-(1 → 4)-*O*-(2,3-di-*O*-methyl-α-L-rhamnopyranosyl)-(1 → 2)-*O*-(3-*O*-methyl-α-L-rhamnopyranosyl)-(1 → 4′)-oxy-(3-phenylpropionyl)-BSA (NT-P-BSA). These protracted syntheses are not described here. However, the coupling protocols have been modifed since the acyl azide method of Lemieux *et al.*,[13,14] although very efficient, generates a large change in the charge of the protein which may have deleterious effects for other, nonserological uses of these NGCs. The following modified method has proved effective for the final synthesis of NT-P-BSA.

The trisaccharide *p*-(2-methoxycarbonylethyl)phenyl *O*-(3,6-di-*O*-methyl-β-D-glucopyranosyl)-(1 → 4)-*O*(2,3-di-*O*-methyl-α-L-rhamnopyranosyl)-(1→2)-*O*-3-*O*-methyl-α-L-rhamnopyranoside[19] (59 mg, 85.7 μmol) is suspended in 4 ml of 80% hydrazine monohydrate and stirred overnight at room temperature. It is then diluted with ethanol and evaporated. The residue is redissolved in toluene, evaporated to remove the residual hydrazine, and extracted with chloroform. The extract is evaporated and dried over phosphorus pentoxide *in vacuo* to give the hydrazide (57.2 mg, 81.2 μmol, 94%) which is dissolved in 1.8 ml of dry dimethylformamide (DMF) and chilled to −25° to −30°. To the solution are added hydrogen chloride in dry dioxane (4 *M*, 0.12 ml) and *tert*-butyl nitrite in DMF (100 mg/ml, 0.17 ml). The mixture is stirred for 30 min at −25° to −30°. Sulfamic acid (70 mg/ml in DMF, 0.17 ml) is then added. After 15 min at −25° to −30°, the mixture is added dropwise to 3.25 ml of BSA (25 mg/ml) in

80 mM $Na_2B_2O_7$–350 mM $KHCO_3$ (pH 9.2) in an ice–water bath with continuous stirring. The pH is kept between pH 9.0 and 9.3 by adding 0.5 M hydrogen chloride or sodium hydroxide. The mixture is stirred for 2 hr, dialyzed against double-distilled water 8 times, and lysophilized to give the sugar–BSA conjugate (NT-P-BSA, 86 mg).

Purification of the sugar hydrazide is desirable but not necessary. The hydrogen chloride/dioxane solution can be stored for about 2 months in a sealed tube in the cold. This batch of NT-P-BSA contains 40 mol of hapten per mole of BSA. The sugar content generally ranges from 35 to 40 mol per mole of BSA but can be altered by changing the ratio of sugar to protein in the coupling reaction.

In case the sugar is not soluble in DMF, it can be effectively coupled to protein by the carbodiimide method[21] as follows. The NT-P (53 mg, 78.4 μmol) is stirred in 1.5 ml of sodium hydroxide to saponify the ester group of NT-P at room temperature (1.5 hr). The solution is adjusted to pH 4.75 with 0.1 M hydrogen chloride, and then BSA (56 mg in 1.5 ml of water) is added to the solution. A solution of 1-(3-dimethylaminopropyl)-3-ethylcarbodiimide methiodide (154 mg in 0.7 ml of water) is added dropwise at room temperature, the solution being maintained at pH 4.75 during the course of the reaction. After stirring for an additional 3 hr at room temperature, the solution is dialyzed against double-distilled water 8 times. Purification of the product by gel filtration is not necessary, but when conducted on a column (80 × 1.6 cm) of Sephadex G-75 in phosphate-buffered saline (PBS), the NGC solution is lyophilized in aliquots of 0.5–1.0 ml and stored at −10°. Lyophilization of the solution gives the sugar–protein conjugate NT-P-BSA (59 mg). The sugar content of this batch is 37.8 mol per mole of BSA and generally ranges from 33 to 41 mol per mole of BSA.

Seroreactivity of Leprosy-Specific Neoglycoconjugates

Indirect ELISA, modified from Voller *et al.*,[22] is the method of choice for the application of the NGCs to the detection of anti-PGL-I antibodies [mostly immunoglobulin M (IgM)] in the serodiagnosis of leprosy.[7,16,17] The NGCs are dissolved in carbonate–bicarbonate buffer, pH 9.6, to a concentration of 50–100 ng of rhamnose equivalent. The NGC solutions (50 μl) are applied to wells of U-bottomed microtiter plates (Dynatech

[21] J. Lonngren, I. J. Goldstein, and J. E. Neiderhuber, *Arch. Biochem. Biophys.* **175,** 661 (1976).

[22] A. Voller, D. E. Bidwell, and A. Bartlett, *in* "The Enzyme-Linked Immunosorbent Assay (ELISA)," p. 23. Dynatech Laboratories, Alexandria, Virginia, 1979.

Laboratories, Alexandria, VA). Equivalent amounts of BSA are applied to control plates. Plates are then placed in a moist chamber and incubated overnight at 37°. Unbound antigen is then removed by suction, and the wells are washed with PBS, pH 7.4, containing 0.05% (v/v) Tween 20 (PBST), and incubated with 100 μl of blocking solution composed of PBST containing 5% BSA (w/v) at 37° for 1 hr. After removal of the blocking solution, 50 μl of human serum (from normal controls, known lepromatous and tuberculoid leprosy patients, and suspects) diluted 1 : 300 in PBST containing 25% (v/v) normal goat serum (GIBCO Laboratories, Grand Island, NY) (goat serum-PBST) is added to wells, and these are incubated for 1 hr at 37°. The wells are washed 4 times with PBST, followed by incubation with 50 μl of horseradish peroxidase-conjugated, goat anti-human IgM (Cappel-Organon Teknika Corp., Durham, NC), diluted 1 : 1000 in goat serum-PBST, at 37° for 1 hr. After final washing of wells 5 times with PBST, 50 μl of the substrate solution composed of *o*-phenylenediamine (0.4 mg/ml) and 30% (v/v) H_2O_2 (0.4 μl/ml) in citrate–phosphate buffer, pH 5.0, is added, and the plates are incubated for 30 min at room temperature. Reactions are stopped by adding 50 μl of 2.5 *N* H_2SO_4, and the absorbance is read at 490 nm. On account of batch-to-batch variation in the degree of hapten substitution on NGCs, the working concentration of individual antigens is adjusted by block titration against a positive control serum from a lepromatous leprosy patient, to obtain an absorbance value (A_{490}) of 1.200.

In an initial study, the seroreactivity of those NGCs prepared by reductive amination, and containing the intact 3,6-di-*O*-methyl-α-D-glucopyranoside, has been compared to the native PGL-I against serum samples from leprosy patients representing the entire disease spectrum and healthy controls.[23] In general, the seroreactivity of PGL-I and the NGC is well correlated [correlation coefficient (r) of 0.84], and the agreement rate in determining seropositivity and seronegativity in leprosy patients and controls is over 90%. However, there are some serum samples which give much higher reactivity against the native PGL-I, suggesting that a minority of anti-PGL-I antibodies recognize the intact penultimate or reducing-end sugar of the full trisaccharide.

The results of the application of ND-O-BSA[16] to the detection of anti-PGL-I IgM in sera from select groups of patients showing the different manifestations of leprosy are summarized in Table I. The results paralleled those obtained earlier with PGL-I and demonstrated a very high positivity for patients with active multibacillary (lepromatous) leprosy with or without various reactional states (erythema nodosum leprosum, Lucio reac-

[23] S.-N. Cho, T. Fujiwara, S. W. Hunter, T. H. Rea, R. H. Gelber, and P. J. Brennan, *J. Infect. Dis.* **150,** 311 (1984).

TABLE I
ESTIMATION OF ANTIGLYCOLIPID IMMUNOGLOBULIN M ACTIVITY IN SERA FROM GROUPS OF LEPROSY PATIENTS USING SYNTHETIC NEOGLYCOCONJUGATE

Patient classification[b]	Number of sera assayed	ND-O-BSA[a] Number positive (%)	A_{240} (mean ± SD)
Lepromatous leprosy (active)	14	14 (100)	1.433 ± 0.650
Lepromatous leprosy (inactive)	18	11 (61)	0.277 ± 0.258
Erythema nodosum leprosum (ENL)	16	15 (94)	0.867 ± 0.719
Lucio reaction	6	6 (100)	1.307 ± 0.666
Reversal reaction	9	8 (89)	1.067 ± 0.788
Tuberculoid leprosy (active)	17	5 (29)	0.165 ± 0.240
Tuberculoid leprosy (inactive)	8	2 (25)	0.075 ± 0.079
Borderline lepromatous	6	5 (83)	0.526 ± 0.366

[a] *O*-(3,6-Di-*O*-methyl-β-D-glucopyranosyl)-(1 → 4)-*O*-(2,3-di-*O*-methyl-α-L-rhamnopyranosyl)-(1 → 9)-oxynonanoyl-BSA.

[b] Patient classification system has been described by D. S. Ridley, and W. H. Jopling, *Int. J. Lepr.* **34,** 255 (1966).

tion, reversal reaction). However, as with PGL-I, patients with paucibacillary (tuberculoid) leprosy showed a poor antibody response.

The results of a comparison between some of the various available NGCs, among each other and with PGL-I, are summarized in Table II. There was excellent correlation in seroreactivity between ND-O-BSA and NT-O-BSA, between ND-O-BSA and NT-P-BSA, and between NT-O-

TABLE II
PAIRWISE COMPARISON OF PHENOLIC GLYCOLIPID I AND NEOANTIGENS IN DETECTION OF ANTIGLYCOLIPID IMMUNOGLOBULIN M IN SERA FROM LEPROSY PATIENTS[a]

Antigens compared	Correlation coefficient (*r*)	Agreement[b] (%)
PGL-I versus ND-O-BSA	0.884	91.5
PGL-I versus NT-O-BSA	0.886	91.5
PGL-I versus NT-P-BSA	0.889	90.5
ND-O-BSA versus NT-O-BSA	0.988	100.0
ND-O-BSA versus NT-P-BSA	0.984	99.0
NT-O-BSA versus NT-P-BSA	0.990	99.0

[a] Total number of sera tested was 199.

[b] Calculated as [(number of sera positive to both antigens + number of sera negative to both antigens)/(total number of sera)] × 100.

BSA and NT-P-BSA, with correlation coefficients of over 0.98. The agreement rates between the NGCs in determining seropositivity and seronegativity from leprosy patients and controls were also very high, reaching almost 100%. When the absorbance values of the individual sera against PGL-I, ND-O-BSA, NT-O-BSA, and NT-P-BSA were plotted in various combinations, the degree of correspondence between the antigens was nearly perfect, demonstrating that ND-O-BSA and ND-P-BSA were as effective as NT-O-BSA and NT-P-BSA in reacting with anti-PGL-I antibodies. Thus, the inner 3-*O*-methyl-α-L-rhamnopyranosyl unit does not appear to contribute to human leprosy immunogenicity or antigenicity. Likewise, the presence of a phenyl substituent in NT-P-BSA, further emulating the natural glycolipid, neither resulted in increased reactivity with leprosy sera nor conferred cross-reactivity against control sera.

Role of Leprosy-Specific Neoglycoconjugate in Leprosy Elimination Programs

Leprosy, unlike tuberculosis, is on the wane, although it is still a substantial problem. Prevalence has been steadily revised downward to a present-day worldwide figure of about 3 million registered cases and 5.5 million estimated cases, owing in part to an effective multiple drug regimen.[24] The cases are concentrated mostly in India, Brazil, Myanmar/Burma, Indonesia, and Nigeria. The World Health Assembly (of the WHO) has dedicated itself to the reduction of leprosy to a prevalence rate of less than 1 per 10,000 population by the year 2000. There is a strong conviction that the goal can be achieved through aggressive case finding, reliance on conventional clinical diagnosis, and supervised administration of the WHO-recommended multiple drug regimen.

Most investigators involved in leprosy control believe that those goals can be accomplished without recourse to new methods for the detection of asymptomatic leprosy, especially serology. Anti-PGL-I antibodies are found in virtually all multibacillary lepromatous leprosy patients. However, the number of false-positive response is about 4% in nonendemic populations, and, also, the responses of paucibacillary tuberculoid leprosy patients and contacts is disappointingly low,[7] thus limiting the use of this type of serology for epidemiological purposes or for identification of patients with subclinical leprosy. Nevertheless, the ready availability of NGCs has ensured that this form of serology has a place in leprosy management programs, especially in monitoring the efficacy of chemotherapy. More importantly, this simple technology, readily implemented in endemic

[24] S. K. Noordeen, L. Lopez-Brazo, and T. K. Sundaresan, *Bull. WHO* **70,** 7 (1992).

regions, has stimulated interest in leprosy research among laboratory scientists, physicians, and other health care workers. This interest within the research community is proving to be an important factor in the ongoing successful elimination program.

Acknowledgments

Much of the research reported by the authors was funded by the National Institute of Allergy and Infectious Diseases, National Institutes of Health Contract NO1-AI-05074, and fellowships from the Heiser Program for Research on Leprosy.

[4] Detection and Quantification of Carbohydrate-Binding Sites on Cell Surfaces and in Tissue Sections by Neoglycoproteins

By HANS-JOACHIM GABIUS, SABINE ANDRÉ, ANDRÉ DANGUY, KLAUS KAYSER, and SIGRUN GABIUS

Introduction

Plant and invertebrate lectins are popular tools for the isolation, characterization, and localization of glycoconjugates that represent a lectin-reactive determinant. In the tissue from which a lectin is isolated, it is possible to detect protein–carbohydrate interactions involving the lectin. In this approach, the histochemically employed lectin searches explicitly for potential endogenous ligands of physiological significance.[1]

Conversely, chemically tailored carbohydrate structures can be used to trace cell and tissue sites with corresponding specificity. This approach establishes reverse lectin cyto- and histochemistry in relation to the common morphological research with lectins as markers.[2,3] Availability of carrier-immobilized carbohydrates with designed structures is the essential requirement to embark on this approach of lectin detection.[4–7] Fluores-

[1] H.-J. Gabius and A. Bardosi, *Prog. Histochem. Cytochem.* **22**(3), 1 (1991).

[2] H.-J. Gabius and S. Gabius (eds.), "Lectins and Cancer." Springer-Verlag, Heidelberg, 1991.

[3] H.-J. Gabius, S. Gabius, T. V. Zemlyanukhina, N. V. Bovin, U. Brinck, A. Danguy, S. S. Joshi, K. Kayser, J. Schottelius, F. Sinowatz, L. F. Tietze, F. Vidal-Vanaclocha, and J.-P. Zanetta, *Histol. Histopathol.* **8,** 369 (1993).

[4] C. P. Stowell and Y. C. Lee, *Adv. Carbohydr. Chem. Biochem.* **37,** 225 (1980).

[5] Y. C. Lee and R. T. Lee, this series, Vol. 179, p. 253.

[6] M.-C. Shao, L.-M. Chen, and F. Wold, this series, Vol. 184, p. 653.

Copyright © 1994 by Academic Press, Inc.
All rights of reproduction in any form reserved.

cent and radioactive labeling of the carrier or adsorption to gold granules are practical procedures to gain access to tools for glycohistochemical application.[8-11] Enzyme–neoglycoprotein conjugates and neoglycoenzymes, exhibiting enzymatic activity as natural label, or biotinylated neoglycoproteins constitute highly sensitive, nonradioactive probes for cell and tissue receptors with specificity for the exploited ligand structure.[1-3,12-14] Their application is increasingly gaining attention in cell biology and histochemistry.

Preparation of Neoglycoprotein–Enzyme Conjugates

Availability of sensitive substrates and easy access are key factors in choosing the enzyme for conjugation. The selection of the coupling agent depends on the presence of suitable functional groups on the enzyme and the neoglycoprotein. When the chosen enzyme itself or the carrier for the carbohydrate ligands are glycoproteins, the "glyco" part must be removed either by extensive enzymatic treatment with glycosidases or by chemical oxidation with periodate.

Oxidation of Horseradish Peroxidase. Horseradish peroxidase (70 mg), a mannose-rich glycoprotein, is dissolved in 12 ml of 5 m*M* sodium acetate buffer, pH 4.5, and a freshly prepared 0.1 *M* sodium periodate solution in acetate buffer (4.5 ml) is added. Following 2 hr at room temperature with gentle stirring, 2–3 ml of a 10% glycerol solution in acetate buffer, and, after a further 30 min, 1.7 ml of a freshly prepared sodium cyanoborohydride solution (12 mg/ml) are added. After incubation for 30 min at room temperature the periodate-treated protein is separated from the reagents by gel filtration, dialyzed against water, and lyophilized.

Conjugation of Peroxidase with Neoglycoprotein. The neoglycoprotein used here is bovine serum albumin modified with a specific carbohydrate group. First, the periodate-oxidized peroxidase and the albumin are each treated with *N*-succinimidyl-3-(2-pyridyldithio)propionate (SPDP). A solu-

---

[7] Y. C. Lee, *FASEB J.* **6,** 3193 (1992).

[8] C. Kieda, A. C. Roche, F. Delmotte, and M. Monsigny, *FEBS Lett.* **99,** 329 (1979).

[9] S. Kojima and H.-J. Gabius, *J. Cancer Res. Clin. Oncol.* **114,** 468 (1988).

[10] V. Kolb-Bachofen, this series, Vol. 179, p. 111.

[11] C. Rushfeldt and B. Smedsrod, *Cancer Res.* **53,** 658 (1993).

[12] H.-J. Gabius, R. Engelhardt, K. P. Hellmann, T. Hellmann, and A. Ochsenfahrt, *Anal. Biochem.* **165,** 349 (1987).

[13] S. Gabius, K. P. Hellmann, T. Hellmann, U. Brinck, and H.-J. Gabius, *Anal. Biochem.* **182,** 447 (1989).

[14] H.-J. Gabius and S. Gabius (eds.), "Lectins and Glycobiology." Springer-Verlag, Heidelberg, 1993.

tion of SPDP (840 μg in 1.2 ml ethanol or 630 μg in 0.9 ml ethanol) is added to a solution of peroxidase (9 mg) or albumin (15 mg), respectively, dissolved in 1.8 ml of 100 m*M* sodium phosphate buffer, pH 7.5, containing 0.9% NaCl. After 30 min at room temperature, excess reagents are removed from proteins by gel filtration. The modified enzyme is then treated with 25 m*M* dithiothreitol at pH 4.5 to reduce the newly incorporated disulfide bridges. Excess dithiothreitol and released pyridine-2-thione are again removed by gel filtration, and the protein-containing fraction is concentrated. The activated enzyme and the 2-pyridyl disulfide-containing neoglycoprotein are then mixed in a final volume of 3.5 ml and incubated for 22 hr at room temperature. Fractionation of the mixture on an Ultrogel AcA-44 column (0.9 × 100 cm), equilibrated with 100 m*M* sodium phosphate buffer (pH 7.5) containing 0.9% NaCl, at a flow rate of 6.5 ml/hr (fraction size: 1.5 ml) and gel electrophoretic analysis of the fractions in the presence and absence of reducing agent indicate the location of the monoconjugate, as shown in Fig. 1.

Preparation of Neoglycoenzymes

Several enzymes like horseradish peroxidase are glycoproteins with known oligosaccharides. The rather limited panel of natural glycoenzymes

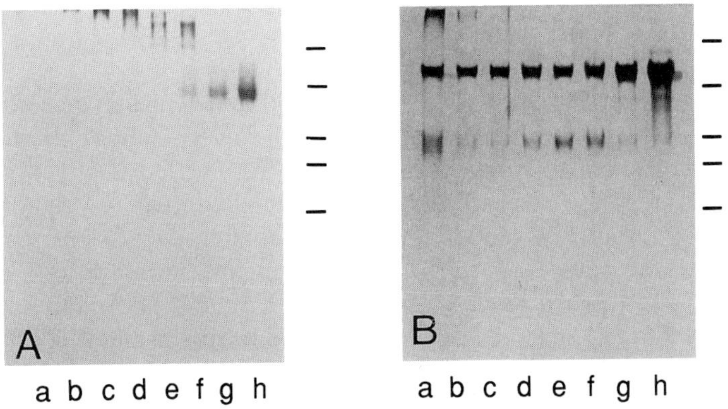

FIG. 1. Sodium dodecyl sulfate–polyacrylamide gel electrophoresis in the absence (A) and presence (B) of reducing agent of fractions from gel filtration that separates peroxidase, lactosylated bovine serum albumin, and their conjugates, especially the monoconjugate in fractions e and f, after SPDP-mediated attachment. Standards for molecular mass designation, given by bars, are phosphorylase *b* (97 kDa), bovine serum albumin (66 kDa), egg albumin (44 kDa), glyceraldehyde-3-phosphate dehydrogenase (36 kDa), and carbonate dehydratase (29 kDa).

can conveniently be enlarged by chemical glycosylation of an enzyme as an alternative to preparation of neoglycoprotein–enzyme conjugates. To generate an appropriate density of carbohydrate ligands on the enzyme surface, any coupling procedure is feasible, unless the enzymatic activity is harmed by the chemical procedure. Mild, activity-preserving conditions are maintained during carbodiimide-mediated coupling of *p*-aminophenyl glycosides. The enzyme, for example, 8 mg *Escherichia coli* β-galactosidase, is dissolved in 1 ml of 20 m*M* phosphate-buffered saline (pH 7.4) and reacted with a mixture of 16 μmol of *p*-aminophenyl glycoside and 32 μmol of 1-ethyl-3-(3-dimethylaminopropyl)carbodiimide for 16 hr at 4° prior to dialysis against the buffer and isolation of the glycosylated enzyme by gel filtration. Any enzyme oligomers formed are discarded. The active fractions are pooled, dialyzed against buffer containing 60% glycerol, and stored at −20°.

The sugar content of the glycosylated enzyme is assayed by a resorcinol–sulfuric acid micromethod.[15] Twenty microliters of the neoglycoenzyme solution, containing 1–100 nmol of neutral sugar, is placed into a microtiter plate well. To this are added 20 μl of resorcinol solution (6 mg/ml), 100 μl of 75% sulfuric acid, and finally 50 μl of pristane, and the plate is cautiously shaken and heated to 90° for 30 min. After keeping the plate for 30 min in the dark, the absorbance is read at 430 or 480 nm. Under these conditions, each subunit of the tetrameric *E. coli* β-galactosidase will incorporate 25 ± 5 sugar moieties.

Cell Surface Binding of Neoglycoproteins

Neoglycoprotein–enzyme conjugates and neoglycoenzymes are used to detect and to quantify cell surface sites with corresponding carbohydrate specificity in a two-step procedure, whereas biotinylated neoglycoproteins require a three-step method and a commercial reagent. To remove any potentially inhibitory serum or medium components such as glycoproteins or glucose, the cells ($1–20 \times 10^4$ cells per assay), grown in suspension, are cautiously washed by centrifugation with 10 m*M* phosphate-buffered saline (pH 7.2) or Hanks' balanced salt solution with 20 m*M* *N*-2-hydroxyethylpiperazine-*N'*-2-ethanesulfonic acid (HEPES, pH 7.5) that contains 0.1–0.5% (w/v) carbohydrate-free bovine serum albumin.

Cells are then incubated with 400 μl of a buffer solution that contains a neoglycoconjugate for a period of time, usually 5–240 min, to achieve plateau binding at 4° with constant shaking. These parameters are determined in preliminary experiments. Adhesive cells, grown in 24-well culture

[15] M. Monsigny, C. Petit, and A. C. Roche, *Anal. Biochem.* **175,** 525 (1988).

plates for at least 24 hr after trypsinization, are similarly incubated with 200 μl of probe-containing buffer. To remove the unbound probe, the cells are carefully and rapidly washed three times to keep the effect on the established equilibrium as low as possible (800 μl for cells in suspension, 300 μl for adhesive cells per step). When β-galactosidase is used as the enzyme part, 200 μl of freshly prepared substrate solution (100 m*M* HEPES buffer, pH 7.0, containing 0.5% Triton X-100, 150 m*M* NaCl, 2 m*M* $MgCl_2$, 0.1% NaN_3, 0.1% albumin, and 1.5 m*M* chlorophenolred-β-D-galactopyranoside) is added. The enzymatic reaction is stopped by the addition of 200 μl of 0.2 *M* glycine–NaOH (pH 10.5), and the degree of product formation is assessed spectrophotometrically at 590 nm. The activity of each enzyme batch is calculated graphically, transforming the measured absorbance into femtomoles of enzyme.

When biotinylated markers are used, a further incubation step with a commercially available streptavidin–enzyme conjugate needs to be incorporated into the procedure (e.g., with 5 μg/ml of the conjugate for 90 min at 4°). The unbound conjugate is thoroughly removed by wash steps. Solid-phase assays in microtiter plate wells with the biotinylated neoglycoproteins yield the equivalent of activity graphs for neoglycoenzymes. These are required for calculation of the bound quantity of probe on the basis of the measured optical density.

Each series of measurements with cells is accompanied by assays in the presence of an excess of carbohydrate inhibitor to assess the extent of noninhibitable binding, referred to as nonspecific (Fig. 2). This assay enables comparative quantification of cell surface expression of sites specific for the chosen carbohydrate structure, for example, structures implicated in the metastatic capacity of tumor cells, stages of differentiation, and oncogene expression and structures associated with parasitic organisms.[3,16–19]

Visualization of Carbohydrate-Binding Sites in Fixed Cells

Owing to the availability of commercial ABC (avidin–biotin complexes) kit reagents, which contain complexes of avidin and biotinylated enzymes with the capacity to bind to biotin of the applied probe that has

[16] S. Gabius, V. Schirrmacher, H. Franz, S. S. Joshi, and H.-J. Gabius, *Int. J. Cancer* **46,** 500 (1990).

[17] S. Gabius, N. Yamazaki, W. Hanewacker, and H.-J. Gabius, *Anticancer Res.* **10,** 1005 (1990).

[18] H.-J. Gabius, S. Gabius, M. Fritsche, and G. Brandner, *Naturwissenschaften* **78,** 230 (1991).

[19] J. Schottelius and H.-J. Gabius, *Parasitol. Res.* **78,** 529 (1992).

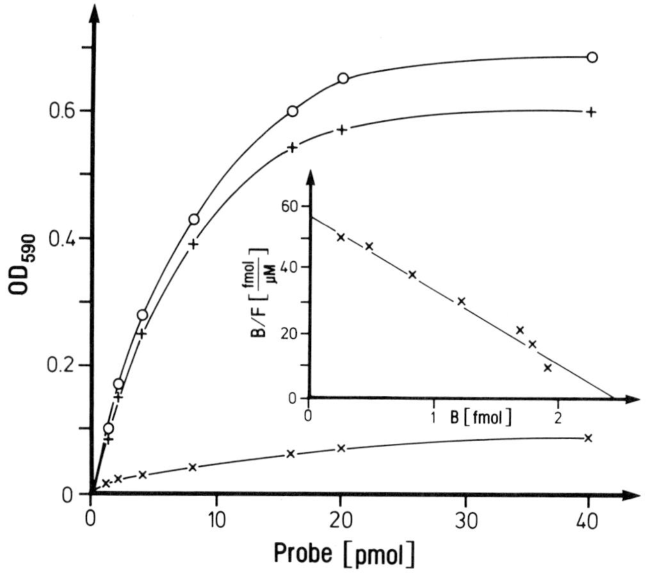

FIG. 2. Determination of specific binding (+) of melibiosylated *E. coli* β-galactosidase to methylcholanthrene-induced ER-15 fibrosarcoma cells with high capacity for liver colonization (8 × 10^4 cells per assay; original cell batch kindly provided by Prof. Dr. G. Edel, St. Franziskus-Hospital, Münster) at 4° as a function of ligand concentration. Total binding (○) was reduced by the extent of nonspecific binding (×). Scatchard plot analysis (inset) is consistent with the presence of a single, noninteracting class of binding sites (K_D = 43 n*M*; B_{max} = 1.8 × 10^4 enzymes/cell).

specifically bound in the section (e.g., Camon, Wiesbaden, Germany), glycocytological staining of cells by biotinylated markers can be amplified relative to application of neoglycoenzymes. These kits therefore provide a high level of sensitivity to visualize cellular binding sites. Carefully washed cells, either grown on coverslips or prepared as cytospins, are dried, fixed in a mild fixative such as 80% ice-cold acetone for 5–10 min, and washed with phosphate-buffered saline containing 0.1% carbohydrate-free bovine serum albumin. Protein-binding sites are saturated by incubation with buffer containing 0.1–2% albumin for 15–60 min at room temperature, and the solution is carefully blotted off with filter paper. When the presence of Ca^{2+} is required for binding, 20 m*M* HEPES buffer is used in subsequent steps. The cells are incubated with 5–200 μg/ml biotinylated neoglycoprotein, dissolved in the appropriate buffer containing 0.1–0.5% carbohydrate-free albumin, for 1–12 hr at 4°, room temperature, or 37° in an incubator. In parallel, control slides are processed with the same amount of nonglycosylated but biotinylated carrier protein and with an excess of carbohydrate inhibitor to exclude binding of the probes by protein–protein interaction. Following three washes with buffer, endogenous peroxidase activity is abolished by treatment with 0.23% periodic acid in buffer for 45 sec. Successive incubation steps with ABC kit reagents and substrate solution will lead to the generation of staining.

Visualization of Carbohydrate-Binding Sites in Tissue Sections

The following steps lead to detection of accessible binding sites for the carbohydrate part of neoglycoproteins in tissue sections. Either frozen or fixed tissue specimen can be used. Fixation is preferably performed with Bouin's mixture (750 ml saturated aqueous picric acid, 250 ml of 36–40% (w/v) formalin, and 50 ml glacial acetic acid). Subsequently, the tissue is dehydrated in graded ethanol solutions and then embedded in Paraplast at a temperature not higher than 56°. Sections for electron microscopical analysis can be fixed in 0.2 *M* phosphate-buffered saline (pH 7.2), containing 0.2% glutaraldehyde and 2% paraformaldehyde, or in 0.1 *M* cacodylate buffer (pH 7.2) with 5% glutaraldehyde, employing a post fixation step with 1% osmium tetroxide in this buffer.[20–21]

The sections are cut at an appropriate thickness, transferred to glass slides, deparaffinated in xylene or toluene, and rehydrated in graded alcohol–water mixtures with decreasing alcohol content. The endogenous peroxidase activity is blocked by incubation with methanolic hydrogen peroxide (0.3–2%) for 30 min at room temperature. After thorough washes with buffer, protein-binding sites are saturated by incubation with 0.1–2% bovine serum albumin in buffer for 15–60 min at room temperature. Incubation steps with the labeled neoglycoprotein, ABC kit reagents, and an adequate substrate solution as well as the necessary control reactions are carried out, as described in the previous section.

When the staining procedure is completed, the intensity of the developed signal can be evaluated either by independent observers or by a color video camera, which is connected to the microscope. A frame grabber converts the video signal to a binary image matrix. After interactively defining the boundaries for intensity classification, the different compartments are analyzed according to the defined scheme. Any cell type or histological area can be pinpointed, and the relative intensity in graded gray to black values can be displayed, as shown in Fig. 3. In addition to quantification of cell classes with different staining intensities and evaluation of their corresponding features (size or DNA characteristics such as cell cycle phases) the spatial configuration can be analyzed by application

[20] A. Bardosi, T. Dimitri, and H.-J. Gabius, *J. Submicrosc. Cytol. Pathol.* **21,** 229 (1989).
[21] S. Kuchler, J.-P. Zanetta, G. Vincendon, and H.-J.Gabius, *Eur. J. Cell Biol.* **59,** 373 (1992).

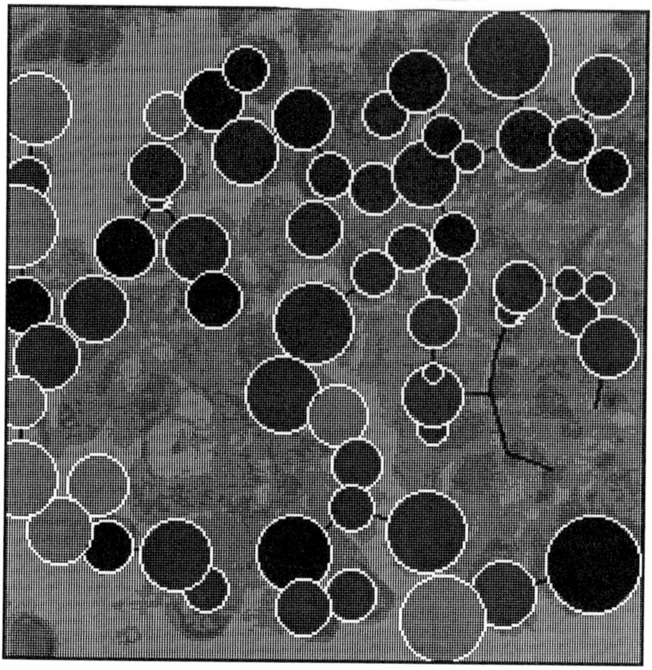

FIG. 3. Semiquantitative image analysis of a section of a human mesothelioma, processed by successive incubation steps with biotinylated neoglycoprotein, containing the covalently attached lysoganglioside part of G_{M1}, ABC kit reagents, and the chromogenic substrates diaminobenzidine/H_2O_2.

of graph theory technique or syntactic structure analysis. This technique allows the assessment of cell clusters with different staining intensities.[22,23] It should be noted that the staining intensity not only is a function of the receptor density, but also can be influenced by tissue processing, thickness and age of section, as well as the concentration and type of kit reagents and chromogenic substrate.

Comments

The procedures described here will define the presence of carbohydrate-binding sites that are neither occupied by endogenous high-affinity ligands nor impaired by processing of the specimen. Several points warrant

[22] K. Kayser, *NATO ASI Ser.* **45,** 115 (1988).

[23] K. Kayser, K. Sandau, G. Boehm, K. D. Kunze, and J. Paul, *Anal. Quant. Cytol. Histol.* **13,** 329 (1991).

attention to ensure adequate interpretation of the data. It is imperative that the carrier does not display inherent receptor properties. When albumin is used, the presence of albumin-binding proteins and of a macrophage migration inhibitory factor-binding subfraction in human but not in bovine albumin should not be neglected.[24–26] Preincubation steps are therefore indispensable. Because any chemical modification of amino or carboxyl groups for carbohydrate coupling alters the isoelectric point of the carrier protein, the contribution of ionic interactions to probe binding must be assessed. For that reason the biotinylated, carbohydrate-free carrier and an array of probes, modified to the same extent with active as well as inert sugar structures with the same type of linkage region, are required to exclude the possibility that properties of the carrier or the linker can account for the observed staining. Moreover, the omission of the probe during otherwise identical processing has to ensure as a further control that no component of the kit reagents is responsible for signal generation. Endogenous biotin or mannose-binding sites that react with, for example, avidin or horseradish peroxidase can cause false-positive results.

Competitive inhibition with the respective ligand structure is necessary to ascertain unequivocally the implied involvement of specific protein–carbohydrate interaction. Enhancement of the affinity of carbohydrate derivatives with hydrophobic groups such as *p*-aminophenyl glycosides, caused by the presence of hydrophobic sites on the lectin in the vicinity of the carbohydrate-binding pocket, may explain the necessity for rather high concentrations of inhibitory sugars. Similarly, the multivalent ligand capacity of the applied probe should be kept in mind. Neoglycoconjugates will bind to more than one cell surface receptor. Consequently, inhibition calls for high-affinity structures, and the measured number of bound enzymes per cell will not necessarily reflect the total number of binding sites actually present. When the degree of sugar incorporation or the spatial extension of the individual marker is lowered, for example, by substituting the tetrameric *E. coli* β-galactosidase with the monomeric *Aspergillus oryzae* enzyme, the calculated dissociation constant of cell binding will increase. The mentioned change of the enzyme has been shown to increase concomitantly the computed number of detectable binding sites.[16,17]

As soon as specific binding with the employed carbohydrate structure is ascertained, its structural complexity can be custom-made. In addition

[24] J. Moroianu, A. Hillebrand, and M. Simionescu, *Eur. J. Cell Biol.* **53,** 20 (1990).

[25] J. E. Schnitzer, *Am. J. Physiol.* **262,** H246 (1992).

[26] F.-Y. Zeng, H. Kratzin, and H.-J. Gabius, *Biol. Chem. Hoppe-Seyler,* in press.

to being used in further binding studies, these probes can serve as valuable tools to inhibit cellular functions or to mediate adhesive contacts in the quest to define cell adhesion molecules and to delineate functions of lectins. In more general terms, similar studies are feasible with any type of ligand by application of neoligandoconjugates.

[5] Chemical Glycosylation of Recombinant Interleukin 2

By SUBRAMANIAM SABESAN and T. JUHANI LINNA

Introduction

Interleukin 2 (IL-2) is a glycoprotein, with an estimated molecular weight of about 16,500, and contains carbohydrates O-linked to the threonine, the third amino acid from the amino-terminal end.[1] It is secreted by lymphocytes and belongs to the lymphokine class of immune modulating substances.[2] Interleukin 2 has been shown to modulate a number of immunological activities of lymphoid cells including T-cell activation, activation of natural killer (NK) cells, activation of B cells, and generation of lymphokine-activated killer (LAK) cells (cells that kill fresh tumor cells but not normal cells).[3–5] Treatment of cancer patients by administration of bacterial recombinant IL-2 (rIL-2) and autologous LAK cells has demonstrated the potential use of rIL-2 as an immunotherapeutic agent.[6] In many patients, however, the administration of therapy has been limited by the toxicity of rIL-2.[7,8] Greater antitumor effects might occur if larger doses of rIL-2 and cells could be administered. Recombinant IL-2

[1] H. S. Conrad, R. Geyer, J. Hoppe, L. Grotjahn, A. Plessing, and H. Mohr, *Eur. J. Biochem.* **153,** 255 (1985).
[2] K. A. Smith, *Science* **240,** 1169 (1988).
[3] E. A. Grim, A. Mazumder, H. Z. Zhang, and S. A. Rosenberg, *J. Exp. Med.* **155,** 1823 (1982).
[4] A. Mazumder and S. A. Rosenberg, *J. Exp. Med.* **159,** 495 (1984).
[5] E. A. Grim and S. A. Resenberg, *Lymphokines* **9,** 279 (1984).
[6] S. A. Rosenberg, M. T. Lotze, M. D. Linda, M. Muul, S. Leitman, A. E. Chang, S. E. Ettinghausen, Y. L. Matory, J. M. Skibber, E. Shiloni, J. T. Vetto, C. A. Seipp, C. Simpson, and C. M. Reichert, *N. Engl. J. Med.* **313,** 1485 (1985).
[7] J. M. Mier, F. R. Aronson, R. P. Numerof, G. Vachino, and M. B. Atkins, *Pathol. Immunopathol. Res.* **7,** 459 (1988).
[8] T. Hamblin, *Chem. Ind.,* 663 (1993).

Copyright © 1994 by Academic Press, Inc.
All rights of reproduction in any form reserved.

has been proposed to cause some of the side effects by possibly stimulating helper T cells to secrete other lymphokines that may be toxic.[6,9]

The gene responsible for the synthesis of human IL-2 has been cloned and sequenced.[10,11] The large quantities of IL-2 that are required for various clinical trials are produced as a result of cloning the gene for IL-2 and expressing it in *Escherichia coli*.[12,13] Even though the bacterially produced recombinant material lacks carbohydrates that are present in the natural material, it is functionally active. However, some of the physical properties of the bacterially produced material (rIL-2) are different from those of the native IL-2. The rIL-2 is produced as insoluble refractile bodies within the bacteria, and therefore denaturants are required during its purification. In the absence of a detergent, purified IL-2 has very limited solubility at neutral pH. Owing to rapid clearance, it also has a short circulatory half-life when administered to animals.[14,15] To overcome the limited solubility at neutral pH and the short circulatory half-life of rIL-2 purified from *E. coli*, Katre *et al.*[16] have modified the rIL-2 by conjugating it with monomethoxypolyethylene glycol. The modified rIL-2 had enhanced solubility, decreased plasma clearance, and increased antitumor potency in a particular animal tumor model.

In this chapter we describe the chemical glycosylation of the amino group of one or more of the 11 lysines in rIL-2 with synthetic oligosaccharides. Although a major objective was to render rIL-2 soluble by the addition of carbohydrates, it was found that several glycosylated rIL-2 preparations lost most of their T lymphocyte activating ability (CTLL activation), while retaining most or all of their ability to enhance NK cell and LAK cell activities. The glycosylated rIL-2 preparations were more readily soluble in water, as anticipated. One biologically active glycosylated IL-2 preparation was found to be thermally more stable than

[9] S. A. Rosenberg, M. T. Lotze, L. M. Muul, A. E. Chang, F. P. Avis, S. Leitman, W. M. Linehan, C. N. Robertson, E. E. Lee, J. T. Rubin, C. A. Seipp, C. G. Simpson, and D. E. White, *N. Engl. J. Med.* **316,** 889 (1987).

[10] T. Taniguchi, H. Matsui, T. Fujita, C. Takaoka, N. Kashima, R. Yoshomoto, and J. Hamaro, *Nature* (*London*) **302,** 305 (1983).

[11] R. Devos, G. Plaetinck, H. Cheroutre, G. Simon, W. Degrave, J. Tavernier, E. Remaut, and W. Fiers, *Nucleic Acids Res.* **11,** 4307 (1983).

[12] G. Ju, L. Collins, K. L. Kaffka, W. H. Tsien, R. Chizzonite, R. Crowl, R. Bhatt, and P. L. Kilian, *J. Biol. Chem.* **262,** 5723 (1987).

[13] D. F. Mark, M. V. Doyle, and K. Koths, *in* "Recombinant Lymphokines and Their Receptors" (S. Gillis, ed.), p. 1. Dekker, New York, 1987.

[14] J. H. Donohue and S. A. Rosenberg, *J. Immunol.* **130,** 2203 (1983).

[15] M. T. Lotze, L. W. Frana, S. O. Sharrow, R. J. Robb, and S. A. Rosenberg, *J. Immunol.* **134,** 157 (1985).

[16] N. V. Katre, M. J. Knauf, and W. J. Laird, *Proc. Natl. Acad. Sci. U.S.A.* **84,** 1487 (1987).

rIL-2 when heated to 90°. Three carbohydrates, namely, βDGal-OR **(1)**, βDGal1,3βDGlcNAc-OR **(2)**, and βDGalNAc1,4βDGal1,4βDGlc-OR (asialo-G_{M2} hapten, **3**), were chosen as representatives of mono- to trisaccharides for conjugation to rIL-2 and for evaluation of the influence of the size of the carbohydrates on the physical and biological properties of glycosylated rIL-2. Of these, the asialo-G_{M2}-conjugated rIL-2 was used for evaluation of the thermal stability of chemically glycosylated rIL-2.

Procedures

General Methods

The carbohydrates βDGal-OR **(1a)**, βDGal1,3βDGlcNAc-OR **(2a)**, and βDGalNAc1,4βDGal1,4βDGlc-OR (asialo-G_{M2} hapten, **3a**) [**1a–3a**, R = $(CH_2)_5COOCH_3$] are prepared according to published methods,[17,18] except that 5-methoxycarbonylpentanol[19] instead of 8-methoxycarbonyloctanol is used. These are converted to the acyl hydrazides [**1b**, **2b**, and **3b**, R = $(CH_2)_5CONHNH_2$] by refluxing with hydrazine in methanol as described by Lemieux and co-workers.[20] Recombinant human IL-2 with the natural amino acid sequence manufactured by Du Pont (Wilmington, DE) is used throughout the studies.[21] Following conjugation to carbohydrates, the products are purified by high-performance liquid chromatography (HPLC). Purification of the lysine-glycosylated rIL-2 is done using a Vydac C_4 column (The Nest Group, Southboro, MA; 4.6 mm × 15 cm, 5 μm particle size) using a Waters HPLC system (Waters, Milford, MA) equipped with a 600E pump, Wisp autoinjector, Rheodyne 7125 manual injector (Altech Associates, Deerfield, IL), and 990 photodiode array detector (Waters). HPLC-grade water, acetonitrile, and trifluoroacetic acid (Pierce Chemical Co., Rockford, IL) are used for elution of the protein. The elution solvents consist of solvent A [water–acetonitrile–trifluoroacetic acid, 85 : 15 : 0.1 (v/v/v)] and solvent B (water–acetonitrile–trifluoroacetic acid, 10 : 90 : 0.1 (v/v/v)]. Estimations of glycosylated rIL-2 concentration are done by injecting a known volume of a standardized solution of rIL-2 and the unknown glycosylated rIL-2 into a reversed-phase column

[17] R. U. Lemieux, D. R. Bundle, and D. A. Baker, *J. Am. Chem. Soc.* **97,** 4076 (1975).

[18] S. Sabesan and R. U. Lemieux, *Can. J. Chem.* **62,** 644 (1984).

[19] S. Sabesan and J. C. Paulson, *J. Am. Chem. Soc.* **108,** 2068 (1986).

[20] R. U. Lemieux, D. A. Baker, and D. R. Bundle, *Can. J. Biochem.* **55,** 507 (1977).

[21] T. J. Linna, M. Moke, and H. W. Chen, *in* "Drugs of Abuse, Immunity and Immunodeficiency" (H. Friedman, S. Specter, and T. W. Klein, eds.), p. 278. Plenum, New York, 1991.

(4.6 mm × 5 cm) and eluting with a linear gradient of solvent A (100% at 0 min) and solvent B (100% at 6 min). The absorptions of the eluted rIL-2 and the glycosylated rIL-2 peaks at 280 nm are compared, and from the integrals of the eluted peaks, the concentration of the glycosylated rIL-2 was calculated.

Conjugation of Carbohydrates to Recombinant Interleukin 2. The conjugation of oligosaccharides to rIL-2 is carried out either in aqueous dimethylformamide (Procedure A) or in water (Procedure B). The activation of the acylhydrazide in **1b**, **2b**, and **3b** [R = $(CH_2)_5CONHNH_2$] to the acyl azide [**1c**, **2c**, and **3c**, R = $(CH_2)_5CON_3$] is done according to the procedure described by Lemieux *et al.*,[20] except that in Procedure B *tert*-butyl nitrite is replaced by sodium nitrite. The activated oligosaccharide solutions are immediately used for coupling to rIL-2.

A solution of rIL-2 in 0.3 *M* mannitol or glucose (5.0 ml, 1.0 mg/ml) is lyophilized. The resulting powder is suspended in 5 ml of a buffer (pH 9.0) containing sodium borate (80 m*M*) and potassium bicarbonate (350 m*M*). The pH of the solution is then raised to 9.0 by the addition of 0.75 *M* potassium hydroxide. The solution of rIL-2 is cooled in an ice bath with stirring, and the solution of activated mono- or oligosaccharide-acylazide [R = $(CH_2)_5CON_3$] is added in drops. The resulting solution is stirred at 4° for 24 h. The glycosylated rIL-2 is recovered by means of high-performance liquid chromatography.

Chromatographic Purification of Lysine-Glycosylated Recombinant Interleukin 2. The following solvent mixtures are used for HPLC purification: solvent A [acetonitrile–water–trifluoroacetic acid, 85 : 15 : 0.1 (v/v/v)] and solvent B [acetonitrile–water–trifluoroacetic acid, 10 : 90 : 0.1 (v/v/v)]. The glycosylation reaction mixture is loaded onto a C_4 reversed-phase column (typically 1.5 to 2 mg of protein) and washed with solvent A to remove all the UV-active materials (normally 2 min). The solvent composition is gradually changed over 5.5 min to contain 45% of solvent B and then maintained at this composition for 2 min, during which all the aggregated protein impurities elute (Fig. 1). Following this, the solvent is changed to 100% B over a period of 2.5 min and maintained at this solvent composition until all the glycosylated rIL-2 elutes as a single peak (Fig. 1, GIL-2). The product is collected in 17 × 100 mm sterile polypropylene tubes, frozen, and lyophilized. Finally, the glycosylated rIL-2 is dissolved in 0.3 *M* glucose solution and stored at −78°.

Determination of Increased Solubility of Glycosylated Recombinant Interleukin 2. The increased solubility of glycosylated IL-2 over that of unglycosylated rIL-2 is determined by visually comparing the ease of dissolution of 1 mg of the glycosylated rIL-2 with that of the unglycosylated rIL-2 in aqueous buffers and in the absence of any detergent. In all exam-

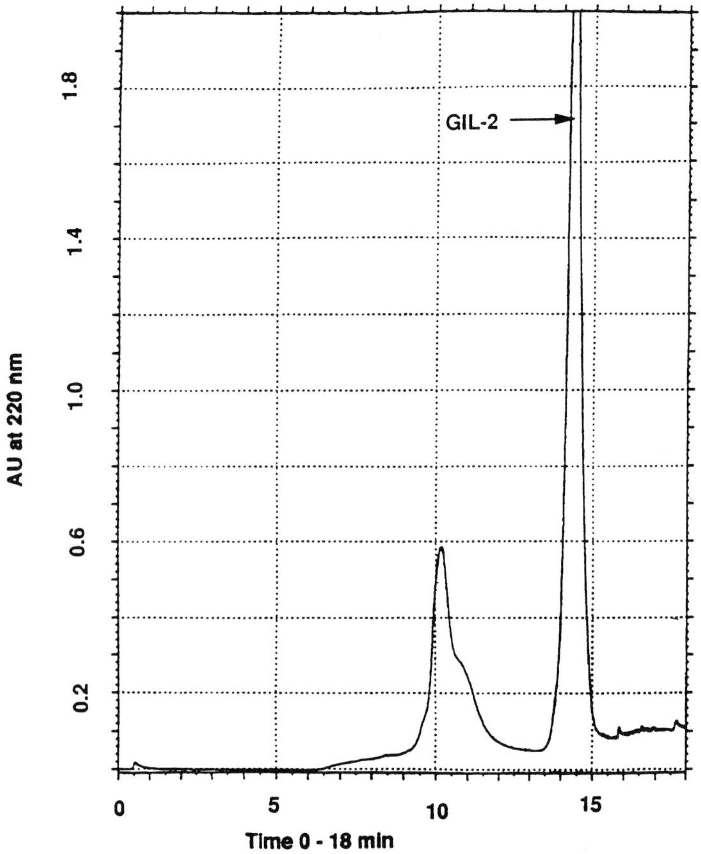

FIG. 1. Purification of rIL-2 glycosylated with βDGal1,3βDGlcNAc on a Vydac C_4 reversed-phase column. The HPLC eluant composition was changed from 100% solvent A [acetonitrile–water–trifluoroacetic acid, 15 : 85 : 0.1 (v/v/v)] at the beginning to solvent B [acetonitrile–water–trifluoroacetic acid, 90 : 10 : 0.1 (v/v/v)] over a period of 12 min.

ples, the fluid in the tubes to which the glycosylated IL-2 is added rapidly becomes clear, and there is no discernible residue in the bottom of the tube on standing. In contrast, the solution containing unglycosylated rIL-2 has a noticeable residue in the bottom of the tube and is not completely clear when agitated.

Determination of Increased Stability of Glycosylated Recombinant Interleukin 2. The increased stability of the glycosylated rIL-2 relative to unglycosylated rIL-2 solutions (1 mg/ml) is established by circular dichroism measurements at room temperature (Fig. 2) as well as at 90° (Fig. 3).

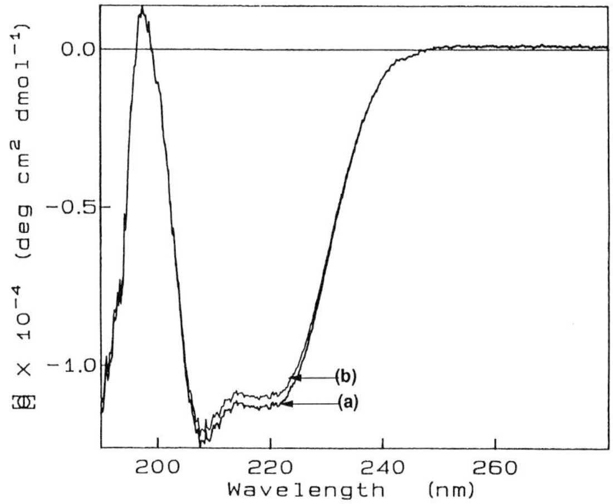

FIG. 2. Comparison of circular dichroism spectra of rIL-2 (a) and glycosylated (asialo-G_{M2}) rIL-2 (b).

Enhancement of Natural Killer Cell Activity. Circulating normal lymphocytes exhibit the ability to kill certain cultured tumor cell lines.[22] This ability is markedly enhanced following prior incubation with IL-2.[23] Nonadherent human peripheral blood lymphocytes are obtained after Ficoll–Hypaque centrifugation and plastic adherence, using procedures described by Boyum.[24] Preparations of such cells are incubated for 3 days in the presence of the different preparations of glycosylated rIL-2 and unglycosylated rIL2. The concentrations are adjusted so that cells are incubated with 5 units of rIL-2 and an equivalent of 10, 5, 2, and 1 units of glycosylated rIL-2. Under standard conditions, the equivalent units of glycosylated rIL-2 are determined by knowing the specific activity of the rIL-2 prior to its glycosylation and assuming no loss of activity during glycosylation. The lymphocytes are used as effector cells in a short-term (3.5 hr) cytotoxicity assay against ^{51}Cr-labeled K562 erythroleukemia target cells [American Type Culture Collection (ATCC), Rockville, MD, CCL-243] essentially as described by Linna *et al.*[25] Effector cells are added to target cells at varying ratios, and the number of target cells required to release 50% of the ^{51}Cr label is determined for each effector

[22] R. B. Herberman and J. R. Ortaldo, *Science* **214,** 24 (1981).
[23] G. Trinchieri, M. M. Kobayashi, S. C. Clark, J. Seehra, L. London, and B. Perussia, *J. Exp. Med.* **160,** 1147 (1984).
[24] A. Boyum, *Scand. J. Clin. Lab. Invest.* **21**(Suppl. 97), 77 (1968).
[25] T. J. Linna, H. D. Engers, J. C. Cerottini, and K. T. Brunner, *J. Immunol.* **120,** 1544 (1978).

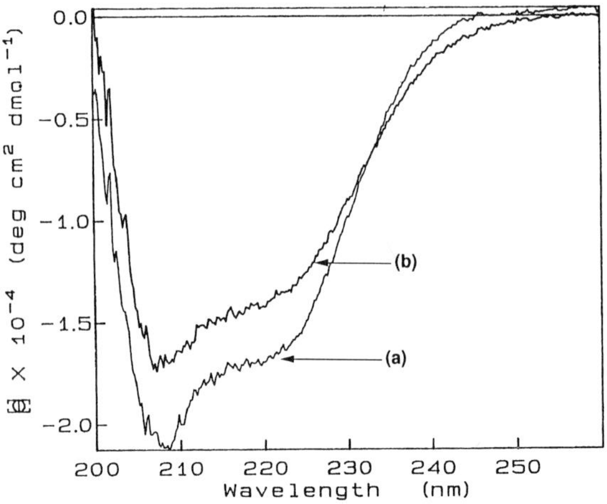

FIG. 3. Circular dichroism spectra of glycosylated (asialo-G_{M2}) rIL-2 at 20° (a) and at 90° (b).

cell population. By comparing the number of effector cells required to bring about 50% lysis, the relative activities of the different effector cell populations are determined. The activity of the glycosylated rIL-2 is reported as a percentage of that of the equivalent amount of unglycosylated rIL-2.

Generation of Lymphokine-Activated Killer Cell Activity in Human Peripheral Blood Lymphocytes. The procedure to generate LAK cell activity from preparations of human peripheral lymphocytes is the same as that used to enhance NK cell activity. The assay for LAK cell activity is done like the assay for NK cell activity except that the target cells are NK-resistant Raji cells. Raji cells are obtained from the ATCC, bearing ATCC accession number CCL-86. Assay results are determined as described for the NK cells, and the activity of the glycosylated rIL-2 is reported as a percentage of that of the equivalent amount of unglycosylated rIL-2.

Assay for Stimulation of Lymphocyte Proliferation by Recombinant Interleukin 2. The stimulation of lymphocyte proliferation by rIL-2 is determined by measuring the IL-2 concentration-dependent incorporation of [^{3}H]thymidine by a cloned IL-2-dependent murine cell line (CTLL) as described by Gillis *et al.*[26] Comparison of the incorporation induced by an unknown sample with that of a laboratory standard yields the relative bioactivity of the sample. The laboratory standard contains 34 BRMP units/ml (units defined by the Biological Response Modifiers Program, National Institutes of Health, Bethesda, MD). Two-fold dilutions of the

[26] S. Gillis, M. M. Ferm, W. Ou, and K. A. Smith, *J. Immunol.* **120**, 2027 (1978).

standard and unknown samples are made and appropriate controls included on a 96-well microtiter plate. Murine IL-2-dependent CTLL cells are added to each well (4000 cells/well). The plate is incubated overnight at 37° (5% CO_2, 95% humidity), 50 μl of [^{3}H]thymidine deoxyriboside with a specific activity 6.7 Ci/mmol at a concentration of 10 μCi/ml is added, and each plate is incubated at 37° for an additional 4 hr. The cells are harvested on glass fiber filter paper using the PHD cell harvester (Cambridge Technology, Inc., Cambridge, MA) and the amount of ^{3}H radioisotope on each filter disk determined.

For standardization, the incorporation of ^{3}H of each sample is compared to the maximum incorporation achieved with the IL-2 standard. The values are calculated using weighted linear regression through least squares analysis, wherein each data point is weighed according to an appropriate measure of accuracy. The fitted lines generated for both the standard and unknown samples are forced to be parallel prior to calculating the rIL-2 activity. The rIL-2 activity is calculated from the distance of the $\log_2$ dilution axis between the parallel lines. Because of assay variability and calculation methodology, some of the sample values may exceed 100%.

Conjugation of βDGal-O-$(CH_2)_5CONHNH_2$ to Recombinant Interleukin 2 (Procedure A)

The βDGal-O-$(CH_2)_5CONHNH_2$ (**1b**, 8.18 mg) in dimethylformamide (DMF; 1.9 ml) is converted to the acyl azide **1c** by reacting with *tert*-butyl nitrite (5.0 μl) and 4 *M* hydrochloric acid in dioxane (40 μl) as described under General Methods followed by quenching with sulfamic acid (1 mg in 20.5 μl of DMF). The resulting solution is reacted with a rIL-2 (12.5 mg) solution as described for Procedure A. The ratio of the monosaccharide to rIL-2 is 0.65 to 1 by weight. The weight of the HPLC-purified glycosylated protein recovered is 4.34 mg.

The rIL-2 glycosylated with βDGal-O-$(CH_2)_5CONHNH_2$ is tested for its solubility and its ability to enhance NK cell activity, generate LAK activity, and stimulate lymphocyte proliferation. The recovered glycosylated rIL-2 is significantly more soluble than native rIL-2. The NK-enhancing activity is 104%, the LAK-generating activity is 111%, and the ability to stimulate the proliferation of lymphocytes is 45%.

Conjugation of βDGal-O-$(CH_2)_5CONHNH_2$ to Recombinant Interleukin 2 (Procedure B)

The βDGal-O-$(CH_2)_5CONHNH_2$ (**1b**, 8.50 mg) is reacted with an aqueous solution (1 ml) of sodium nitrite (8.8 mg) and 4 *M* hydrochloric acid in dioxane (40 μl) and then with a rIL-2 (12.5 mg) solution is described for Procedure B. The ratio of the monosaccharide to rIL-2 is 0.68 to 1 by weight. The amount of glycosylated rIL-2 recovered by HPLC is 4.37 mg. The NK-enhancing activity is 104%, the LAK-generating activity is 99%, and the ability to stimulate the proliferation of lymphocytes is 40%.

Conjugation of βDGal1,3βDGlcNAc-O-$(CH_2)_5CONHNH_2$ to Recombinant Interleukin 2 (Procedure B)

Two sets of reactions are carried out (sample 1 and sample 2) to study the effect of increased activated carbohydrate to protein ratios on the biological activity. βDGal1,3βGlcNAc-O-$(CH_2)_5CONHNH_2$ (**2b**, 39.0 mg) is dissolved in 1 ml of water and cooled in an ice bath. A solution of hydrochloric acid (4 *M*) in dioxane (600 μl) and a solution of aqueous sodium nitrite (450 m*M*, 250 μl) are added, and the solution is stirred at ice temperature for 30 min. This is followed by the addition of a solution of sulfamic acid in water (250 μl of a stock solution prepared by dissolving 14.2 mg of sulfamic acid in 1 ml water), and the solution is stirred in an ice bath for 10 min. The product acylazide **2c** is used subsequently.

Two solutions, each containing 5.8 mg of rIL-2 in sodium borate buffer (see above) are prepared, labeled sample 1 and sample 2, then treated with 700 and 1400 μl of the solution of azide **2c**, respectively, and stirred in a cold room for 24 hr. The resulting glycosylated rIL-2 is purified as described under General Methods. The yields are 2.06 mg sample 1 and 1.64 mg sample 2. For sample 1, the NK-enhancing activity is 81%, the LAK-generating activity is 86%, and the ability to stimulate the proliferation of lymphocytes is 15%. For sample 2, the NK-enhancing activity is 73%, the LAK-generating activity is 67%, and the ability to stimulate the proliferation of lymphocytes is 11%.

Conjugation of βDGal1,3βDGlcNAc-O-$(CH_2)_5CONHNH_2$ to Recombinant Interleukin 2 (Procedure A)

Asialo-G_{M2} hydrazide (**3b**, 8.55 mg) is dissolved in DMF (800 μl) and the solution is cooled to −20°. Hydrochloric acid (4 *M*) in dioxane (13 μl) and *tert*-butyl nitrite (2.4 μl) are added, and the solution is stirred at −20° for 30 min. Sulfamic acid (0.49 mg in 10 μl DMF) is added, and the solution is cooled to −30°. After 10 min, this is added to a solution of rIL-2 (5.8 mg) in sodium borate buffer (prepared as described above). After 24 hr the product is purified by HPLC as described under General Methods. The yield of the protein is 2.37 mg. The NK-enhancing activity is 103%, the LAK-generating activity is 123%, and the ability to stimulate the proliferation of lymphocytes is 5%.

Conjugation of βDGal1,3βDGlcNAc-O-$(CH_2)_5CONHNH_2$ to Recombinant Interleukin 2 (Procedure B).

The preparation is the same as above except the asialo-G_{M2} is dissolved in 800 μl of water and sodium nitrite is used instead of *tert*-butyl nitrite. Also, the sulfamic acid solution is made in water. The yield is 2.67 mg. The NK-enhancing activity is 76%, the LAK-generating activity is 89%, and the ability to stimulate the proliferation of lymphocytes is 8%.

Discussion

Carbohydrates are extremely hydrophilic and readily soluble in aqueous buffers. This property of the oligosaccharides has been used to modify the physical properties of a very hydrophobic rIL-2, and as anticipated the chemical glycosylation renders rIL-2 more soluble in water. In general rIL-2 molecules glycosylated with di- or trisaccharides were more readily soluble than those glycosylated with monosaccharide. The increased ratio of activated carbohydrate–protein results in the loss of NK and LAK activities as well. In contrast to the large size of the polyethylene glycols, the oligosaccharides described in this chapter are much smaller in size and can be made synthetically with defined structure. However, as in the case with any other chemical conjugate methodology, the product obtained is a heterogeneous mixture containing proteins that have different degrees of oligosaccharide conjugation to lysines. The near identical circular dichroism spectrum of rIL-2 and glycosylated rIL-2 (Fig. 2) shows that the chemical glycosylation is mild enough not to perturb the secondary structure. The fact that the glycosylated rIL-2 maintained its secondary structure up to 90° (Fig. 3; unglycosylated rIL-2 irreversibly denatures under these conditions) clearly demonstrates the stabilizing effect of carbohydrates on protein structure.

In view of the results reported for the polyethylene glycol-modified rIL-2, the dramatic reduction on chemical glycosylation in the CTLL activity, mostly without loss of the NK and LAK activities, was unexpected. Because the reactivities of the activated polyethylene glycol and carbohydrate are different, it is likely that the lysines that reacted in these two methods may also be different. One possible explanation for this relative deficiency may be that the p70 receptors on the NK and LAK cells[27] may still be able to interact with the glycosylated rIL-2 and allow for postreceptor signaling and initiating of a cytolytic cascade. Chemical glycosylation should not have a detrimental effect in this regard, since

[27] J. P. Siegel, M. Sharon, P. L. Smith, and W. J. Leonard, *Science* **238,** 75 (1987).

the most critical peptide region of IL-2 required for p70 binding appears to be between amino acid residues 10 and 27 and this region is devoid of any lysines. On the other hand, the peptide epitope of IL-2 required for binding to the p56 receptor, which together with p70 mediates T cell activation, has a number of lysines that can be expected to become glycosylated under the chemical conditions used here. Consequently, the basic ε-amino groups of lysines in the key peptide region can be acylated. This could result in a decreased ability to bind to the critical p56 receptor or to the receptor complex and consequently the diminution of T cell activation as measured by proliferation.

It has been proposed that the toxic effects of IL-2 treatment may arise from the activation of helper T cells and the associated triggering of a cascade of other lymphokines,[6] and so it may be desirable to have IL-2 be capable of more selective NK and LAK cell activation, ultimately with a view to immunotherapy of some malignancies. Chemical glycosylation of rIL-2 provides such an opportunity, and these results could be expanded to the exploration of other recombinant proteins.

[6] Neoglycoprotein–Liposome and Lectin–Liposome Conjugates as Tools for Carbohydrate Recognition Research

By NOBORU YAMAZAKI, MAKOTO KODAMA, and HANS-JOACHIM GABIUS

Introduction

Specific interactions between carbohydrate ligands and lectin receptors on membrane surfaces play an important role in various biological processes such as cell-to-cell recognition, adhesion, and communication. For carbohydrate recognition research as well as for applied areas such as cell type-specific targeting, it is essential to provide a basic understanding of underlying mechanisms of the multivalent carbohydrate–protein interactions on membrane surfaces. To evaluate these interactions in *in vitro* and *in vivo* systems, model systems such as liposomes whose surfaces are modified by chemical conjugation of neoglycoproteins or lectins can provide appropriate tools.[1-4]

[1] N. Yamazaki, *in* "Advances in Chromatography 1986, Part II" (A. Zlatkis, ed.), p. 371. Elsevier, Amsterdam, 1987.

[2] N. Yamazaki, *J. Membr. Sci.* **41,** 249 (1989).

[3] N. Yamazaki, S. Kojima, S. Gabius, and H.-J. Gabius, *Int. J. Biochem.* **24,** 99 (1992).

Copyright © 1994 by Academic Press, Inc.
All rights of reproduction in any form reserved.

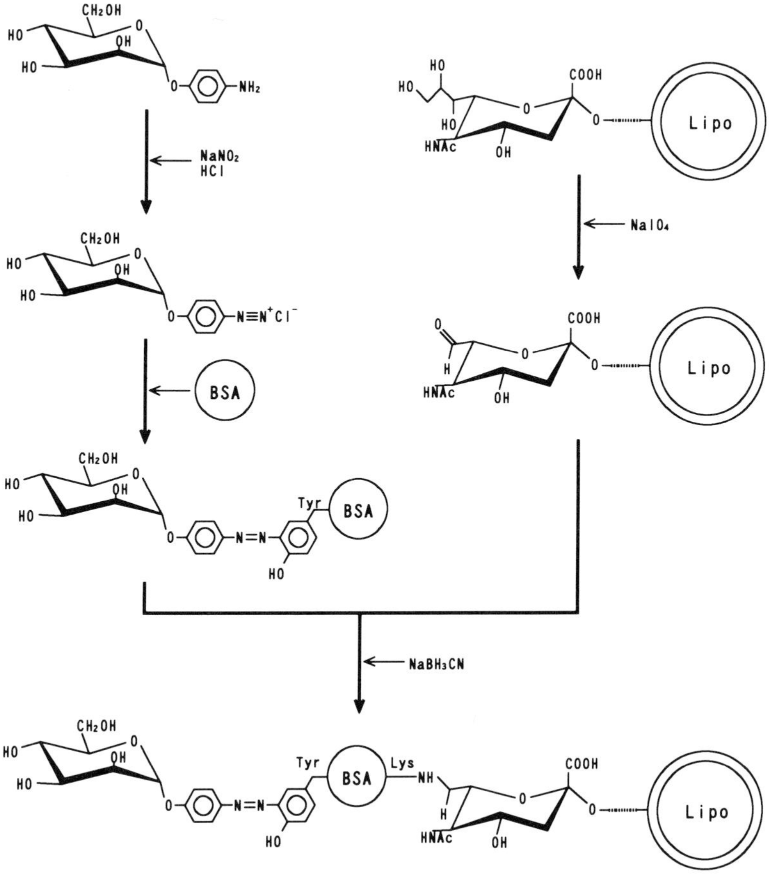

FIG. 1. Outline of the reaction sequence for neoglycoprotein–liposome conjugation. Preparation of mannosylated bovine serum albumin (BSA)-coupled liposomes (Lipo) was chosen as an example.

This chapter describes procedures for the preparation and characterization of one type of versatile and stable liposomes, as well as procedures for covalent conjugation of neoglycoproteins and lectins to the liposomes and characterization of the conjugates. A scheme outlining the procedure described in this chapter for neoglycoprotein–liposome conjugation is represented in Fig. 1, and the procedure for lectin–liposome conjugation

[4] N. Yamazaki, S. Gabius, S. Kojima, and H.-J. Gabius, *in* "Lectins and Glycobiology" (H.-J. Gabius and S. Gabius, eds.), p. 319. Springer-Verlag, Heidelberg, 1993.

is the same except that unmodified lectins, instead of neoglycoproteins, are coupled to oxidized liposomes.

Materials

All chemicals mentioned here, obtained from the indicated sources, are of analytical reagent grade and are used without further purification: L-α-dipalmitoylphosphatidylcholine, cholesterol, dicetyl phosphate, gangliosides (type III from bovine brain), and sodium cholate (Sigma Chemical Co., St. Louis, MO), and lectins of the highest available level of purity (Honen Corporation, Tokyo, Japan). Neoglycoproteins are prepared as follows: bovine serum albumin (BSA) is modified with the diazo derivatives of *p*-aminophenyl glycosides, which results in yields of 10 ± 2 carbohydrate moieties per carrier protein, as described in detail elsewhere.[5]

Preparation of Liposomes

The following procedure for preparing a type of liposomes is a modified version of the published method,[1-4] which is an adaptation of the controlled cholate dialysis method of Zumbuehl and Weder.[6] The procedure consists of two steps, namely, preparation of lipid–detergent mixed micelles and flow-through dialysis of the mixed micelles. The lipid composition of the liposomes, namely, L-α-dipalmitoylphosphatidylcholine (DPPC), cholesterol (Chol), dicetyl phosphate (DCP), and gangliosides at a molar ratio of 35 : 45 : 5 : 15, has been designed for achieving covalent coupling of liposome-bound gangliosides to proteins via Schiff base formation, and for yielding homogeneous and stable liposomes.

The lipids DPPC (17 mg), Chol (12 mg), DCP (2 mg), and gangliosides (15 mg) are mixed with sodium cholate (48 mg) in a 10-ml round-bottomed flask, suitable for a standard rotary evaporator, to give a lipid to detergent molar ratio of 0.6, and the solids are dissolved in 3 ml of chloroform/methanol (1 : 1, v/v) to make a clear solution. The solvent is evaporated under a stream of nitrogen on a rotating evaporator at about 30° in a water bath. The flask with the deposited lipid–detergent mixture is kept for about 1 hr under vacuum in a desiccator equipped with a high vacuum pump and stored overnight *in vacuo* to ensure complete removal of organic solvents and to give a completely dry lipid film on the walls of the flask. The lipid film is dissolved in 3 ml of tris(hydroxymethyl)methylaminopropanesulfonic acid (TAPS)-buffered saline [10 m*M* TAPS buffer (pH 8.4)

[5] C. R. McBroom, C. H. Samanen, and I. J. Goldstein, this series, Vol. 28, p. 212.

[6] O. Zumbuehl and H. G. Weder, *Biochim. Biophys. Acta* **640,** 252 (1981).

containing 150 m*M* NaCl] with gentle stirring for 1 hr in the flask, during which time the air is replaced by N_2. The homogeneous lipid–detergent suspension is sonicated in a bath-type ultrasonicator equipped with time regulator for 1 hr with an intermission of 3 sec every 2 min until a homogeneous suspension of mixed micelles is formed.

To prepare homogeneous unilamellar liposomes, the lipid–detergent mixed micelle suspension obtained should be as transparent as possible before starting the following dialysis. For flow-through dialysis of the mixed micelles, a 10-ml ultrafiltration cell (Model 8010; Amicon Division, W. R. Grace & Co., Danvers, MA) is fitted with an Amicon Diaflo PM10 membrane and equipped with a concentration/dialysis selector (Amicon Model CDS-10) connected to an 800-ml reservoir. Then 3 ml of the mixed micelle suspension is added and diafiltered with approximately 100 ml of TAPS-buffered saline (pH 8.4) for about 24 hr at room temperature, by applying about 2 atm N_2 pressure. The volume of the prepared liposome suspension will increase after the diafiltration step, and it can be between 5 and 7 ml. The liposome suspension is filtered through a 0.45-μm Millex-HV membrane (Millipore, Bedford, MA) and stored in a glass tube with a Teflon-lined screw cap in a refrigerator at 4°–7°.

Characterization of Liposomes

To analyze the homogeneity and stability of liposomes, size distribution is an important characteristic among several physical properties. Gel-permeation chromatography (GPC), electron microscopy (EM), and dynamic light scattering (DLS) can be applied for size characterization of liposomes. In the following, a procedure for GPC analysis is described, and a typical GPC result is compared with analyses by EM and DLS.

Analysis by Gel-Permeation Chromatography. A precise analysis can be performed by using a high-performance liquid chromatography (HPLC) system, which is equipped with an on-line degasser, a computer-controlled pump, a sample injector, and a data processor. A water-jacketed, thick-walled glass column (100 cm × 1 cm i.d.) is packed with Sephacryl S1000 Superfine (Pharmacia Fine Chemicals, Uppsala, Sweden), equilibrated, and eluted at 0.1 ml/min with phosphate-buffered saline [i.e., 10 m*M* phosphate buffer (pH 7.2) containing 150 m*M* NaCl]. The column is kept at 25° by circulating thermostatted water in the jacket. Column effluent is monitored by two detectors connected in series: a UV detector and a fluorometer for measuring absorbance at 280 nm and 90° light scattering at 633 nm, respectively. The measurement of light scattering permits estimation of the liposome content in the effluent with such high sensitivity that a 50-μl injection of liposome suspension gives a well-defined elution

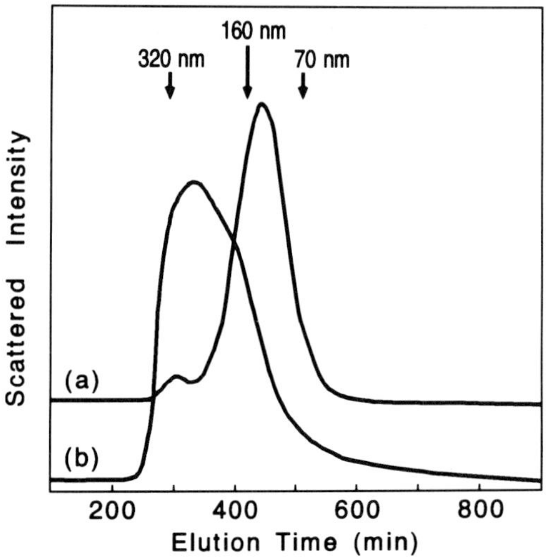

FIG. 2. Gel-permeation chromatography on a Sephacryl S1000 Superfine column. Elution profiles of (a) one type of liposome which is described in this chapter and (b) another type of liposome which differs in lipid composition are shown. The column was calibrated, as depicted by arrows, with three sizes of polystyrene latex particles: 320, 160, and 70 nm.

profile, as shown in Fig. 2. For calibration of the column with polystyrene latex particles (Polysciences, Warrington, PA), a Triton-containing buffer (phosphate-buffered saline, pH 7.2, containing 0.5% Triton X-100) is used to suspend the latex particles as well as to preequilibrate and elute the column. The use of Triton X-100 prevents aggregation of the latex particles.

The result of GPC (Fig. 2a) is in good agreement with the results of EM (Fig. 3) and DLS (Fig. 4), and this agreement clearly supports the accuracy of the method described above. As shown in Fig. 2, two types of liposome preparations can be distinguished from one another by using the GPC analysis, which is convenient for routine size characterization of liposomes.

Conjugation of Neoglycoproteins or Lectins to Liposomes

The following procedure is a modified version of the published method[1-4] for conjugation of proteins to liposomes, which is an adaptation of the covalent coupling method of Heath *et al.*[7] As outlined in Fig. 1, the

[7] T. D. Heath, B. A. Macher, and D. Papahadjopoulos, *Biochim. Biophys. Acta* **640,** 66 (1981).

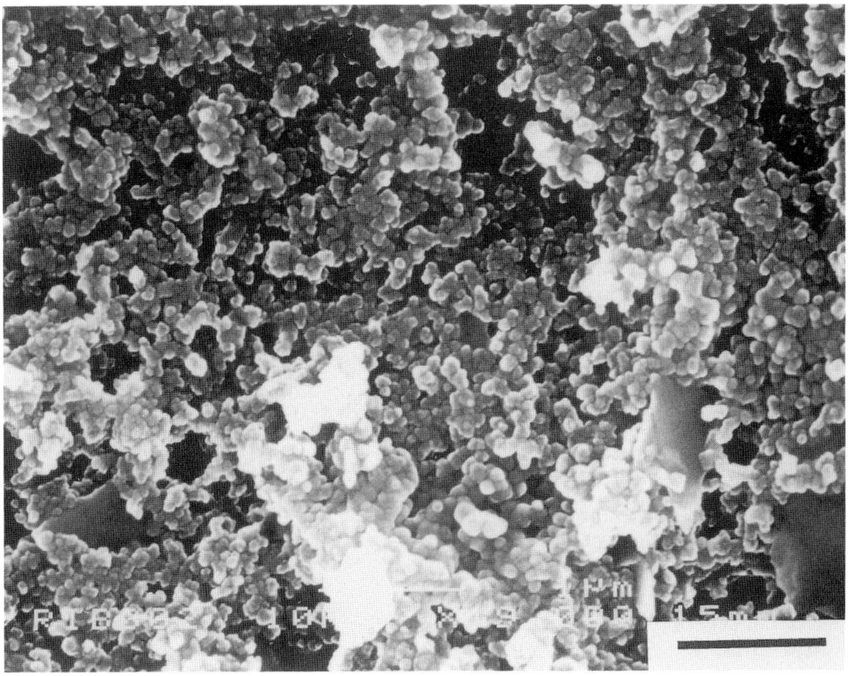

FIG. 3. Scanning electron micrograph of the same liposome preparation as used in Fig. 2a. Magnification: ×19,000. Bar: 1 μm.

procedure for neoglycoprotein–liposome conjugation can be performed in a simple two-step reaction involving periodate oxidation of gangliosides in the liposome membrane and coupling of neoglycoproteins to oxidized liposomes by reductive amination. The periodate treatment at pH 8.4 oxidizes a large portion of the external gangliosides of the liposomes without oxidizing the internal gangliosides. This method provides a useful basis for covalent coupling of biologically active proteins to the outer surface of membranes without destroying the integrity of the liposomes.

Periodate Oxidation. The liposome suspension described above, which contains approximately 46 mg of total lipid in 5–7 ml of TAPS-buffered saline (pH 8.4) in a glass tube, is mixed with 0.5–0.7 ml of 0.2 *M* sodium periodate in TAPS-buffered saline (pH 8.4) to obtain a final periodate concentration of 20 m*M* in the reaction mixture. The mixture is incubated at room temperature in the dark for 2 hr with gentle stirring. To separate residual periodate from the liposomes and to change the buffer, a 10-ml ultrafiltration cell is fitted with a 0.03-μm polycarbonate membrane (Nuclepore, Pleasanton, CA) and equipped with a concentration/dialysis

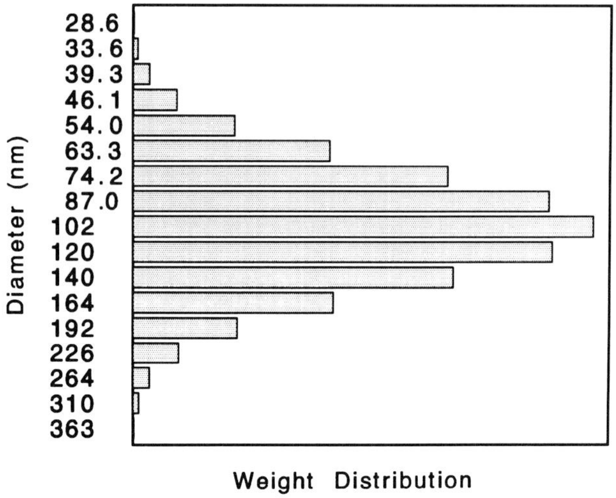

FIG. 4. Bar histogram of dynamic light scattering of the same liposome preparation as shown in Figs. 2a and 3.

selector connected to an 800-ml reservoir which contains phosphate-buffered saline [i.e., 10 m*M* phosphate buffer (pH 8.0) and 150 m*M* NaCl]. Then the reaction mixture is added and diafiltered overnight with approximately 100 ml of phosphate-buffered saline (pH 8.0) at about 10° by applying 1 atm N_2 pressure. The oxidized liposomes obtained, the volume of the suspension being between 5 and 7 ml, can be stored at about 4° for at least several months under a nitrogen atmosphere in a tightly closed glass tube with Teflon-lined screw cap.

Reductive Amination. The second step for coupling proteins to liposomes is usually performed by using one-tenth volume of the suspension of oxidized liposomes described above. The following procedure is applied to coupling of neoglycoproteins as well as lectins. In a 2-ml glass tube with a Teflon-coated micro stirring bar, 1–2 mg of solid neoglycoproteins or 2–3 mg of solid lectins are mixed with 0.5–0.7 ml of the oxidized liposome suspension, containing about 4.5 mg of total lipid in phosphate-buffered saline (pH 8.0). The mixture is incubated for 2 hr at room temperature with gentle stirring, and 5–7 μl of 2 *M* sodium cyanoborohydride in phosphate-buffered saline (pH 8.0) is added to reach a final concentration of 20 m*M* in cyanoborohydride. The mixture is further incubated overnight at about 10° with gentle stirring. To separate unbound neoglycoproteins or lectins from the liposomes and to change the buffer and pH environment, a 10-ml ultrafiltration cell is fitted with a 0.03-μm polycarbonate membrane and equipped with a concentration/dialysis selector connected to an 800-

ml reservoir, containing phosphate-buffered saline [i.e., 10 m*M* phosphate buffer (pH 7.2) and 150 m*M* NaCl]. The reaction mixture is adjusted with phosphate-buffered saline (pH 8.0) until the total volume reaches 2 ml, then transferred to the cell and diafiltered overnight with approximately 100 ml of phosphate-buffered saline (pH 7.2) at about 10° by applying 1 atm N_2 pressure. The suspension of neoglycoprotein–liposome conjugates or lectin–liposome conjugates obtained is filtered through a 0.2-μm Millex-HV membrane, and stored under a nitrogen atmosphere in a tightly closed glass tube with Teflon-lined screw cap in a refrigerator at 4°–7°.

Characterization of Neoglycoprotein–Liposome and Lectin–Liposome Conjugates

Analysis by Gel-Permeation Chromatography. To analyze the purity and stability of neoglycoprotein-coupled liposomes or lectin-coupled liposomes, GPC analysis is routinely carried out by using the same HPLC system as described above. Figure 5a shows the purity of a preparation of mannosylated BSA-coupled liposomes. The size distribution of this type of neoglycoprotein–liposome conjugate is identical with that of non-

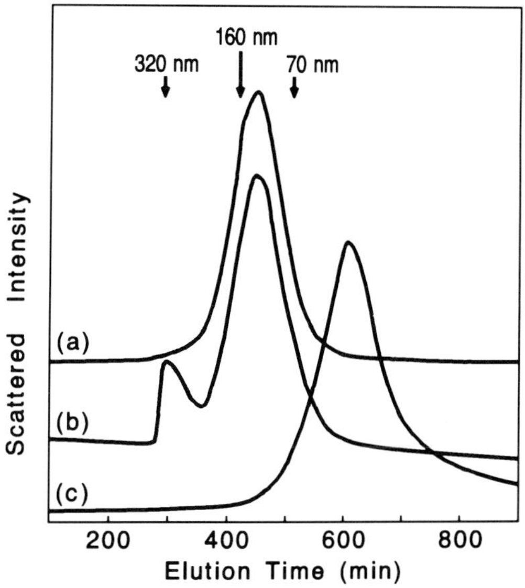

FIG. 5. Gel-permeation chromatography on a Sephacryl S1000 Superfine column. Elution profiles of (a) mannosylated BSA-coupled liposomes, (b) N-acetylgalactosaminylated BSA-coupled liposomes, which were stored for about 2 years at 4°, and (c) mannosylated BSA are shown. The column was calibrated as in Fig. 2.

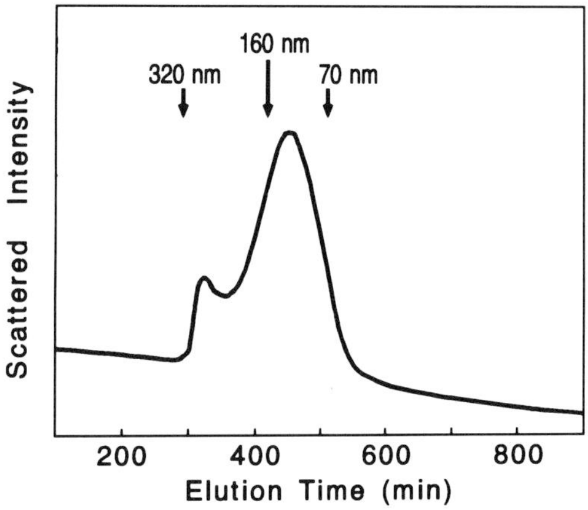

FIG. 6. Gel-permeation chromatography on a Sephacryl S1000 Superfine column. An elution profile of a mistletoe lectin-coupled liposome preparation, which had been stored for 3 weeks at 4°, is shown. The column was calibrated as in Fig. 2.

coupled liposomes, as compared with Fig. 2a, and the conjugate suspension does not contain free neoglycoproteins, as illustrated in Fig. 5a, c. This type of neoglycoprotein–liposome conjugate is stable and does not undergo changes in physical properties after storage at about 4° for several months, and it is noteworthy that one preparation of N-acetylgalactosaminylated BSA–liposome conjugates has been kept in a refrigerator for about 2 years without substantial property changes, as shown in Fig. 5b. Figure 6 shows the size distribution of mistletoe lectin-coupled liposomes. Some lectin–liposome conjugates tend to form liposome aggregates, which can be removed by filtration through a 0.22-μm Millex-HV membrane prior to use.

Composition Analysis. To analyze the yield and quality of neoglycoprotein-coupled liposomes or lectin-coupled liposomes, protein and lipid compositions in each preparation are routinely determined, and coupled protein/total lipid ratios are calculated. The protein content of neoglycoprotein–liposome and lectin–liposome conjugates is estimated either by employing the modified Lowry method in the presence of 1% sodium dodecyl sulfate[8] or by using a commercial micro-BSA protein assay kit (Pierce Chemical Co., Rockford, IL) in the presence of 1% sodum dodecyl sulfate. For determination of the lipid concentration in the preparation, the cholesterol content of the liposome suspension is first analyzed by

[8] M. A. K. Markwell, S. M. Haas, L. L. Bieber, and N. E. Tolbert, *Anal. Biochem.* **87**, 206 (1978).

using a commercial kit for total-cholesterol determination, Determiner TC"555" (Kyowa Medex, Tokyo, Japan), in the presence of 0.5% Triton X-100. The amount of lipids in the liposome preparation is then estimated using the molar ratio of lipids, namely, DPPC, Chol, DCP, and gangliosides (35 : 45 : 5 : 15). Results by this method are consistent with results by microanalysis of phosphorus using the Bartlett assay. Coupled protein/total lipid ratios in the final products are usually between 0.1 and 0.2 g/g for neoglycoprotein–liposome conjugates, and between 0.1 and 0.4 g/g for lectin–liposome conjugates.

[7] Modification of Proteins with Polyethylene Glycol Derivatives

By YUJI INADA, AYAKO MATSUSHIMA, MISAO HIROTO, HIROYUKI NISHIMURA, and YOH KODERA

Introduction

Chemical modification of proteins became commonplace in the late 1950s: the techniques were developed to aid in the structural analysis of protein molecules. The intention of such a modification was to develop reagents which would specifically react with amino acid side chains in the protein molecule to discriminate the state of amino acid residues and to identify the amino acid involved in particular protein functions. From the 1970s, chemical modification of proteins by conjugation with synthetic or natural macromolecules has been performed. The purposes of these modifications include alteration of immunoreactivity, immunogenicity, and suppression of immunoglobulin E production, or making enzymes soluble and active in organic solvents.[1,2,2a] This chapter deals with the chemical modification of proteins with synthetic macromolecules, polyethylene glycol derivatives.

Since polyethylene glycol [PEG; general formula $HO(CH_2CH_2O)_nH$] has been synthesized, many industrial and biochemical applications in the areas of pharmaceutics, cosmetics, and textiles have been developed to

[1] Y. Inada, A. Matsushima, Y. Kodera, and H. Nishimura, *J. Bioact. Compat. Polym.* **5**, 343 (1990).

[2] Y. Inada, Y. Kodera, A. Matsushima, and H. Nishimura, *in* "Synthesis of Biocomposite Materials" (Y. Imanishi, ed.), p. 85. CRC Press, Boca Raton, Florida, 1992.

[2a] Y. Inada, M. Furukawa, H. Sasaki, Y. Kodera, M. Hiroto, H. Nishimura, and A. Matsushima, *Trends Biotechnol.*, in press.

Copyright © 1994 by Academic Press, Inc.
All rights of reproduction in any form reserved.

utilize its unique properties.[3] No significant toxicity is observed when polyethylene glycol (molecular weight 4000) is injected repeatedly into dogs.[4]

Activation of Polyethylene Glycol

Proteins can be modified with an activated polyethylene glycol derivative. The methods of activation of polyethylene glycol have been summarized by Harris.[5] The modifier is usually synthesized from monomethoxypolyethylene glycol, which has a hydroxy group at one end of the molecule amenable to manipulation. In this section, syntheses of the modifiers are described. Most of the modifiers have a chain-shaped form, such as 2,4-bis(*O*-methoxypolyethylene glycol)-6-chloro-*s*-triazine, abbreviated as "activated PEG_2." We have developed a new type of modifier with a comb-shaped form: a copolymer of maleic anhydride and a monomethoxypolyethylene glycol derivative, abbreviated as "activated PM." Each modifier reacts mainly with the ε-amino group of lysine residues and/or the N-terminal amino group.

Synthesis of Activated PEG_2

Activated PEG_2, 2,4-bis(*O*-methoxypolyethylene glycol)-6-chloro-*s*-triazine [Eq. (1)], was synthesized by Matsushima *et al.*[6] from monomethoxypolyethylene glycol and cyanuric chloride in benzene using sodium carbonate as an acid scavenger. Ono *et al.*[7] obtained the activated PEG_2 in a homogeneous state by using zinc oxide in lieu of Na_2CO_3 without formation of by-products such as 2-(*O*-methoxypolyethylene glycol)-4,6-dichloro-*s*-triazine (activated PEG_1) and its polymerized macromolecules. Because the modifier has two polyethylene glycol chains in the molecule, two PEG chains can be attached to one amino group. Therefore, a more effective modification of protein with activated PEG_2 is achieved, in comparison with activated PEG_1 [Eq. (2)] which has one PEG chain in the molecule. Activated PEG_2 is available from Seikagaku Kogyo Co., Ltd. (Tokyo, Japan).

[3] H. C. Schultze, *in* "Glycols" (G. O. Curme, Jr., ed.), p. 153. Reinhold, New York, 1952.

[4] C. P. Carpenter, M. D. Woodside, E. R. Kinkead, J. M. King, and L. J. Sullivan, *Toxicol. Appl. Pharmacol.* **18,** 35 (1971).

[5] J. M. Harris, *Macromol. Chem. Phys.* **C25,** 325 (1985).

[6] A. Matsushima, H. Nishimura, Y. Ashihara, Y. Yokota, and Y. Inada, *Chem. Lett.,* 773 (1980).

[7] K. Ono, Y. Kai, H. Maeda, F. Samizo, K. Sakurai, H. Nishimura, and Y. Inada, *J. Biomater. Sci. Polym. Ed.* **2,** 61 (1991).

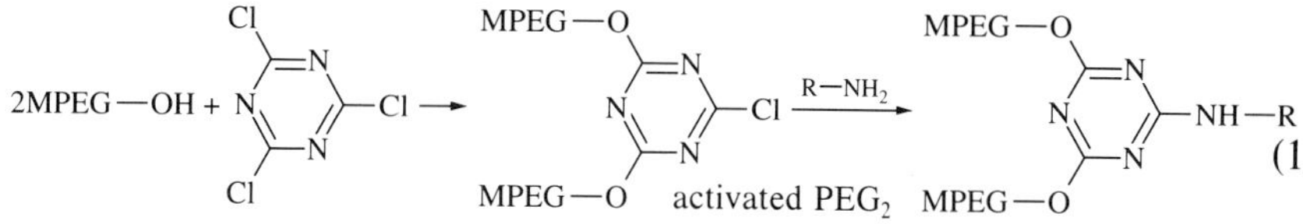

(1)

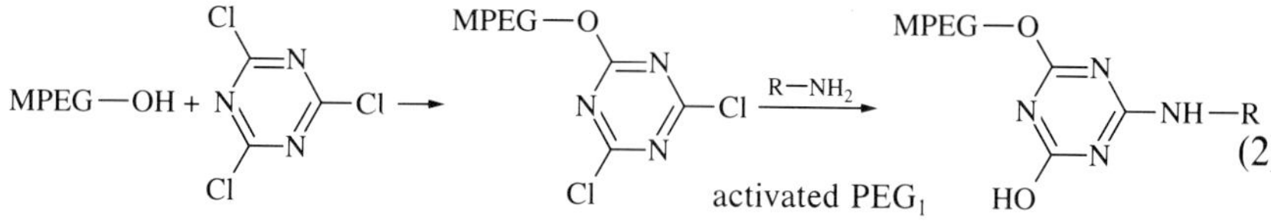

(2)

MPEG = methoxypolyethylene glycol, $CH_3—(OCH_2CH_2)_n—$

Reagents

Monomethoxypolyethylene glycol (average molecular weight 5000) is available from Chemiscience (Tokyo, Japan), Nippon Oil and Fat, Inc. (Tokyo, Japan), or Sigma Chemical Co. (St. Louis, MO)

Cyanuric chloride [2,4,6-trichloro-1,3,5-triazine, Wako Pure Chemical Industries (Osaka, Japan)] is recrystallized in benzene prior to use

Other chemicals used are of reagent grade.

Procedures. Into 500 ml of anhydrous benzene is dissolved 110 g of monomethoxypolyethylene glycol (molecular weight 5000). The solution is refluxed in the presence of 25 g of powder Molecular Sieves 4A (Nacalai Tesque, Kyoto, Japan) for 6 hr to remove water. After cooling, 50 g of zinc oxide and 1.85 g of cyanuric chloride are added, and the mixture is refluxed for 53 hr. The resulting mixture is diluted with 500 ml of benzene and filtered. The product is precipitated by adding a 2-fold volume of petroleum ether followed by drying under vacuum to obtain 108 g of activated PEG$_2$.

Figure 1 shows the chromatographic pattern during the course of activated PEG$_2$ synthesis using ZnO. After forming activated PEG$_1$ initially, the reaction leads to selective synthesis of activated PEG$_2$ without any by-products. Because the binding of amino groups in a protein with activated PEG$_2$ is a substitution reaction through the triazine ring, PEG-modified proteins are rather stable *in vitro* and *in vivo* and are not readily degraded either chemically or enzymatically.

Synthesis of Other Activated Polyethylene Glycols

Other modifiers to introduce polyethylene glycol into a protein molecule have been also developed. A variety of modifiers, having chain-

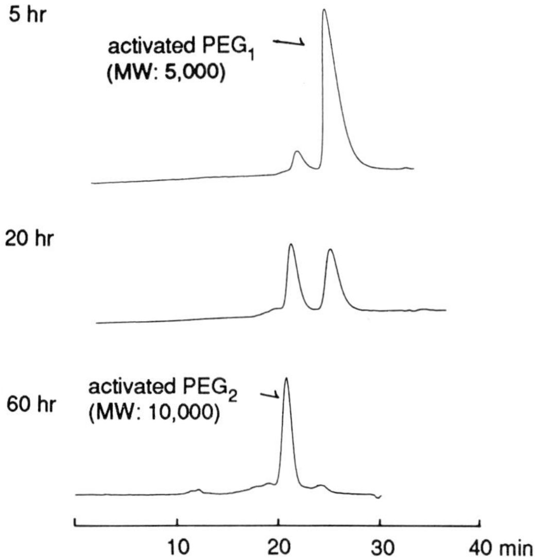

FIG. 1. Time course of activated PEG_2 synthesis using zinc oxide. Gel filtration was performed with a high-performance liquid chromatography system equipped with a TSK-gel G3000SW column (7.5 mm × 60 cm, Tosoh Co. Ltd., Tokyo, Japan) at a flow rate of 0.7 ml/min. The eluent was a 1 : 19 (v/v) mixture of ethanol and 10 m*M* phosphate buffer (pH 7.0) containing 0.2 *M* sodium chloride. The eluate was monitored by the absorbance at 254 nm. (From Ono *et al.*[7])

shaped forms with one polyethylene glycol chain and one reactive site, are readily synthesized from methoxypolyethylene glycol with amine or carboxyl groups (available from Nippon Oil and Fat and from Sigma).

Activated PEG_1. 2-(*O*-Methoxypolyethylene glycol)-4,6-dichloro-*s*-triazine, activated PEG_1 shown in Eq. (2), is prepared by the method of Abuchowski *et al.*[8] from methoxypolyethylene glycol and cyanuric chloride using anhydrous sodium carbonate. This modifier reacts with amino groups at pH 9.2, which is slightly lower than the pH required for activated PEG_2 (pH 9.5–10.0).

PEG Succinimide Preparation. PEG succinimide is prepared from methoxypolyethylene glycol carboxylic acid by activation with *N*-hydroxysuccinimide (NHS) and dicyclohexylcarbodiimide [Eq. (3)].[9] α-Carboxy-

[8] A. Abuchowski, T. van Es, N. C. Palczuk, and F. F. Davis, *J. Biol. Chem.* **252,** 3578 (1977).

[9] G. W. Anderson, J. E. Zimmerman, and F. M. Callahan, *J. Am. Chem. Soc.* **86,** 1839 (1964).

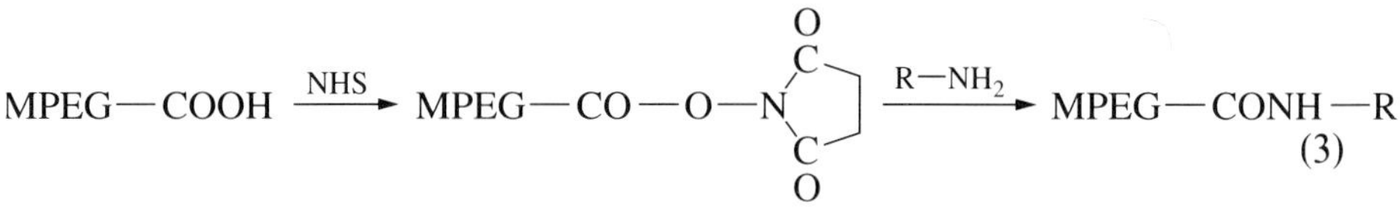

(3)

methyl-ω-methoxypoly(oxyethylene) (average molecular weight 4500) is obtained from Nippon Oil and Fat. All solvents used in the protocol are dried by adding Molecular Sieves 3A.

α-Carboxymethyl-ω-methoxypoly(oxyethylene) (25 g) is dissolved in 80 ml of dimethylformamide, and 0.75 g of *N*-hydroxysuccinimide (10% molar excess) and 5 g of Molecular Sieves 3A are added. To the mixture is added dropwise 1.35 g of dicyclohexylcarbodiimide dissolved in 5 ml of dry dimethylformamide, and the reaction mixture is kept at room temperature for 24 hr. The mixture is then diluted with 50 ml of cold benzene, and the activated PEG is precipitated by adding 100 ml of cold petroleum ether (or hexane). The precipitate is collected on a sintered glass filter with suction. The dissolution and the precipitation procedure is then repeated three times, and the final precipitate is dried under vacuum.

This type of modifier can also be prepared as follows.[10] First, methoxypolyethylene glycol is coupled with succinic anhydride to introduce a carboxylic acid, and then it is activated as described above. Methoxypolyethylene glycol succinimidyl succinate thus synthesized is available from Sigma. After coupling with protein, ester and amide bonds will be formed between the modifier and amino groups in the protein.

Phenylchloroformate Activation. Methoxypolyethylene glycol can be activated using *p*-nitrophenyl chloroformate as shown in Eq. (4).[11] The

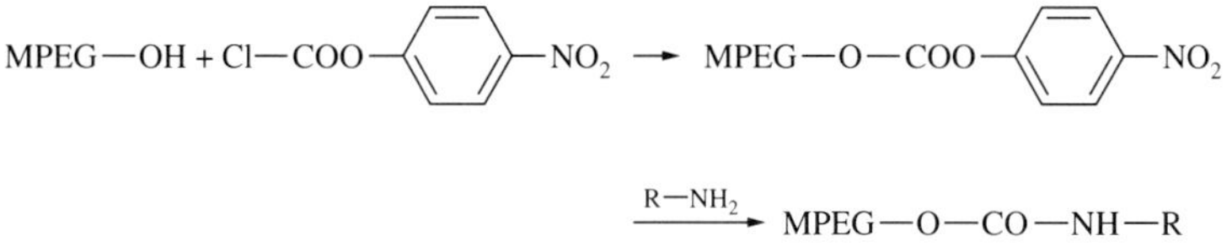

(4)

modifier, methoxypolyethylene glycol *p*-nitrophenyl carbonate, reacts with the amino groups of a protein between pH 8 and 9 to form a stable urethane linkage.

Synthesis of Activated PM

Activated PM, a copolymer of maleic anhydride and a derivative of polyethylene glycol, has a comb-shaped form with multivalent reactive

[10] A. Abuchowski, G. M. Kazo, C. R. Verhoest, Jr., T. van Es, D. Kafkewitz, M. L. Nucci, A. T. Viau, and F. F. Davis, *Cancer Biochem. Biophys.* **7,** 175 (1984).

[11] F. M. Veronese, R. Largajolli, E. Boccu, C. A. Benassi, and O. Schiavon, *Appl. Biochem. Biotechnol.* **11,** 141 (1985).

groups as shown in Eq. (5). Two kinds of activated PM copolymers have been prepared: activated PM_{13} [molecular weight 13,000; $m \approx 8$, $n \approx 33$, R = H, Eq. (5)] and activated PM_{100} [molecular weight 100,000; $m \approx 50$, $n \approx 40$, R = CH_3, Eq. (5)]. Amino groups in a protein are directly coupled with maleic anhydride in the PM modifier to form amide bonds. These comb-shaped modifiers possess unique properties, covering a protein molecule with PM molecules and/or placing anionic groups (—COOH) on the surface of a protein molecule.

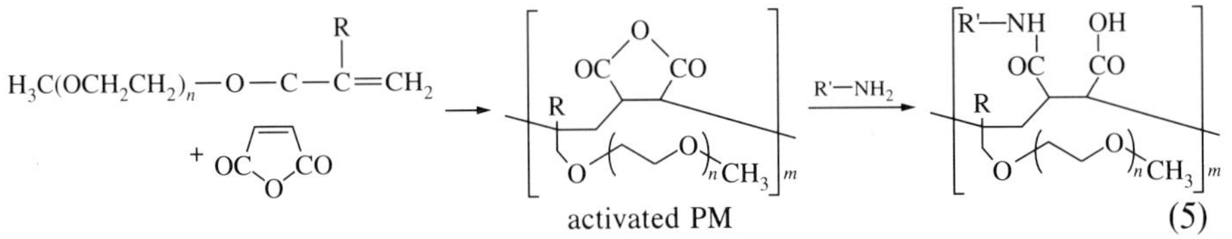

(5)

Reagents. Polyoxyethylene allyl methyl diether and polyoxyethylene (2-methyl-2-propenyl) methyl diether are obtained from Nippon Oil and Fat. Benzoyl peroxide, lauroyl peroxide, and maleic anhydride are of analytical grade. Bio-Beads S-X1, a preparation of porous polystyrene beads for gel-permeation chromatography, is available from Bio-Rad Laboratories (Richmond, CA). The gel swollen with tetrahydrofuran is poured into an organic solvent-resistant column assembly (SR25) equipped with gel bed supports and with a four-port valve (SRV-4) as a sample injector (Pharmacia, Uppsala, Sweden). The system is eluted with chloroform by gravity.

Procedure. To 1520 g of polyoxyethylene allyl methyl diether [molecular weight 1500, $n \approx 33$, Eq. (5)], dissolved in 1 liter of toluene, are added 103 g of maleic anhydride and 12 g of benzoyl peroxide. The mixture is kept for 7 hr at 80° under nitrogen. After evaporation, 1450 g of the crude copolymer is obtained. The crude copolymer dissolved in chloroform is subjected to gel-permeation chromatography with Bio-Beads S-X1. The eluate at the void volume is pooled and evaporated. The molecular weight of the eluate is approximately 13,000 as measured by a gel-permeation chromatography system equipped with Tosoh TSK-Gel $G4000H_{HR}$ and $G3000H_{HR}$ columns serially connected. The degree of polymerization [m in Eq. (5)] is calculated to be 8, based on the measured molecular weight of activated PM_{13} (molecular weight 13,000) and of PEG monomer (molecular weight 1500).

Activated PM_{100} may also be synthesized by the same manner described above: a mixture of polyoxyethylene (2-methyl-2-propenyl) methyl diether (1850 g) and maleic anhydride (103 g) in radically polymerized with lauroyl peroxide (8 g) as a catalyst. The molecular weight of activated PM_{100} is approximately 100,000. Activated PM_{13} and PM_{100} are available from Nippon Oil and Fat.

Reactivity of Activated PM. Before application of activated PM copolymers in protein modifications, the reactivity is checked, using benzoylglycyl-L-lysine (Bz-Gly-Lys; Peptide Institute, Inc., Osaka, Japan) as a model compound.[12] The degree of modification is determined by measuring amino groups with trinitrobenzenesulfonate.[13] The coupling reaction of peptide with activated PM_{13} in 0.5 *M* sodium borate buffer (pH 8.5) is performed at 22° for 1 hr. The degree of protein modification is enhanced by increasing the molar ratio of activated PM_{13} to Bz-Gly-Lys, and it tends to approach a constant level. At a molar ratio of 10, the ε-amino group of the lysine residue in the dipeptide is completely modified between pH 7 and 10 and between 0° and 37°.

Modification of Hydrolytic Enzymes

Principle

In 1984, Inada and co-workers[14,15] demonstrated that enzymes modified with a polyethylene glycol derivative become soluble and remain active in organic solvents. Because PEG is an amphipathic macromolecule, its hydrophilic nature makes it possible to modify enzymes in aqueous solution, and its hydrophobic nature would make modified enzymes soluble in hydrophobic environments. In fact, modified enzymes such as catalase[14] and peroxidase[15] have markedly high activities in organic solvents. Furthermore, PEG-modified hydrolytic enzymes such as lipase catalyze effectively the reverse reaction of hydrolysis in hydrophobic media, namely, ester synthesis and ester exchange reactions in transparent organic solvents [Eqs. (6) and (7)].[15a]

$$R^1COOH + R^2OH \rightarrow R^1COOR^2 + H_2O \quad (6)$$

$$R^1COOR^2 + R^3COOR^4 \rightarrow R^1COOR^4 + R^3COOR^2 \quad (7)$$

[12] M. Hiroto, A. Matsushima, Y. Kodera, Y. Shibata, and Y. Inada, *Biotechnol. Lett.* **14,** 559 (1992).

[13] R. Fields, this series, Vol. 25, p. 464.

[14] K. Takahashi, A. Ajima, T. Yoshimoto, and Y. Inada, *Biochem. Biophys. Res. Commun.* **125,** 761 (1984).

[15] K. Takahashi, H. Nishimura, T. Yoshimoto, Y. Saito, and Y. Inada, *Biochem. Biophys. Res. Commun.* **121,** 261 (1984).

[15a] Y. Kodera, H. Nishimura, A. Matsushima, M. Hiroto, and Y. Inada, *J. Am. Oil Chem. Soc.* **71,** 335 (1994).

In the case of proteases such as papain[16] and chymotrypsin,[17] amide bond formation also proceeds in organic solvents.

$$R^1COOH + R^2NH_2 \rightarrow R^1CONHR^2 + H_2O \quad (8)$$

The likelihood of the modified enzymes being rendered soluble and remaining active in organic solvents depends on the degree of modification of amino groups in each enzyme molecule with an activated modifier. Undermodification does not accomplish the desired solubilization in organic solvents, whereas overmodification causes a reduction in enzymatic activity because of the perturbation of protein conformation. The degree of modification of amino groups in the protein molecule can be varied by changing the amount of the modifier added to the reaction mixture.

Analytical Methods

Determination of Protein and Amino Groups. The protein concentration is determined by the biuret method.[18] The degree of modification of amino groups in the modified lipase is determined with trinitrobenzenesulfonate.[13] To 250 μl of lipase solution (0.5 mg protein/ml) are added 250 μl of 0.5 *M* sodium bicarbonate buffer (pH 8.5) and 250 μl of a 0.1% aqueous solution of sodium trinitrobenzenesulfonate. The reaction mixture is incubated at 40° for 2 hr to complete the reaction. After adding 250 μl of 10% sodium dodecyl sulfate (SDS) and 125 μl of 1 *M* hydrochloric acid, the absorbance at 335 nm is measured using quartz cuvettes with a light path of 3-mm. A standard curve is obtained with known concentrations of glycine (0–1 m*M*) instead of a protein solution. The degree of modification is obtained by dividing the number of remaining free amino groups by the total number of amino groups in the lipase molecule.

A fluorometric method for determining amino groups in a protein is as follows.[19] A mixture of 50 μl of sample solution containing approximately 1 mg protein/ml lipase and 50 μl of 10% SDS is incubated at 37° for 10 min. To the sample solution is added 2 ml of 0.1 *M* sodium borate buffer (pH 8.0) containing 2% SDS followed by the addition of 50 μl of 0.17% fluorescamine (in dry dioxane) under vigorous stirring. The fluorescence intensity is measured at an excitation of 390 nm and emission of 480 nm.

[16] H. Lee, K. Takahashi, Y. Kodera, K. Ohwada, T. Tsuzuki, A. Matsushima, and Y. Inada, *Biotechnol. Lett.* **10,** 403 (1988).

[17] A. Matsushima, M. Okada, and Y. Inada, *FEBS Lett.* **178,** 275 (1984).

[18] E. Layne, this series, Vol. 3, p. 450.

[19] S. Udenfriend, S. Stein, P. Bohlen, W. Dairman, W. Leimgruber, and M. Weigle, *Science* **178,** 871 (1972).

Hydrolytic Activity. The esterase activity of the lipase is measured in an emulsified system composed of polyvinyl alcohol and olive oil.[20] The olive oil emulsion is prepared by successive sonication of a mixture of 10 ml of olive oil and 30 ml of 2% polyvinyl alcohol [9 : 1 mixture of Poval #117 (degree of polymerization 1750) and #205 (degree of polymerization 550), which are available from Kurare (Kurashiki, Japan)]. To a mixture of 0.5 ml of emulsified olive oil and 0.4 ml of 0.1 *M* phosphate buffer (pH 7.0) are added 0.1 ml of lipase solution (about 10 U). In the case of *Pseudomonas fragi* lipase, 0.2 *M* Tris-HCl buffer containing 20 m*M* $CaCl_2$, pH 9.0, is used instead of phosphate buffer. The reaction mixture is incubated at 37° for 20 min, at which point 2 ml of a mixture of acetone and ethanol (1 : 1, v/v) is added to stop the reaction. The sample solution is titrated with 0.05 *N* NaOH in the presence of a few drops of phenolphthalein dissolved in ethanol. One unit of enzyme is defined as the amount which liberates 1 μmol of fatty acid per minute.

Ester Synthesis Activity. Water-saturated organic solvents are prepared by mixing the solvents with water at room temperature. The concentration of water in water-saturated benzene is determined to be 30 m*M* by the Karl Fischer titration.[20a]

Ester synthesis with the modified lipase in benzene is determined as follows.[21] To 300 μl of benzene containing two substrates, a carboxylic acid and an alcohol, are added 100 μl of the modified lipase (~0.5 U) in benzene, and the reaction mixture is incubated at 25°. A set of solutions in benzene containing a known concentration of a corresponding ester (0–30 m*M*) is prepared to make a standard curve. At an appropriate time, the reaction is stopped by the addition of 100 μl of 0.2 *N* sulfuric acid in benzene. To the mixture are added 2 ml of 0.25 *N* sodium hydroxide, 2 ml of petroleum ether, and 0.8 ml of methanol. Then the mixture is vigorously shaken and centrifuged at 2500 rpm for 10 min. The amount of ester extracted into the upper layer is colorimetrically determined by the method of Hill.[22] To 1 ml of the upper organic layer containing the extracted ester are added 300 μl of 2.5% sodium hydroxide in 95% ethanol saturated with sodium carbonate and 300 μl of 2.5% hydroxylamine hydrochloride in 95% ethanol. After the solvents are evaporated off in a boiling water bath, the residue is dissolved with 2 ml of the 5% ferrous reagent (see below) and the absorbance at 520 nm measured.

The ferrous reagent is prepared as follows: 17.5 ml of 60% perchloric acid is added to 1.94 g of ferrous chloride dissolved in 20 ml of nitric acid

[20] K. Yamada, Y. Ota, and H. Machida, *J. Agric. Chem. Soc. Jpn.* **36,** 860 (1962).
[20a] D. M. Smith, W. H. D. Bryant, and J. Mitchell, Jr., *J. Am. Chem. Soc.* **61,** 2407 (1939).
[21] K. Takahashi, T. Yoshimoto, A. Ajima, Y. Tamaura, and Y. Inada, *Enzyme* **32,** 235 (1984).
[22] U. T. Hill, *Anal. Chem.* **19,** 932 (1947).

(10 *M*). The mixture is heated while being gently stirred on a water bath in a hood until a vigorous generation of yellow gas is observed. After cooling, 35 ml of water and 10 ml of concentrated nitric acid are carefully added. The mixture is made up to 100 ml with 60% perchloric acid (~17.5 ml). The reagent is diluted with 19-fold of 95% ethanol prior to use.

For an ester exchange reaction, the products are determined by high-performance liquid chromatography (HPLC) or gas chromatography. The protein concentration in benzene is determined by the biuret method after removing the solvent by evaporation.

Modification of Lipase from Pseudomonas Species with Activated PEG_2

Reagents

A homogenous lipase of *Pseudomonas fluorescens*[23] or of *P. fragi* 22.39B[24] is purchased from Amano Pharmaceutical Co. (Nagoya, Japan) and from Sapporo Breweries Ltd. (Yaizu, Japan), respectively. Each lipase has a single peptide chain with a molecular weight of 33,000, and catalyzes hydrolysis of triglycerides (about 2000 and 3600 U/mg of protein, respectively) in an emulsified aqueous system

Activated PEG_2 is prepared as described previously [Eq. (1)]

Sodium borate buffers (0.2 *M*, pH 9.5, and 0.1 *M*, pH 8.0) are prepared from boric acid and sodium hydroxide.

Modification Procedure. To 100 ml of 0.2 *M* sodium borate buffer (pH 9.5) containing 1 g of lipase is added stepwise 9.1 g of activated PEG_2 under stirring. The reaction mixture is stirred at 25° for 5 hr and is thoroughly ultrafiltered against 0.1 *M* borate buffer (pH 8.0) using an Amicon (Danvers, MA) Diaflo YM100 membrane to remove unreacted activated PEG_2 and its hydrolyzate. After dialysis against water and lyophilization, PEG_2–lipase is obtained with about 90% recovery by mass. Absence of the unmodified enzyme in the preparation is confirmed by electrophoresis on a 5% polyacrylamide gel containing 0.1% sodium dodecyl sulfate and 4.5 *M* urea. In the case of *P. fragi* lipase,[25] the degree of modification of amino groups in the molecule is 49%, and 43% of the hydrolytic activity determined with olive oil emulsion is retained. The ester synthesis activity of PEG_2–lipase in benzene is 13.6 μmol/min/mg of protein using lauryl alcohol and lauric acid as substrates.

[23] M. Sugiura, T. Oikawa, K. Hirano, and T. Inukai, *Biochim. Biophys. Acta* **488,** 353 (1977).

[24] T. Nishio, T. Chikano, and M. Kamimura, *Agric. Biol. Chem.* **51,** 181 (1987).

[25] T. Nishio, K. Takahashi, T. Yoshimoto, Y. Kodera, Y. Saito, and Y. Inada, *Biotechnol. Lett.* **9,** 187 (1987).

Effect of Solvents. The enzymatic activity of PEG_2–lipase as well as catalase and peroxidase is generally exhibited in water-immiscible organic solvents such as benzene, toluene, chloroform, and 1,1,1-trichloroethane. The lipase activity in 1,1,1-trichloroethane is almost 3.6 times higher than that in benzene or toluene.[26] However, a trace amount of water in the water-immiscible organic solvents is needed for these activities. The water molecules may adsorb at the surface of the protein molecule, so that the protein conformation could be maintained.[27] Therefore, use of organic solvents saturated with water is recommended even in reverse reactions of hydrolysis with PEG_2–lipase. Little activity is shown in water-miscible solvents such as acetone and dimethylformamide.

Stability and Recovery. The modified lipase has a high stability in benzene.[28] The PEG_2–lipase still retains about 50% of the original activity for amyl laurate synthesis after 3 months of storage at room temperature, and about 40% activity remains even after 140 days.

The PEG-modified enzymes can be recovered from the reaction mixture. To 1 ml of benzene solution containing PEG_2–lipase (0.31 mg protein/ml) is added 2 ml of *n*-hexane or petroleum ether, and the precipitate formed is collected by centrifugation at 1500 *g* for 10 min. The precipitate is redissolved in 1 ml of benzene, and the enzymatic activity is shown to be well retained.

Retinyl Ester Synthesis. Ester exchange reactions between retinyl acetate and palmitic acid or oleic acid are performed with PEG_2–lipase in benzene.[29] To 15 ml of benzene containing 0.2 *M* palmitic acid or oleic acid and 20 m*M* retinyl acetate is added 5 ml of water-saturated benzene containing 5 mg (as protein) of PEG_2–lipase. The reaction is carried out under nitrogen gas at 25°. For comparison, conventional organic synthesis of retinyl esters is performed by using *p*-toluenesulfonic acid under nitrogen gas and refluxing for 3 hr.

The products, retinyl palmitate and retinyl oleate, are analyzed by high-performance liquid chromatography using a Merck (Darmstadt, Germany) LiChrosorb RP-18 column (5 μm, 4.0 × 250 mm) monitoring the effluent at 325 nm. The elution is performed with a mixture of acetone, acetonitrile, and chloroform (5 : 4 : 1, v/v/v) at a flow rate of 1.0 ml/min. The peroxide

[26] K. Takahashi, A. Ajima, T. Yoshimoto, M. Okada, A. Matsushima, Y. Tamaura, and Y. Inada. *J. Org. Chem.* **50,** 3414 (1985).

[27] K. Takahashi, H. Nishimura, T. Yoshimoto, M. Okada, A. Ajima, A. Matsushima, Y. Tamaura, Y. Saito, and Y. Inada, *Biotechnol. Lett.* **6,** 765 (1984).

[28] T. Yoshimoto, K. Takahashi, H. Nishimura, A. Ajima, Y. Tamaura, and Y. Inada, *Biotechnol. Lett.* **6,** 337 (1984).

[29] A. Ajima, K. Takahashi, A. Matsushima, Y. Saito, and Y. Inada, *Biotechnol. Lett.* **8,** 547 (1986).

values (*POV*) of the products isolated by the HPLC are measured by the official method of the American Oil Chemists' Society.[29a]

As much as 85% of the retinyl acetate is converted to retinyl palmitate at 25° with PEG_2–lipase. Its *POV* value, 2.5 mEq/kg, is almost 17 times lower than the 43 mEq/kg obtained by organic synthesis (about 30% yield). A similar result is obtained for retinyl oleate synthesis in benzene: 9.0 mEq/kg for the enzyme reaction and 200 mEq/kg for organic synthesis. Obviously, for ester synthesis starting from two substrates with double bonds, PEG–enzyme is preferable to the organic synthesis.

Ester Synthesis Reactions in Neat Substrates. The PEG–lipase is soluble not only in organic solvents but also in hydrophobic substrates. Therefore, ester synthesis and ester exchange reactions with hydrophobic substrates via PEG–lipase proceeds without organic solvents.[30] For example, the ester exchange reaction with PEG_2–lipase proceeds in neat liquid substrates, without organic solvents, as shown in Eq. (9).

$$\text{Amyl laurate} + \text{lauryl alcohol} \rightarrow \text{lauryl laurate} + \text{amyl alcohol} \quad (9)$$

A mixture (1.2 ml) of two liquid substrates is added to 1 mg of PEG_2–lipase with a trace amount of water (about 1 μl). The transparent sample is incubated at 55°–70°. At suitable times, an aliquot is taken out and diluted with acetonitrile to stop the reaction. The products are analyzed by high-performance liquid chromatography using a Merck LiChrosorb RP-8 column (5 μm, 4 × 250 mm) equipped with a refractive index detector. The elution is carried out with acetonitrile with a flow rate of 1.0 ml/min.

In the mixture of two substrates containing 1.35 *M* amyl laurate and 2.7 *M* lauryl alcohol, the apparent specific activity of the ester exchange reaction with PEG_2–lipase is 2.6 μmol/min/mg of protein at the optimum temperature of 65°. The optimum temperature for the hydrolysis reaction in an aqueous emulsified state with unmodified lipase is 45°. The higher optimum temperature shown by PEG_2–lipase may be due to the modification with PEG and/or to the alteration of the reaction environment. The PEG_2–lipase efficiently catalyzes reactions not only between an ester and an alcohol but also between an ester and a fatty acid and between two esters containing triglycerides in neat liquid substrates. The versatile activity and heat-stable nature of the modified enzyme, as well as the solubility in neat hydrophobic substrate, make the PEG_2–lipase extremely useful for many practical applications such as altering the properties of fats and oils.[31]

[29a] American Oil Chemists' Society, Official Method, Cd 8–53.

[30] K. Takahashi, Y. Kodera, T. Yoshimoto, A. Ajima, A. Matsushima, and Y. Inada, *Biochem. Biophys. Res. Commun.* **131,** 532 (1985).

[31] A. Matsushima, Y. Kodera, K. Takahashi, Y. Saito, and Y. Inada, *Biotechnol. Lett.* **8,** 73 (1986).

Kinetic Study on Substrate Specificity. Because the reactions proceed in a homogeneous benzene solution containing substrates and the modified lipase, the Michaelis equation can be applied to the reaction system. The enzymatic activity of an ester synthesis reaction is enhanced with increasing the substrate concentration and tends to reach a constant level. Plotting the reciprocal of the substrate concentration ($1/S$) against the reciprocal of the reaction rate ($1/v$) gives a straight line, from which apparent V_{max} and K_m values together with K_i values of inhibitors can be obtained.

Table I shows the K_m, K_i, and V_{max} values for various alcohols and fatty acids.[21] The V_{max} value is enhanced by increasing the carbon number of either the fatty acid or alcohol, whereas the K_m value is little affected. Fatty acids with a branching carbon chain at the position neighboring the carboxyl group act as competitive inhibitors of ester synthesis from pentyl

TABLE I
K_m AND V_{max} FOR ESTER SYNTHESIS BY PEG_2–LIPASE FROM *Pseudomonas fluorescens* AND K_i OF INHIBITORS[a]

Substrates	K_m (*M*)	K_i (*M*)	V_{max} (μmol/min/mg)
Pentyl alcohol (0.75 *M*)			
Pentanoic acid	0.17		2.8
Hexanoic acid	0.22		8.0
Octanoic acid	0.14		8.1
Dodecanoic acid	0.31		10.3
2-Methylpentanoic acid[b]	—	0.22	—
3-Methylpentanoic acid[b]	—	0.27	—
4-Methylpentanoic acid	0.19		0.2
Benzoic acid[b]	—	0.10	—
3-Phenylpropionic acid	0.18		7.0
4-Phenylbutyric acid	0.15		7.3
Pentanoic acid (0.50 *M*)			
Pentyl alcohol	0.23		2.6
Hexyl alcohol	0.25		2.2
Octyl alcohol	0.24		5.4
Dodecyl alcohol	0.33		8.0
2-Methylpentyl alcohol	0.21		2.5
3-Methylpentyl alcohol	0.62		2.0
4-Methylpentyl alcohol	0.19		1.0
1-Methylbutyl alcohol	0.31		0.1
1,1-Dimethylpropyl alcohol[c]	—	—	—

[a] Reactions were performed in benzene at 20°. Reprinted from Takahashi *et al.*[21]

[b] Acted as inhibitors, not as substrates.

[c] Acted neither as an inhibitor nor as a substrate.

TABLE II
K_m AND V_{max} FOR ESTERIFICATION OF SECONDARY ALCOHOLS[a]

	K_m (M)		V_{max} (μmol/min/mg)		V_{max}/K_m	
Alcohol	*R*	*S*	*R*	*S*	*R*	*S*
2-Butanol	0.43	0.43	1.18	1.23	2.7	2.9
2-Pentanol	0.43	1.34	1.25	0.45	2.9	0.34
2-Octanol	0.54	1.50	1.68	0.42	3.1	0.28
2-Nonanol	0.50	1.66	1.37	0.30	2.7	0.18
α-Phenylethanol	0.47	—[b]	0.90	—[b]	1.9	—[b]

[a] Reactions were conducted using 0.5 *M* dodecanoic acid in benzene with PEG_2–lipase from *P. fragi*. Adapted from Kikkawa *et al.*[32]
[b] Did not serve as a substrate.

alcohol and pentanoic acid. For alcohol substrates the substrate specificity is not as restrictive as for fatty acids, but secondary and tertiary alcohols did not serve as substrates. The results shown here are obtained from *P. fluorescens*, and kinetic properties may differ for lipases from different sources.

Based on a kinetic study of the esterification of chiral secondary alcohols shown in Table II.[32] PEG_2–lipase from *P. fragi* exhibits higher stereoselectivity for chiral secondary alcohols with a longer carbon chain or a phenyl group. For example, optical resolution of (*RS*)-α-phenylethanol is performed with *n*-dodecanoic acid and PEG_2–lipase in 1,1,1-trichloroethane, and only the (*R*)-isomer is esterified with PEG_2–lipase. The optical purity of the unreacted alcohol, (*S*)-α-phenylethanol, reaches 99% enantiomeric excess in 7 hr.

Determination of K_m Value for Water. Kinetic study on indoxyl acetate hydrolysis [Eq. (10)] is performed in transparent benzene solutions in

$$\text{Indoxyl acetate} + H_2O \rightarrow \text{3-hydroxyindole} + \text{acetic acid} \quad (10)$$

order to obtain the K_m value for H_2O.[33] To 2.8 ml of benzene containing indoxyl acetate (3–100 m*M*) and water (0–30 m*M*) is added 0.2 ml of PEG_2–lipase (0–20 μg protein/ml). The transparent reaction mixture is incubated at 37°. During the course of the reaction, the amount of 3-hydroxyindole is spectrophotometrically determined by measuring the

[32] S. Kikkawa, K. Takahashi, T. Katada, and Y. Inada, *Biochem. Int.* **19,** 1125 (1989).
[33] A. Matsushima, A. Okada, K. Takahashi, T. Yoshimoto, and Y. Inada, *Biochem. Int.* **11,** 551 (1985).

absorbance assuming the molar extinction coefficient to be 2100 M^{-1} cm^{-1} at 385 nm.

The double-reciprocal plots of the rate ($1/v$) and water concentration ($1/S$) at fixed concentrations of indoxyl acetate give parallel straight lines, indicating that the hydrolysis takes place as a double-displacement reaction (Ping-Pong).[34] The K_m value is calculated to be 7×10^{-2} M for H_2O and 1.6×10^{-1} M for indoxyl acetate, and V_{max} is 4,700 μmol/min/mg of protein.

Modification of Candida Lipase with PEG Succinimide

Reagents

A crude lipase preparation (EC 3.1.1.3, triacylglycerol lipase) from *Candida cylindracea*[35] is purchased from Meito Sangyo Co., Ltd. (Nagoya, Japan). The *Candida* lipase is unstable in alkaline environment so that the modification should proceed in a neutral solution

α-Carboxymethyl-ω-methoxypoly(oxyethylene) (average molecular weight 4500) is activated with *N*-hydroxysuccinimide as shown in Eq. (3)

Phosphate-buffered saline contains 8 g NaCl, 0.2 g KCl, 2.9 g $Na_2HPO_4 \cdot 12H_2O$, and 0.2 g KH_2PO_4 in 1 liter of water, pH 7.0

Modification Procedure. Lipase from *Candida cylindracea* (560 mg) dissolved in 8 ml of phosphate-buffered saline (pH 7.0) is dialyzed against the same buffer followed by centrifugation at 10,000 rpm for 10 min to remove insoluble materials.[36] To 10 ml of the enzyme solution (190 mg of protein) is slowly added 200 mg of PEG succinimide. The mixture is incubated for 15 min at 25° under gentle stirring. The PEG–lipase is obtained after the mixture is dialyzed against water and is lyophilized. The degree of modification is 47%, and 56% of the original hydrolytic activity is retained using emulsified olive oil as the substrate.

Substrate Specificity. To a benzene solution containing a carboxylic acid (0.6 M) and an alcohol (0.6 M) is added PEG–lipase (final concentration: 0.5 mg/ml). The substrate specificity for ester synthesis is tested.[36] By varying the carbon number (1–12) of the alcohol, the activity of PEG–lipase is found to be relatively high with short-chain alcohols but drops sharply when the carbon number of the alcohol exceeds 5. By varying the carbon number (4–12) of the carboxylic acid, it is shown that the

[34] W. W. Cleland, *Biochim. Biophys. Acta* **67,** 104 (1963).

[35] N. Tomizuka, Y. Ota, and K. Yamada, *Agric. Biol. Chem.* **30,** 576 (1966).

[36] Y. Kodera, K. Takahashi, H. Nishimura, A. Matsushima, Y. Saito, and Y. Inada, *Biotechnol. Lett.* **8,** 881 (1986).

activity of the PEG–lipase is high for butyric, octanoic, and lauric acids but is quite low for pentanoic and hexanoic acids. In the case of branched carboxylic acids, 2-, or 3-methylpentanoic acid can be esterified with the PEG-modified *Candida* lipase, but not with the PEG-modified *Pseudomonas* lipase. The substrate specificity of *Candida* lipase is markedly different from that of *Pseudomonas* lipase (Table I).

Modification of Lipase with Activated PM

Reagents

A homogeneous lipase preparation (EC 3.1.1.3, triacylglycerol lipase) from *Pseudomonas fluorescens* is purchased from Amano Pharmaceutical Ltd. The total number of amino groups in the lipase molecule was reported[37] to be 7

Copolymer (molecular weight 13,000) of polyoxyethylene allyl methyl diether and maleic anhydride, activated PM_{13} shown in Eq. (5), is purchased from Nippon Oil and Fat

Sodium borate buffer (0.5 *M*, pH 8.5). Because of the pH shift caused by hydrolysis of the acid anhydride of activated PM, a strong buffer or a pH-stat should be used for pH control

Sodium borate buffer (50 m*M*, pH 8.5)

Modification Procedure. To 2 ml of lipase (2 mg/ml) in 0.5 *M* borate buffer (pH 8.5) are added 0–150 mg of activated PM_{13}.[12] The mixture is stirred at 4° for 1 hr and is diluted by the addition of 100 ml of 50 m*M* borate buffer (pH 8.5). Unreacted copolymer is removed by ultrafiltration against the buffer solution with an Amicon Diaflo PM30 membrane. The modified lipase is then dialyzed against water and lyophilized.

Enzymatic Activity. Figure 2 shows the degree of modification of lipase with activated PM_{13} on changing the molar ratio of the modifier to amino groups in lipase. The degree of modification is enhanced by increasing the molar ratio and tends to approach a constant level of 60%. This indicates that approximately 4 of the 7 amino groups in the lipase molecule are coupled with activated PM_{13}. Surprisingly, the esterase activity in an emulsified aqueous system is enhanced when amino groups are coupled with activated PM_{13}. The absolute activity of the modified lipase is 1200 μmol/min/mg of protein, which is 1.3 times higher than that of unmodified lipase.

The modified lipase also becomes soluble in organic solvents, so that the reverse of hydrolysis, namely, ester synthesis, proceeds in these solvents. In fact, lauryl stearate is synthesized in a clear benzene solution

[37] M. Sugiura and T. Oikawa, *Biochim. Biophys. Acta* **489,** 262 (1977).

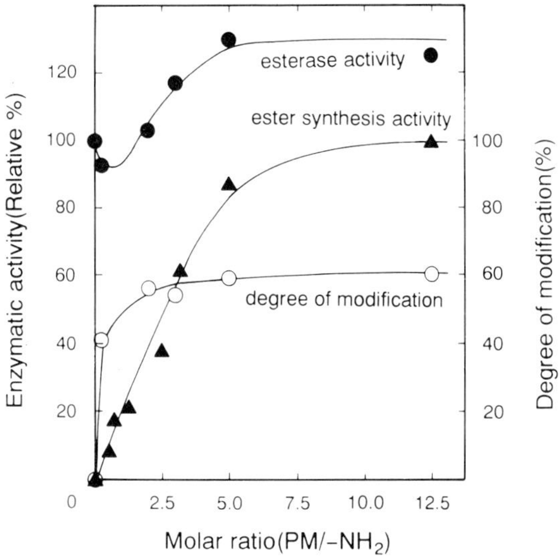

FIG. 2. Dependence of enzymatic activities on the degree of modification of lipase with activated PM. The horizontal axis shows the molar ratio of activated PM to amino groups in the lipase molecule. The lipase modification reaction was performed at 4° for 1 hr. The esterase activity and the ester synthesis activity were determined in emulsified olive oil and in benzene containing lauryl alcohol and stearic acid, respectively. (From Hiroto *et al.*[12])

containing lauryl alcohol and stearic acid with PM–lipase. The unmodified lipase does not catalyze any ester synthesis reaction in benzene. The ester synthesis activity in benzene is enhanced by increasing the degree of modification and tends to approach a constant level (4.7 μmol/min/mg of protein). In these cases, carboxyl groups formed by hydrolysis of activated PM are not esterified in organic solvents, in conformity with its substrate specificity (see Table I).

Heat Stability of PM_{13}–Lipase. The PM_{13}–lipase (molar ratio of PM/NH_2 groups 12.5) or unmodified lipase is incubated at 55°, and the esterase activity in an emulsified system and the ester synthesis activity in benzene are tested. Although the esterase activity of unmodified lipase is completely lost after 150 min of incubation at 55°, the modified lipase retains 60% activity even after 150 min of incubation. Similarly 50% of the ester synthesis activity of PM–lipase in benzene is retained over the same incubation time.

Preparation of Magnetized Lipase

In biotechnological processes, enzyme recovery is often critically important, and immobilized enzymes have been explored to cope with this problem. However, one of the most attractive alternatives is to make use of magnetic force. We describe here two methods to endow enzymes with magnetic properties [Eqs. (11) and (12)].

$$\underset{(Fe_3O_4)}{\text{Magnetite}} + \text{PEG–enzyme} \rightarrow \text{magnetized enzyme} \quad (11)$$

$$\text{Activated magnetite–PEG} + \text{enzyme} \rightarrow \text{magnetized enzyme} \quad (12)$$

Reagents

α,ω-Dicarboxymethylpolyoxyethylene (DCPEG; molecular weight: 2000) is obtained from Nippon Oil and Fat

Lipase from *Pseudomonas fragi* is modified with activated PEG_2 as in the previous section

$FeCl_3 \cdot 6H_2O$, $FeCl_2 \cdot 4H_2O$, 28% ammonium solution, 3 *N* sodium hydroxide, and 0.03% hydrogen peroxide solution are prepared

N-Hydroxysuccinimide, dicyclohexylcarbodiimide, and dioxane are dried with Molecular Sieves 3A

Magnet: magnetic field of 250 or 6000 G (electromagnet). The latter enables the magnetic particles to separate quickly from the reaction mixture

Method 1. To 1.5 ml of a mixture containing PEG_2–lipase (50 mg) and DCPEG (50 mg) is added dropwise a mixture of $FeCl_3 \cdot 6H_2O$ (61 mg) and $FeCl_2 \cdot 4H_2O$ (26 mg) dissolved in 0.2 ml of water, with the reaction mixture being adjusted to pH 8.0–8.5 with 28% ammonium solution under vigorous stirring at room temperature.[38] The reaction mixture is dialyzed against water, followed by lyophilization to obtain the magnetized lipase in a powder form.

The composition of the magnetized lipase is as follows: magnetite 31%, polyethylene glycol 44%, and lipase 25%. The magnetite content is estimated by measuring the amount of iron with a Varian (Palo Alto, CA) AA-875 atomic absorption spectrophotometer, and protein is determined after removing magnetite from magnetized enzyme with concentrated hydrochloric acid. The average size of the magnetized lipase is 120 ± 60 nm as measured by a submicron particle analyzer, model N4 (Coulter Electronics, Inc., Hialeah, FL), at 20° at a concentration of 3 mg/ml. The enzymatic activity is 11.6 μmol/min/mg of protein for lauryl laurate synthesis in 1,1,1-trichloroethane and 250 U/mg of protein for olive oil hydrolysis.

Method 2. To 20 ml of 30% DCPEG in water are added 6 ml of 1% $FeCl_2 \cdot 4H_2O$ under agitation by bubbling with nitrogen gas, followed by a slow addition of 10–20 μl of 0.03% hydrogen peroxide with stirring at

[38] K. Takahashi, Y. Tamaura, Y. Kodera, T. Mihama, Y. Saito, and Y. Inada, *Biochem. Biophys. Res. Commun.* **142,** 291 (1987).

50°.[39] During the process, the pH is kept at 10.0 by adding 3 *N* sodium hydroxide. After the PEG–magnetite is formed, it is recovered by magnetic force and washed several times with water. Once formed, the PEG–magnetite never dissociates into the original components. It disperses well in organic solvents as well as in aqueous solution.

A mixture of vacuum-dried PEG–magnetite (1 g), *N*-hydroxysuccinimide (0.1 g), and Molecular Sieves 3A (1 g) dispersed in dry dioxane (100 ml) is stirred for 1 hr. To the mixture is added dicyclohexylcarbodiimide (0.2 g) in dry dioxane (3 ml), and the reaction is incubated at room temperature for 24 hr with stirring. The activated magnetic modifier is recovered from the reaction mixture with magnetic force and is washed several times with dioxane, then vacuum dried. The average particle size of the activated magnetic modifier thus obtained is approximately 200 nm.

To 2.5 ml of lipase (5 mg) in phosphate-buffered saline (pH 7.0) is added the activated magnetic modifier (100 mg), and the mixture is dispersed by sonication. After stirring at room temperature for 1 hr, the magnetized lipase is recovered by magnetic force and is washed several times with water.

The weight composition of magnetite, polyethylene glycol, and enzyme are 10.2, 85.7, and 4.1%, respectively. The activities of olive oil hydrolysis and lauryl laurate synthesis are 640 U/mg of protein and 14.8 μmol/min/mg of protein, respectively.

Comments. Figure 3 displays the process of magnetic separation after completion of the lauryl laurate synthesis from lauryl alcohol and lauric acid with magnetized lipase prepared by method 1 in benzene. The magnetized lipase particles are completely recovered with a magnet (250 G) in 5 min, and the enzymatic activity was retained completely. Therefore, the magnetized lipase can be used repeatedly without impairing the enzymatic activity. The PEG–lipase is not readily released from magnetite; probably PEG bound to lipase interacts with magnetite via hydrogen bonds and/or coordination bonds. The use of DCPEG is essential for successful dispersion of magnetic particles in benzene.

Method 2 is more flexible in allowing easy preparation of magnetized enzymes from various enzymes. However, the enzymatic activity per weight of magnetized enzyme is lower because a lesser amount of enzyme is coupled to magnetite. Generally, smaller sized magnetite molecules could bind more enzymes, but such magnetized enzymes were difficult to recover by magnetic force. In biotechnology applications, magnetization

[39] T. Yoshimoto, T. Mihama, K. Takahashi, Y. Saito, Y. Tamaura, and Y. Inada, *Biochem. Biophys. Res. Commun.* **145,** 908 (1987).

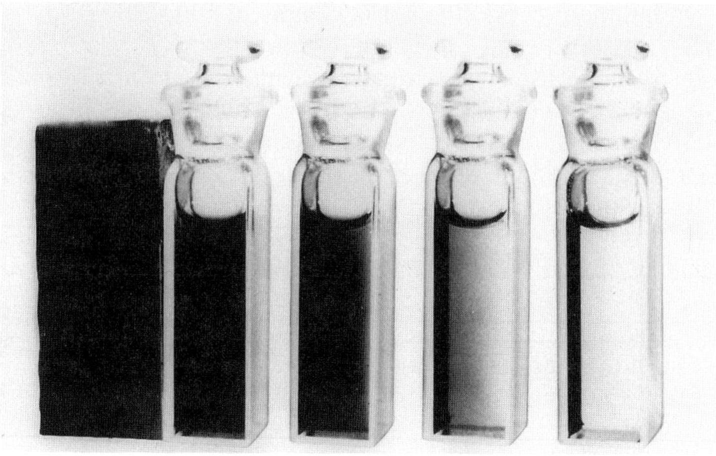

FIG. 3. Magnetic separation of magnetized lipase from benzene after lauryl laurate synthesis. The magnetic separation was accomplished within 5 min when a magnet (250 G) was used. [From Y. Inada, K. Takahashi, T. Yoshimoto, Y. Kodera, A. Matsushima, and Y. Saito, *Trends Biotechnol.* **6,** 131 (1988)].

techniques such as those described here are very valuable. The magnetized enzymes can be easily separated from the reaction mixture. Magnetic delivery of drugs is another biomedical application.[40]

Modification of Protein Drugs

Principle

It is frequently necessary to administer proteins of nonhuman origin to humans as protein drugs, even though foreign proteins have serious immunogenic side effects such as anaphylactic shock induced by frequent injections. Many investigators have attempted to reduce the immunogenicity and immunoreactivity associated with the protein drugs while retaining their biomedical functions. In 1977, Abuchowski *et al.* demonstrated that catalase[41] modified with a polyethylene glycol derivative (activate PEG_1) lost its immunoreactivity toward the antiserum while retaining its enzymatic activity. Polyethylene glycols, which are nontoxic and nonimmunogenic synthetic macromolecules, have been used as precipitants of proteins, agents for cell fusion, and plasma expanders.

[40] T. Yoshimoto, Y. Saito, K. Sugibayashi, Y. Morimoto, T. Tsukada, K. Kanmatsuse, Y. Kodera, A. Matsushima, and Y. Inada, *Drug Delivery Syst.* **4,** 121 (1992).

[41] A. Abuchowski, J. R. McCoy, N. C. Palczuk, T. van Es, and F. F. Davis, *J. Biol. Chem.* **252,** 3582 (1977).

The technique of modification with polyethylene glycol has opened new avenues to reduce immunoreactivity and to prolong the clearance time of foreign protein drugs for biomedical applications. More than 200 reports of such applications have been published. The following proteins have been successfully modified with polyethylene glycol derivatives: (1) antitumor proteins: arginase, asparaginase, glutaminase–asparaginase, γ-interferon, interleukin 2, phenylalanine ammonia-lyase, and tryptophanase; (2) enzymes for inherited deficiency: adenosine deaminase, galactosidase, glucosidase, glucronidase, gulonolactone oxidase, insulin, purine nucleoside phosphorylase, and urokinase; (3) anti-inflammatory enzymes: catalase and superoxide dismutase; (4) allergic proteins: ovalbumin, honeybee venom, ragweed pollen, timothy grass pollen, and mite allergens; (5) antithrombotic enzymes: batroxobin, antithrombin III, elastase, plasminogen, tissue-type plasminogen activator, streptokinase, and urokinase; (6) blood-proteins: albumin, blood coagulation factor VIII, thrombin, hemoglobin, immunoglobulins, and serine protease inhibitors; and (7) others: alkaline phosphatase, antibiotics, Arg-Gly-Asp (fibronectin), bilirubin oxidase, eglin C, glutathione, protein A, ribonuclease, soybean trypsin inhibitor, and substance P.

Modification of L-Asparaginase with Activated PEG_2

L-Asparaginase, which catalyzes the hydrolysis of L-asparagine to L-aspartic acid and ammonia [see Eq. (13)], has been used clinically in

$$\text{Asparagine} + H_2O \rightarrow \text{aspartic acid} + NH_3 \qquad (13)$$

therapy of leukemia and lymphosarcoma.[42] L-Asparaginase from *Escherichia coli* has a molecular weight of 136,000 and four identical subunits. Its administration to patients causes serious immunological side effects owing to its "foreign" origin. To overcome this drawback, Inada *et al.* explored two groups of modifiers: the first group includes chain-form polyethylene glycol derivatives, including activated PEG_1 [Eq. (2)][43] and activated PEG_2 [Eq. (1)],[6] as previously described. Another group includes comb-shaped copolymers of polyethylene glycol derivatives and maleic anhydride, that is, activated PM [Eq. (5)],[44] which also has been described in an earlier section. Each group of modifiers has been used to modify

[42] R. L. Capizzi, *Cancer Treat. Rep.* **65**(suppl. 4), 115 (1981).

[43] Y. Ashihara, T. Kono, S. Yamazaki, and Y. Inada, *Biochem. Biophys. Res. Commun.* **83,** 385 (1978).

[44] Y. Kodera, H. Tanaka, A. Matsushima, and Y. Inada, *Biochem. Biophys. Res. Commun.* **184,** 144 (1992).

L-asparaginase for the purpose of reducing immunoreactivity and prolonging clearance time.

Reagents

Crystalline L-asparaginase (EC 3.5.1.1) from *Escherichia coli* A-1-3[45] is provided from Kyowa Hakko Co., Ltd. (Tokyo, Japan). The asparaginase molecule contains a total of 92 amino groups: four terminal amino groups and 88 lysine ε-amino groups. The specific activity of asparaginase is 200 IU/mg protein by the GOT method (see below)
Activated PEG_2 is prepared as described in Eq. (1)
Sodium borate buffer (0.1 *M*, pH 10.0)
Phosphate-buffered saline (pH 7.0)
Glutamate–oxaloacetate transaminase (GOT) (EC 2.6.1.1, aspartate aminotransferase) from porcine heart and malate dehydrogenase (MDH) (EC 1.1.1.37) from porcine heart are purchased from Boehringer Mannheim GmbH (Mannheim, Germany). NADH and α-ketoglutaric acid are of analytical grade

Modification Procedure. Activated PEG_2 (1 g) is added to 2 ml of asparaginase (5 mg/ml) in 0.1 *M* borate buffer (pH 10.0), and the mixture is stirred at 37° for 1 hr.[6] To the reaction mixture is added 80 ml of cold phosphate-buffered saline to stop the reaction. The reaction mixture is ultrafiltered through a ULVAC Diafilter A-50T (Tokyo, Japan) membrane to remove the remaining activated PEG_2. The degree of modification and the protein concentration are determined by the trinitrobenzenesulfonate method[13] and biuret method,[18] respectively.

Determination of Asparaginase Activity. The enzymatic activity of L-asparaginase is determined by the modified GOT method.[46] The decrease of absorbance of NADH at 340 nm resulting from the formation of malate from L-asparagine by the combined action of glutamate–oxaloacetate transaminase and malate dehydrogenase is measured. To 30 ml of 2.5 m*M* Tris-HCl buffer containing 4.5 mg of L-asparagine, 3.4 mg of NADH, and 19 mg of α-ketoglutaric acid are added 250 μl of GOT (0.5 mg) and 200 μl of MDH (1 mg). The substrate mixture (3 ml) is incubated at 37° for 10 min. To the above mixture in quartz cuvette is added 50 μl of asparaginase solution (about 0.05 mg protein/ml), and the absorbance at 340 nm is measured as a function of time. The activity is reported as micromoles of L-aspartate formed per minute (IU).

Quantitative Precipitin Reaction. The quantitative precipitin reaction

[45] N. Nakamura, Y. Morikawa, T. Fujio, and M. Tanaka, *Agric. Biol. Chem.* **35,** 219 (1971).
[46] D. A. Cooney, R. L. Capizzi, and R. E. Handschumacher, *Cancer Res.* **30,** 929 (1970).

curve is obtained by the method of Kabat.[47] Anti-asparaginase serum is obtained from rabbits immunized three times (3.3 mg × 3) and is stored at −80°. To a known amount of anti-asparaginase serum are added sample solutions, and the mixtures are kept standing at 37° for 1 hr and then 4° for 16 hr to obtain precipitates of the antigen–antibody complex. The amount of protein precipitated is determined by the Lowry method.[18]

Comments. The effect of the modification with polyethylene glycol on the reduction of immunoreactivity depends on the molecular weight of the polyethylene glycol, the degree of modification of amino groups, and the shape of the modifiers (chain form or comb form). If the average molecular weight of polyethylene glycol is lower than 5000, serious reduction in the enzymatic activity results without complete elimination of immunoreactivity. Therefore, polyethylene glycol with a molecular weight of more than 5000 is recommended as a modifier.

The degree of modification of amino groups in the asparaginase molecule is controlled by varying the amount of each modifier. Addition of activated PEG_2 to amino groups in asparaginase with a molar ratio of 15 : 1 gives rise to modification of 52 of the total 92 amino groups (57% modification) in the asparaginase molecule, and binding by anti-asparaginase serum is completely lost, with retention of 11% of the initial enzymatic activity. The pH dependency and K_m value of the modified asparaginase thus prepared are identical with those of the unmodified enzyme.

Prolonging Clearance Time. The clearance times of the modified and unmodified enzymes in blood are studied by injecting the preparations at 80 IU/kg i.p. into a group of five rats (Wistar strain of male albino rats, weighting about 300 g).[48] Blood (~0.25 ml) is taken from the tail vein without anticoagulant and immediately centrifuged to obtain serum. The sample (10–50 μl) is assayed for enzymatic activity. The sample (100 μl) is deproteinized with 15% perchloric acid (50 μl) followed by centrifugation at 1500 *g* for 20 min at 4°. The amino acid composition of the supernatant (75 μl) is analyzed with an Atto MLC-703 (Tokyo, Japan) amino acid analyzer.

As shown in Fig. 4, enzymatic activity appears in the serum after injection of unmodified asparaginase; however, the activity rapidly decreases, and the amount of L-asparagine returns to the normal level after 3 days. On the other hand, injection of PEG_2–asparaginase gives quite a different profile. The enzymatic activity of the modified enzyme is detected in the serum within 30 min, and the peak level persists for about 20 hr. Thereafter, the activity decreases far more slowly than that of the native

[47] E. A. Kabat, this series, Vol. 70, p. 13.

[48] Y. Kamisaki, H. Wada, T. Yagura, A. Matsushima, and Y. Inada, *J. Pharmacol. Exp. Therap.* **216,** 410 (1981).

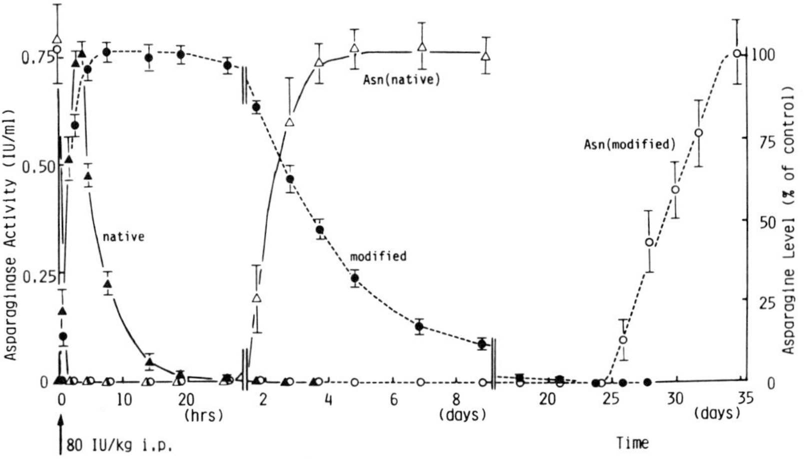

FIG. 4. L-Asparaginase activity and L-asparagine concentration in rat serum after intraperitoneal injection of native or PEG_2-modified enzyme. L-Asparaginase activity is denoted by filled symbols and L-asparagine concentration by open symbols. (From Kamisaki *et al.*[48])

enzyme, and is still detectable after 3 weeks. The half-lives of the native and modified enzymes in the serum are calculated to be 2.9 and 56 hr, respectively. The slower clearance rate of the modified asparaginase may render its antitumor activity more effective.

Preparation of PM–Asparaginase

Modification Procedure. To 1.0 ml of asparaginase (4.0 mg/ml) dissolved in 0.5 *M* borate buffer (pH 8.5) is added 0–200 mg of activated PM_{13}, and the mixture is stirred at 4° for 1 hr.[44] The reaction mixture is diluted with 100 ml of 50 m*M* borate buffer (pH 7.0), and then the uncoupled activated PM_{13} is removed by ultrafiltration with an Amicon Diaflo YM100 membrane to obtain PM_{13}–asparaginase. A similar modification is performed with activated PM_{100} to obtain PM_{100}–asparaginase using an Amicon Diaflo XM300 ultrafiltration membrane.

Reduction of Immunoreactivity. In the case of PM_{13}–asparaginase, the immunoreactivity is completely lost with 50% modification at a PM_{13}/NH_2 group molar ratio of approximately 3. In the case of PM_{100}–asparaginase, the degree of modification reaches a constant level at a PM_{100}/NH_2 group ratio exceeding 0.4, which leads to approximately 34% of the amino groups in the asparaginase molecule being modified with activated PM_{100}. Its immunoreactivity is completely lost, and 85% of the enzymatic activity is preserved, which is approximately 2-fold greater than the value of 46% for PM_{13}–asparaginase.

TABLE III
ACTIVITY AND IMMUNOREACTIVITY OF *Escherichia coli* ASPARAGINASE MODIFIED WITH POLYETHYLENE GLYCOL DERIVATIVES

Modifying reagent	Molecular weight of modifier	Degree of modification (%)	Enzymatic activity (%)	Immunoreactivity (%)	Ref.
Unmodified asparaginase		0 (0)[a]	100[b]	100	—
Activated PEG_1	5000	79 (73)	0.9	0	43
Activated PEG_2	10,000	57 (52)	11	0	6
Activated PM_{13}	13,000	50 (46)	45	0	44
Activated PM_{100}	100,000	34 (31)	85	0	44

[a] In parentheses are given the number of amino groups modified by the PEG reagent. The total number of amino groups in the asparaginase molecule is 92.
[b] Activity of the unmodified enzyme is 200 IU/mg protein by the GOT method.

Table III shows a comparison of the immunoreactivity, enzymatic activity, and degree of modification of L-asparaginase modified with the chain-shaped modifiers (activated PEG_1 and activated PEG_2) and with the comb-shaped modifers (activated PM_{13} and activated PM_{100}). It can be seen clearly that the comb-shaped activated PM_{100} is the superior modifying reagent for asparaginase (molecular weight 136,000) to reduce its immunoreactivity with retention of enzymatic activity. The comb-shaped polymers with many reactive groups of acid anhydrides react directly with ε-amino groups in lysine residues and N-terminal amino groups in the asparaginase molecule. Furthermore, they may cover nearly the entire surface of the asparaginase molecule by forming hydrogen bonds between the side chains of amino acid residues and the oxygen atoms in polyethylene glycol chains. A similar phenomenon was also observed for bovine serum albumin modified with activated PMs.

Conclusion

Polyethylene glycol-modified proteins and enzymes have been demonstrated to be invaluable as catalysts in biotechnological processes and as protein drugs in biomedical processes. The number of reports on chemical modification of proteins with polyethylene glycol derivatives is increasing yearly.

Genetic engineering has also been used in protein modification. Goodson and Katre[49] introduced a single cysteine residue into interleukin 2 by site-directed mutagenesis. Modification of the SH group with a polyethyl-

[49] R. J. Goodson and N. V. Katre, *Bio/Technology* **8,** 343 (1990).
[49a] H. Sasaki, Y. Ohtake, A. Matsushima, M. Hiroto, Y. Kodera, and Y. Inada, *Biochem. Biophys. Res. Commun.* **197,** 287 (1993).

ene glycol derivative gave a homogeneously modified protein. This is an example of "site-specific" modification. Hershfield *et al.*[50] introduced additional amino groups into purine nucleoside phosphorylase by site-directed mutagenesis for attaching polyethylene glycol on the entire enzyme surface and succeeded in the effective reduction of its immunoreactivity.

Adenosine deaminase coupled with a polyethylene glycol derivative, which was constructed by Enzon Inc. (South Plainfield, NJ), gained U.S. Food and Drug Administration approval as an orphan drug in 1990. This is the first PEG-modified enzyme to gain such approval. The PEG modification of proteins is expected to flourish in biotechnology and in medical use.

Acknowledgments

We thank Dr. Hiroshi Wada, professor emeritus of Osaka University, and Dr. Katsukiyo Sakurai, Research Institute of Seikagaku Kogyo Co., Ltd., as well as all of our collaborators at the Tokyo Institute of Technology and Toin University of Yokohama.

[50] M. S. Hershfield, S. Chaffee, L. Koro-Johnson, A. Mary, A. A. Smith, and S. A. Short, *Proc. Natl. Acad. Sci. U.S.A.* **88,** 7185 (1991).

[8] Michael Additions for Syntheses of Neoglycoproteins

By A. Romanowska, S. J. Meunier, F. D. Tropper, C. A. Laferrière, and R. Roy

Introduction

Artificial carbohydrate–protein conjugates (neoglycoproteins) represent useful antigens and immunogens from which immunodiagnostic or therapeutic reagents can be derived.[1,2] They also constitute important tools for receptor binding studies, for the inhibition of adherence of bacteria and viruses, and for cell-targeted drug delivery. To expand further existing methodologies[3] for the conjugation of functionalized carbohydrates to proteins and polymers, we describe a mild and effective procedure for

[1] R. Bell and G. Torrigiani, "Towards Better Carbohydrate Vaccines." Wiley, London, 1987.

[2] W. E. Dick and M. Beurret, *in* "Contributions to Microbiology and Immunology, Conjugate Vaccines" (J. M. Cruse and R. E. Lewis, Jr., eds.), Vol. 10, p. 48. Karger, Basel, 1989.

[3] C. P. Stowell and Y. C. Lee, *Adv. Carbohydr. Chem. Biochem.* **37,** 225 (1980).

Copyright © 1994 by Academic Press, Inc.
All rights of reproduction in any form reserved.

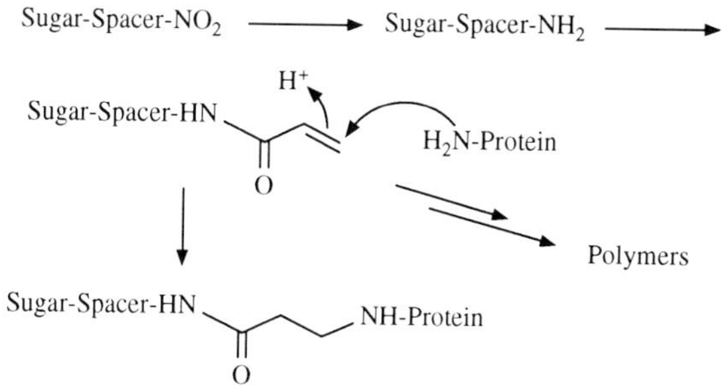

SCHEME 1

the preparation of neoglycoproteins. The key reaction involves a 1,4-conjugate addition (Michael-type addition) of nucleophilic sites on proteins (thiols or lysine ε-NH_2 groups) onto *N*-acrylamido-substituted glycosides (Scheme 1).[4–6]

The first model compound, 4-*N*-acrylamidophenyl β-lactoside (**3**) is easily derived (Scheme 2) from 4-nitrophenyl lactosisde (**1**) obtained by phase-transfer catalysis (PTC).[7] It is of interest to mention that many 4-nitrophenyl glycosides, which constitute useful glycohydrolase substrates, are commercially available; therefore, the strategy described here represents a practical entry to this family of neoglycoproteins. For the model compound, the effects of buffers, pH, and temperature have been thoroughly examined and the optimum conditions described. The influences of pH and time on the degradation of protein (bovine serum albumin, BSA) have also been evaluated using sodium dodecyl sulfate–polyacrylamide gel electrophoresis (SDS–PAGE). The binding properties of the lactosyl conjugate have also been demonstrated using agar gel diffusion with peanut lectin taken as model.

Other important antigenic glycosides have also been utilized to demonstrate the wide range of application of this new strategy. Thus, *N*-acryloylated sialic acid (**5**) and, sialyllactosamine (**7**),[4] as well as the T-antigen derivatives [β-D-Gal-(1 → 3)-α-D-GalNAc] (**11**),[6] have been conjugated to BSA and tetanus toxoid, respectively (Scheme 3). The final example strongly supports the usefulness of utilizing the ε-amino group of lysyl residues to act as efficient nucleophiles in the Michael addition. In this example,

[4] R. Roy and C. A. Laferrière, *J. Chem. Soc., Chem. Commun.* **1709** (1990).

[5] R. Roy, F. D. Tropper, T. Morrison, and J. Boratynski, *J. Chem. Soc., Chem. Commun.* **536** (1991).

[6] R. Roy, F. D. Tropper, A. Romanowska, R. K. Jain, C. F. Piskorz, and K. L. Matta, *Bioorg. Med. Chem. Lett.* **2,** 911 (1992).

[7] R. Roy, F. D. Tropper, and A. Romanowska, *Bioconjugate Chem.* **3,** 256 (1992).

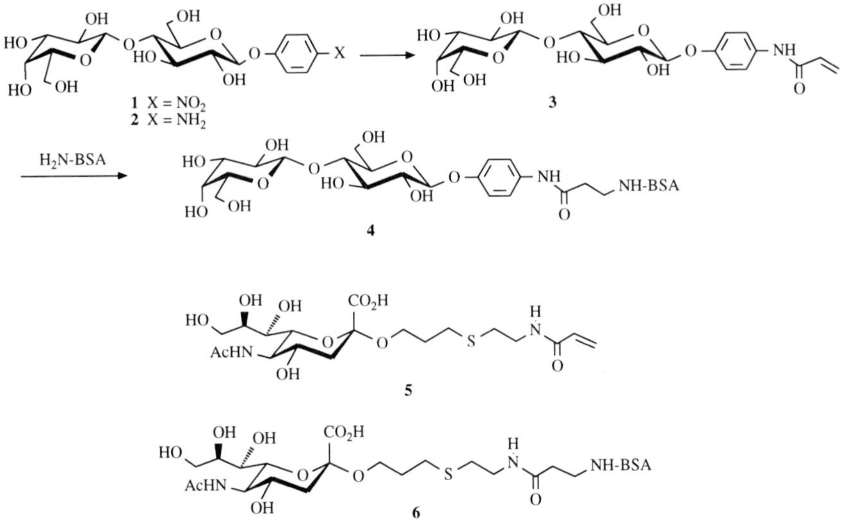

SCHEME 2

poly(L-lysine) is used as carrier peptide[8] for the direct conjugation to the lactoside **3**. The conjugate thus obtained shows a very high level of incorporation.

Materials and Methods

The details for the general procedures are given elsewhere in this volume.[9]

Synthesis of Lactoside Containing N-Acryloylated Aglycon

4-Nitrophenyl 2,3,6,2′,3′,4′,6′-Hepta-O-acetyl-β-lactoside. To a solution of peracetylated lactosyl bromide (acetobromolactose)[10] (5.20 g, 7.43 mmol), tetrabutylammonium hydrogen sulfate (2.50 g, 1 equivalent), and 4-nitrophenol (2.01 g, 14.5 mmol) in dichloromethane (50 ml) is added 1 *M* NaOH (50 ml). The reaction mixture is gently warmed until no solid precipitate appears. The two-phase reaction mixture is then vigorously

[8] R. Roy, R. A. Pon, F. D. Tropper, and F. O. Andersson, *J. Chem. Soc., Chem. Commun.* 264 (1993).

[9] R. Roy, A. Romanowska, and F. O. Andersson, this volume [18].

[10] C. P. Stowell and Y. C. Lee, this series, Vol. 83, p. 282.

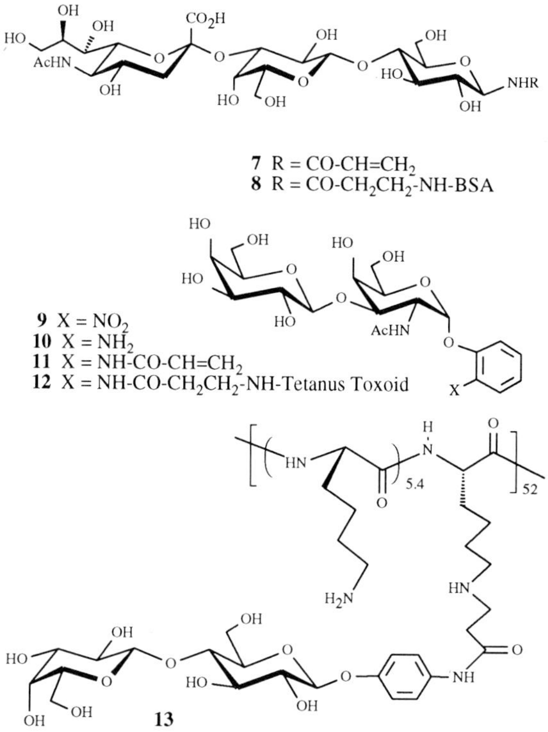

SCHEME 3

stirred at room temperature for 2.5 hr. Thin-layer chromatography (TLC) in hexane/ethyl acetate (1 : 1 v/v), containing 0.5% 2-propanol indicates complete transformation of acetobromolactose into the peracetylated 4-nitrophenyl β-lactoside. The resulting organic phase is successively washed with 1 *M* NaOH and water and dried over sodium sulfate. The residue obtained after filtration and evaporation of the solvent is purified by silica gel column chromatography using the above solvent as eluent. Crystallization of the pooled fractions from ethanol gave the title compound, 3.19 g (57%): mp 133.3°–135.4°, $[\alpha]_D$ −21.1° (*c* 1.0, $CHCl_3$).[7]

*4-Nitrophenyl β-Lactoside (**1**).* The peracetylated lactoside prepared above (1.21 g, 1.60 mmol) is dissolved in warm methanol (30 ml) containing 1 *M* sodium methoxide (40 μl). After a few hours at room temperature, **1** partially crystallizes. To ensure complete de-O-acetylation, more methanol is added (20 ml) and the solution is warmed again. The solution is stirred overnight at room temperature, and ether (25 ml) is added to the cooled solution to ensure complete crystallization. Pure **1** is thus obtained after filtration and drying (733 mg, 99% yield): mp 249.0°–250.2°, $[\alpha]_D$ −39.1° [*c* 1.25, dimethyl sulfoxide (DMSO)].

4-Aminophenyl β-Lactoside (**2**). Compound **1** (178 mg, 0.384 mmol), ammonium formate (120 mg, 5 equivalents), and 10% Pd/C (25 mg) are warmed in methanol (25 ml) containing 0.05% formic acid in a stoppered round-bottomed flask. When TLC in $CHCl_3$/methanol/water (30 : 10 : 1, v/v/v) indicates complete reduction (15 min), the reaction mixture is cooled to room temperature. The solid is filtered through Celite and rinsed with methanol. The filtrate is evaporated to dryness to give **2** still containing ammonium formate. The salt is removed by repeated freeze-drying from water. The homogeneous product (quantitative yield) is crystallized from warm (not hot) ethanol, mp 237.0°–238.5°, or water, mp 270.2°–271.5°; $[\alpha]_D$ −24.2° (*c* 1.26, DMSO).[7,11]

4-N-Acrylamidophenyl β-Lactoside (**3**). To a solution of compound **2** (141 mg, 0.326 mmol) and triethylamine (200 μl) in cold methanol (20 ml, 0°), acryloyl chloride (35 μl) in $CHCl_3$ (4 ml) is added dropwise. After complete addition, the reaction mixture is stirred at room temperature for an additional hour. The ions are removed by successive treatment with anionic (OH^-) and cationic resins (H^+) until the solution reaches neutrality. Compound **3** is obtained in 94% yield after evaporation of the methanolic solution. It is homogeneous by TLC in $CHCl_3$/methanol/water (30 : 10 : 1, v/v/v); mp 252.6°–254.8° (methanol), $[\alpha]_D$ −26.7° (*c* 0.81, DMSO).

Direct Conjugation of 4-N-Acrylamidophenyl Lactoside onto Bovine Serum Albumin

Bovine serum albumin (BSA, Sigma, St. Louis, MO, Fraction V, 50 mg) and 4-*N*-acrylamidophenyl lactoside **3** (44 mg) are incubated at 37° in 0.2 *M* carbonate buffer at pH 8.5, 9.5, or 10.5 (5 ml). The ratio of carbohydrate derivative to ε-amino groups is kept at 2 : 1. Aliquots (1 ml) are withdrawn at time intervals and exhaustively dialyzed against distilled water (3 times 5 liters). The lactose content in the conjugate **4** is estimated by the phenol–sulfuric acid microassay method.[12] The conjugates in which there is no unbound *N*-acrylamidophenyl lactoside detected (TLC and ^1H NMR) are isolated by freeze-drying. The above experiments are repeated with 0.2 *M* sodium carbonate buffer at 30° and with 0.2 *M* phosphate or

[11] F. H. Babers and W. F. Goebel, *J. Biol. Chem.* **105,** 473 (1934).
[12] J. D. Fox and J. F. Robyt, *Anal. Biochem.* **195,** 93 (1991).

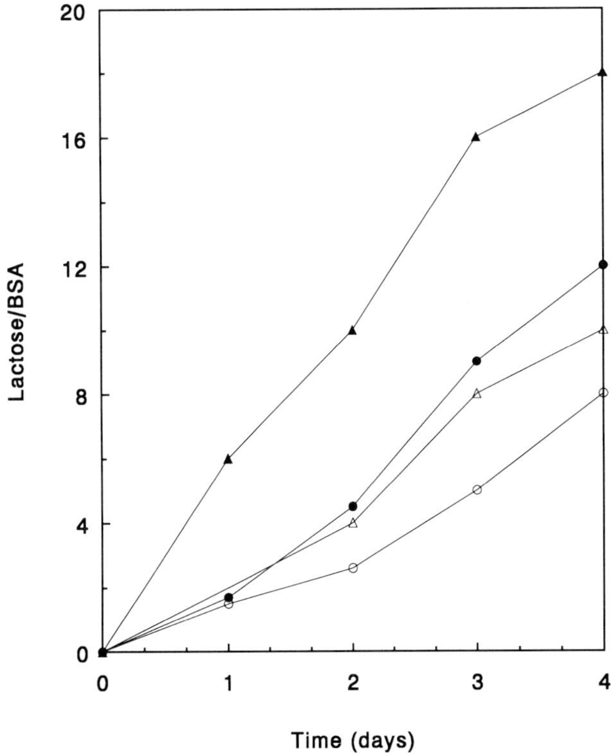

FIG. 1. Time course of Michael additions of BSA to 4-*N*-acrylamidophenyl β-D-lactoside (**3**) in 0.2 *M* phosphate buffer, pH 8.5 (○) or pH 9.5 (△), or 0.2 *M* borate buffer, pH 8.5 (●) or pH 9.5 (▲), at 37°.

borate buffer at 37°. The results of these experiments are illustrated in Figs. 1 and 2.

Electrophoresis of Neoglycoproteins

Sodium dodecyl sulfate–polyacrylamide gel electrophoresis is carried out by the method of Laemmli[13] using a Bio-Rad (Richmond, CA) Mini-PROTEAN II dual-slab cell apparatus. Stock solutions of the conjugates (1 mg/ml) in SDS-reducing buffer are diluted 9-fold with the electrophoretic buffer (0.5 *M* Tris-HCl, pH 6.8, containing 10% (w/v) SDS, 10% (v/v) glycerol, 5% (v/v) 2-mercaptoethanol, and 0.05% (w/v) bromphenol blue). The samples are heated at 95° for 5 min, and 2.5- or 5-μl portions

[13] U. K. Laemmli, *Nature (London)* **227,** 680 (1970).

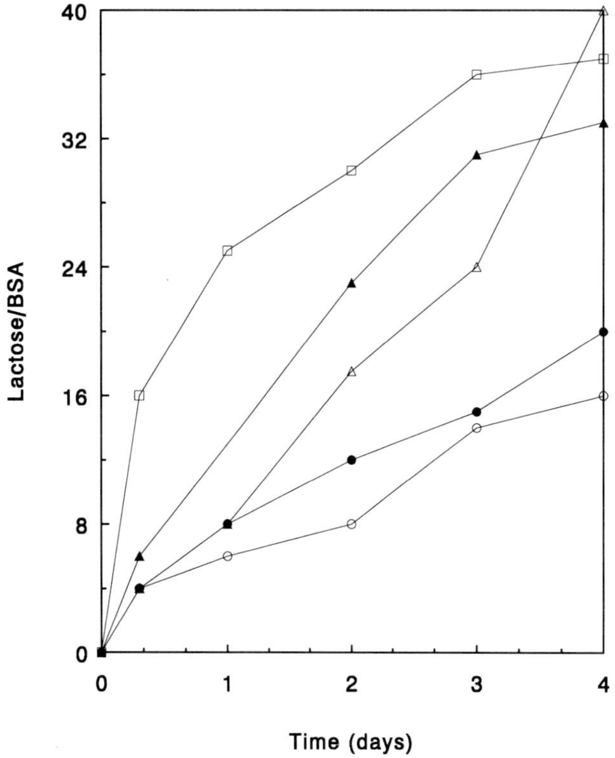

FIG. 2. Time course of Michael additions of BSA to **3** in 0.2 *M* carbonate buffer, pH 8.5, at 30° (○) or 37° (●); pH 9.5, 30° (△) or 37° (▲); and pH 10.5, 30° (□).

are applied to the gel. Electrophoresis is performed in 12% acrylamide as the separating slab gel (0.375 *M* Tris, pH 8.8) and in 4% acrylamide as the stacking gel (0.125 *M* Tris, pH (6.8) with a constant current of 20 mA until the tracking dye reaches the bottom of the gel. The standard protein kit (Bio-Rad, Cat. No. 161-0303) is composed of ovalbumin (M_r 42,697), bovine serum albumin (M_r 66,200), phosphorylase (M_r 97,400), β-galactosidase (M_r 116,250), and myosin (*M*r 200,000). The bands are then stained with 0.1% Coomassie Blue R-250 (10% aqueous acetic acid, 40% aqueous methanol) according to the manufacturer's instruction manual (Bio-Rad). The result is illustrated in Fig. 3.

Agar Gel Diffusion

Agar gel diffusion is performed in 1% agarose containing 2% polyethylene glycol (PEG 8000, Sigma) in phosphate-buffered saline (PBS). The

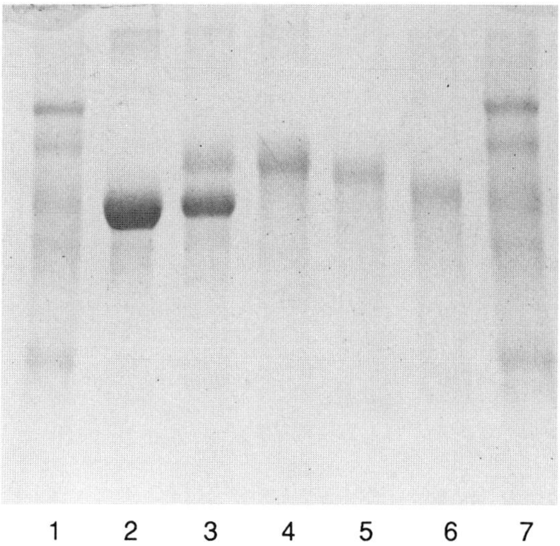

FIG. 3. Electrophoresis of lactose–BSA conjugates (**4**). Lanes 1 and 7 (left to right) contain protein standards: myosin (200 kDa), β-galactosidase (116 kDa), phosphorylase (97.4 kDa), BSA (66.2 kDa), and ovalbumin (42.7 kDa); lane 2, BSA native; lane 3, BSA plus lactosylated BSA (45 : 1); lane 4, lactosylated BSA (45 : 1); lane 5, lactosylated BSA (23 : 1); lane 6, lactosylated BSA (6 : 1).

lactose–BSA conjugates (**4**) and the lectin from *Arachis hypogaea* (peanut lectin, Sigma) are used at a concentration of 1 mg/ml PBS. The precipitin bands are allowed to form overnight at 4°. The result is illustrated in Fig. 4.

Conjugation of N-Acryloylated Sialosides onto Bovine Serum Albumin

To a solution of BSA (50 mg) in 0.1 *M* carbonate buffer, pH 10 (1 ml), is added compound **5** (25 mg.[14,15] The reaction is allowed to proceed for 2 days at 37°, after which time the reaction mixture is dialyzed against distilled water. The solution is then lyophilized to provide **6** as a white powder (50% yield). Based on both the resorcinol colorimetric determination of sialic acid[16] and amino acid analysis (phenylisothiocyanate), the conjugate showed a sialic acid to BSA content of 14 NeuAc/BSA (±2 NeuAc residues). The sialyl lactose derivative **7**[4,17] (2 mg) in water (50 μl) is coupled to BSA (10 mg) in 0.1 *M* carbonate buffer, pH 10.5 (100 μl),

[14] R. Roy and C. A. Laferrière, *Carbohydr. Res.* **177,** C1 (1988).

[15] C. A. Laferrière, F. O. Andersson, and R. Roy, this volume [25].

[16] L. Svennerholm, *Biochim. Biophys. Acta* **24,** 604 (1957).

[17] E. Kallin, H. Lonn, T. Norberg, and M. Elofsson, *J. Carbohydr. Chem.* **8,** 597 (1989).

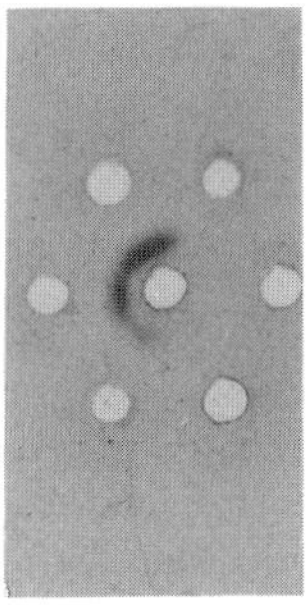

FIG. 4. Agar gel diffusion. The middle well contains peanut lectin (1 mg/ml); wells 1–6 clockwise, lactosylated BSA with 3, 8, 9, 12, 20, and 35 lactosyl residues, respectively.

under the conditions described above. The reaction provides conjugate **8** (5 mg) containing 19 NeuAc/BSA. SDS-PAGE (not shown) indicates an increase of molecular weight of 7 kDa. The time course for conjugation of **5** and **7** onto BSA is illustrated in Fig. 5.

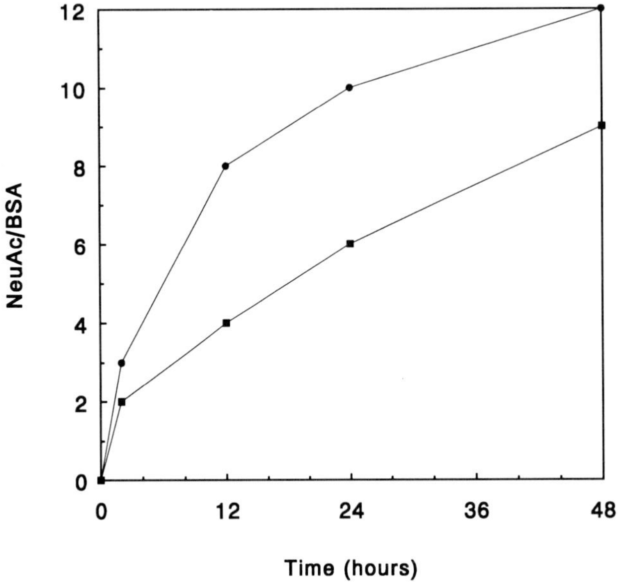

FIG. 5. Time course of Michael additions of BSA (7.5%, w/v) to *N*-acrylamido-substituted sialoside **5,** 25 mg/ml (■), and **7,** 10 mg/ml (●), in 0.1 *M* sodium carbonate buffer, pH 10, at 37°.

T-Antigen Tetanus Toxoid Conjugate

The *o*-nitrophenyl glycoside of the T-antigen is a generous gift of Dr. K. L. Matta (Roswell Park Cancer Institute, Buffalo, NY). The nitrophenyl derivative **9**[6] (7.1 mg, 14.1 μmol), ammonium formate (5 mg, 5 equivalents), and 5% Pd/C (5 mg) are mixed with methanol (2 ml) containing 0.05% formic acid in a small round-bottomed flask. The flask is stoppered and warmed briefly (65°, 1–2 min). While the solution cools to room temperature, TLC [acetonitrile/acetone/10% acetic acid in water, 5:3:1 (v/v/v)] shows only a partial reduction of **9** (R_f 0.32) to the amine **10** (R_f 0.23). The reaction is warmed briefly and allowed to cool a second time. Analysis by TLC then shows that a clean and complete conversion to the amine has been achieved. The solution is filtered through a small bed of Celite using warm methanol for rinsing. After evaporation (≤30°), the resulting clear oil is dissolved in water and repeatedly lyophilized to remove any remaining ammonium formate or formic acid.

The lyophilized amine **10** is then dissolved in methanol (2 ml). Wet anion-exchange (HO^-) resin (Amberlite IRA-400) is added and the solution cooled to 0°. A dilute solution of acryloyl chloride (2 equivalents) in 1,4-dioxane (1%) is slowly added dropwise to a cool, stirring solution until TLC indicates complete conversion of the amine to the acrylamide derivative **11** (R_f 0.33). After warming to room temperature, the resin is filtered, rinsed with warm methanol, and evaporated to a clear oil. The oil is dissolved in water and lyophilized to a white fluffy solid (8.5 mg). The ^{1}H nuclear magnetic resonance (NMR) spectrum of the sample reveals the presence of a 2.5 molar excess of acrylamide to the product **11**. The acrylamide contaminant, produced most likely from acryloylation of incompletely removed ammonium formate in the previous stage, is removed by size-exclusion chromatography on a column of Sephadex LH-20 (1.5 × 85 cm) using methanol/water (9:1, v/v) as eluent (flow rate 32 ml/hr). The separation is followed with a Waters (Milford, MA) 700 series refractive index monitor and recorded. An excellent separation is achieved. The desired fractions (eluted first) are pooled and evaporated. The product **11** is recovered in 95% overall yield (7.1 mg) as a white lyophilized powder.

^{1}H NMR data for **11,** δ (D_2O, referenced to HOD at 4.828 ppm), are as follows: 7.15–7.44 (mult, 4H, aromatic, 7.57 [dd, 1H, J_{cis} = 10.2, J_{trans} = 17.0 Hz, C(O)CH=], 6.40 (dd, 1H, J_{gem} = 1.2 Hz, C=CH_2 *trans*), 5.96 (dd, 1H, C=CH_2 *cis*), 5.71 (d, 1H, $J_{1,2}$ = 3.5 Hz, H1), 4.53 (dd, 1H, $J_{2,3}$ = 11.1 Hz, H2), 4.43 (d, 1H, $J_{1,2}$ = 7.5 Hz, H1′), 4.30 (dd, 1H, $J_{3,4}$ = 2.4 Hz, H4′), 4.03–4.11 (mult, 2H), 3.92 (dd, 1H, $J_{3,4}$ = 3.0, $J_{4,5}$ ≤ 1 Hz, H4′), 3.77–3.51 (mult, 7H), and 2.05 (s, 3H, NAc).

Tetanus Toxoid Protein Purification

A 4.0-ml aliquot of a commercial tetanus toxoid (TT) preparation (Dr. P. Rousseau, Institut Armand Frappier, Laval, Quebec, Canada, lot AT-18) is diluted to 10 ml with distilled water and dialyzed twice against 5-liter volumes of distilled water overnight. After the second dialysis, the precipitated material (27 mg) is removed by centrifugation. The combined supernatants are lyophilized and give 46 mg of soluble TT. The soluble TT fraction is assayed by SDS-PAGE electrophoresis and is shown to contain mainly monomeric TT (150 kDa) with almost no higher molecular weight aggregates present. It is kept as a lyophilized solid at 4° prior to conjugations. This material is always soluble in water and did not denature or aggregate even after 4 days in 0.2 *M* carbonate buffer, pH 10.0, at room temperature.

Conjugation of T-antigen to Tenanus Toxoid

To a solution of **11** (2.0 mg, 4 μmol) in 0.2 *M* carbonate buffer, pH 10.0 (500 μl), is added purified tetanus toxoid (5.1 mg, 2 μmol NH_2 groups). The suspension is homogenized by mixing on a Vortex mixer and left at room temperature for 72 hr. The reaction mixture is diluted to 2.5 ml with water and then centrifuged at 14,000 *g* in a Fisher microcentrifuge Model 235B for 5 min. The homogeneous mixture is then dialyzed against distilled water (4 times 1 liter) and lyophilized to give 5.8 mg of the protein conjugate **12** (82% yield by weight, 39 T-antigen/TT).

Conjugation of 4-N-Acrylamidophenyl β-Lactoside to Poly(L-lysine)

To a small vial containing a micro stir bar is added poly(L-lysine hydrobromide) (24.8 mg, 118 μmol NH_2 groups) [Sigma, P-2636, $M_{W_{(vis)}} = 69{,}000$, $DP_{(vis)} = 333$, $M_W/M_{n_{(sec\text{-}LALLS)}} = 1.1$], 4-*N*-acrylamidophenyl β-D-lactoside (**3**) (9.7 mg, 20 μmol) and 0.2 *M* carbonate buffer, pH 10.0 (2 ml). After stirring for 15 min, the lactoside **3** cannot be properly dissolved. The solution quickly becomes homogeneous after adding DMSO (500 μl). The reaction is stirred for 4.5 days at room temperature, after which a primary dialysis in 25 ml of distilled water is performed. Treatment of the dialyzate with mixed bed resin (H^+/HO^-) followed by lyophilization does not provide any detectable unconjugated **3** (TLC and ^{1}H NMR), indicating that the reaction is essentially complete. The crude glycopolymer **13** is then exhaustively dialyzed against two 5-liter volumes of distilled water and freeze-dried. The glycopolymer **13** is isolated as a white spongy solid in 25 mg yield. ^{1}H NMR (D_2O) analysis of **13** reveals that the ratio of lactose to unmodified lysine residues is 1 : 5.4, which is quite close to the theoretical ratio of 1 : 4.9, had the lactoside **3** been totally incorporated into polylysine.

^{13}C NMR data, $\delta(D_2O$, referenced to internal CH_3OH at 48.97 ppm), are as follows: 173.8 (C=O), 169.9 (C_{ipso}), 132.1 (C_{para}), 123.7 (C_{meta}), 117.2 (C_{ortho}), 103.2 (C_1 Glc), 100.4 (C_1 Gal), 78.3, 75.6, 75.1, 74.4, 72.8 (2C), 71.1, 68.7 (C_2 to C_5 Glc and Gal), 61.2, 60.1 ($2C_6$), 53.6 (C_α lysine), 41.2 (—CH_2—N lactoside), 39.4 (C_ε lysine), 30.7 (CH_2 lysine), 27.3 [CH_2—C(O)N lactoside], 26.8, and 22.5 ($2CH_2$ lysine).

Comments

N-Acryloylated carbohydrate derivatives such as those described in the present study constitute unique precursors for the preparation of both protein and polymer conjugates. On the one hand, they can be used as monomers in copolymerization with acrylamide,[15] and, on the other hand, they can serve as Michael acceptors for the direct protein conjugations.

Under various conditions of pH, buffers, and temperature, the direct coupling of lysyl ε-amino groups of protein was evaluated. The most efficient medium for the Michael additions was the carbonate buffers. There was only very slow coupling in phosphate buffers. Although there was a very high incorporation of carbohydrate residues at pH 10.5 in the carbonate buffer at 37°, the extent of protein degradation was too extensive for the method to be of synthetic utility. The optimum conditions of coupling, while preventing protein degradation under the alkaline pH, were obtained in carbonate buffer, pH 10.5, at 30°, although pH 10.0 at 37° had a similar effect. As previously observed in a similar study involving reductive amination as the key reaction for conjugation,[18] the neoglycoproteins were antigenic in model agglutination assays with peanut lectin beyond a threshold carbohydrate content of 12 mol per mole BSA (agar double diffusion).

[18] R. Roy, F. D. Tropper, A. Romanowska, M. Letellier, L. Cousineau, S. Meunier, and J. Boratynski, *Glycoconjugate J.* **8,** 75 (1991).

[9] Isolation, Modification, and Conjugation of Sialyl α(2→3)-Lactose

By C. A. LAFERRIÈRE and R. ROY

Introduction

The α(2→3) or α(2→6) linkage domains of sialyl lactosides of glycoproteins and glycolipids have been implicated in the fine receptor specificities of influenza virus hemagglutinins.[1] It was therefore of interest to obtain readily available quantities of simple sialyl oligosaccharides of defined linkages for binding and inhibition studies.

Although the literature is well documented with regard to the chemical[2–4] or chemoenzymatic[5–8] syntheses of sialyl oligosaccharides, the cost or procedures involved are sometimes troublesome. As previously demonstrated, bovine colostrum[9,10] or human milk[11] may also constitute appropriate sources for these complex sialyl oligosaccharides. However, none of the isolation procedures are rapid and simple enough for the easy preparation of the α-D-Neu*p*5Ac-(2→3)-β-D-Gal sequence present in the G_{M3} oligosaccharide **(1)**. This chapter describes an isolation protocol which is based on the fact that, among the monosialyl oligosaccharides present in bovine colostrum, **1** is the only major species prone to lactonization.[12] This process transforms **1** into a neutral species **2** which cannot bind to

[1] J. C. Paulson, *in* "The Receptors" (M. Conn, ed.), Academic Press, Orlando, Florida, 1985.

[2] M. Numata, M. Sugimoto, S. Shibayama, and T. Ogawa, *Carbohydr. Res.* **174,** 73 (1988).

[3] H. Paulsen and U. von Deessen, *Carbohydr. Res.* **146,** 147 (1986).

[4] T. Murase, H. Ishida, M. Kiso, and A. Hasegawa, *Carbohydr. Res.* **188,** 71 (1989).

[5] S. Sabesan and J. C. Paulson, *J. Am. Chem. Soc.* **108,** 2068 (1986).

[6] M. M. Palcic, A. P. Venot, R. M. Ratcliffe, and O. Hindsgaul, *Carbohydr. Res.* **190,** 1 (1989).

[7] K. K.-C. Liu and S. J. Danishefsky, *J. Am. Chem. Soc.* **115,** 4933 (1993).

[8] G. F. Herrmann, Y. Ichikawa, C. Wandrey, F. C. A. Gaeta, J. C. Paulson, and C.-H. Wong, *Tetrahedron Lett.* **34,** 309 (1993).

[9] R. W. Veh, J.-C. Michalski, A. P. Corfield, M. Sander-Wewer, and R. Schauer, *J. Chromatogr.* **212,** 313 (1981).

[10] J. Parkkinen and J. Finne, this series, Vol. 138, p. 289.

[11] D. F. Smith, D. A. Zopf, and V. Ginsburg, this series, Vol. 50, p. 221.

[12] R. K. Yu, T. A. W. Koerner, S. Ando, H. C. Yohe, and J. H. Prestegard, *J. Biochem. (Tokyo)* **98,** 1367 (1985).

Copyright © 1994 by Academic Press, Inc.
All rights of reproduction in any form reserved.

anion-exchange resins, thus leading to a simplified purification scheme (Scheme 1).[13]

The small quantities of the analogous *N*-glycolylsialic acid derivative known to be present in bovine colostrum[3–5] were lost in the process [as judged by high-performance liquid chromatography (HPLC) analysis]. Compound **1** was also made multivalent by direct coupling to bovine serum albumin using reductive amination.[14] The aminated glucitol residues in the neoglycoprotein **5** play the role of hydrophilic spacers. Under the improved conditions,[15] 24 sialyl α(2→3)-galactose units/protein could be introduced after only overnight treatment.

Materials

Freshly harvested bovine colostrum is a generous gift from the Experimental Farm, Agriculture Canada (Ottawa, Ontario). All the chemicals and solvents are reagent grade and are used as received (Aldrich Chemical Co., Milwaukee, WI). Sodium cyanoborohydride is kept in the cold under vacuum in a desiccator containing dry calcium sulfate. Bovine serum albumin (BSA) is from Sigma Chemical Co. (St. Louis, MO, No. A-7638). Sephadex DEAE A-25 anion-exchange resin is obtained from Pharmacia (Piscataway, NJ).

General Procedures

Nuclear magnetic resonance spectroscopy (NMR) is run on a Varian (Palo Alto, CA) XL-300 instrument. Fast atom bombardment-mass spectrometry (FAB-MS) is effected on a VG-7070E instrument (Manchester, UK) using xenon as the neutral gas and the samples are dissolved in a glycerol matrix. Infrared spectra are taken on a Perkin-Elmer (Norwalk, CT) 783 infrared spectrophotometer. The samples (1–2 mg) are prepared as KBr pellets (100 mg). Size-exclusion, anion-exchange, or desalting chromatographies are done on 2.5 × 60 cm column. Products are detected with a Waters (Milford, MA) differential refractometer R403 and collected with a LKB (Uppsala, Sweden) 2112 Redirac fraction collector. The HPLC is performed on a Perkin-Elmer Series 10 liquid chromatograph equipped with a LC25 RI detector. Aminex A-28 (Bio-Rad, Richmond, CA) anion-exchange resin is used for HPLC. Thin-layer chromatography (TLC) is performed using precoated 60F-254 silica gel plates eluted with *n*-propanol–water, 8 : 2 (v/v).

[13] R. Roy, C. A. Laferrière, and H. Dettman, *Carbohydr. Res.* **186,** C1 (1989).

[14] G. R. Gray, this series, Vol. 50, p. 155.

[15] R. Roy, E. Katzenellenbogen, and H. J. Jennings, *Can. J. Biochem. Cell Biol.* **62,** 270 (1984).

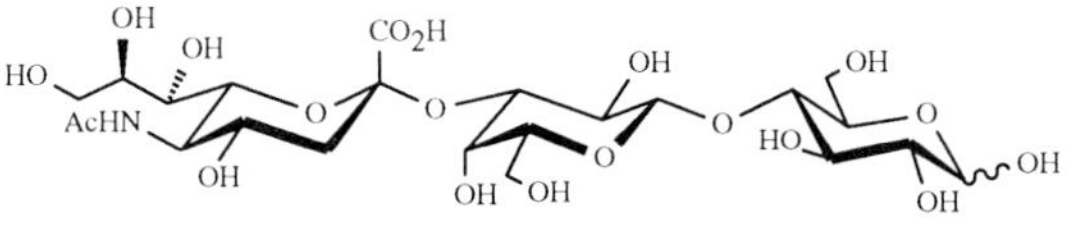

1

α-D-Neu*p*5Ac-(2→3)-β-D-Gal*p*-(1→4)-β-D-Glc*p*

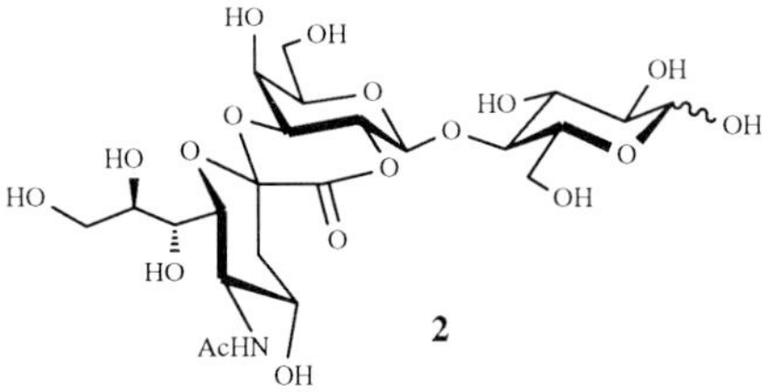

2

α-D-Neu*p*5Ac-(2→3)-lactose lactone

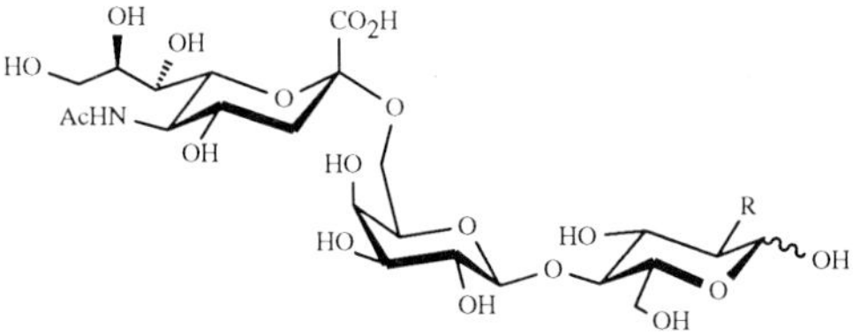

3 R = OH: α-D-Neu*p*5Ac-(2→6)-β-D-Gal*p*-(1→4)-β-D-Glc*p*
3a R = NHAc: α-D-Neu*p*5Ac-(2→6)-β-D-Gal*p*-(1→4)-β-D-Glc*p*NAc

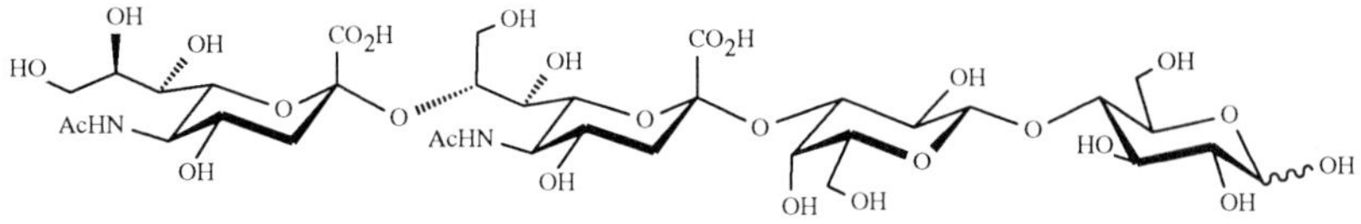

4

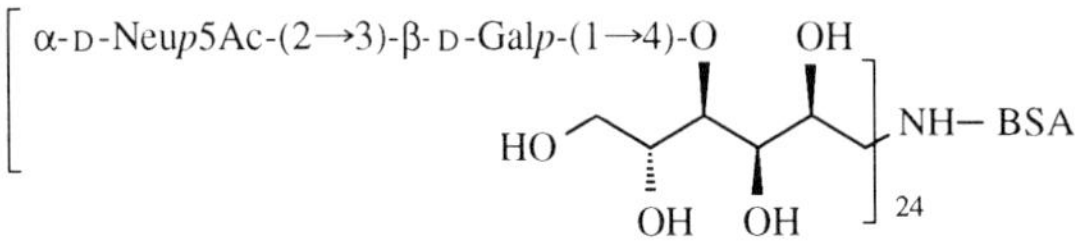

5

Isolation of Sialyl α(2→3)-Lactose

Crude bovine colostrum (1 liter) is extracted with a mixture of chloroform–methanol (3 : 1, v/v, 1 liter) which, after a single centrifigation step at 5000 rpm for 15 min at 4° and phase separation, gives a clear solution.

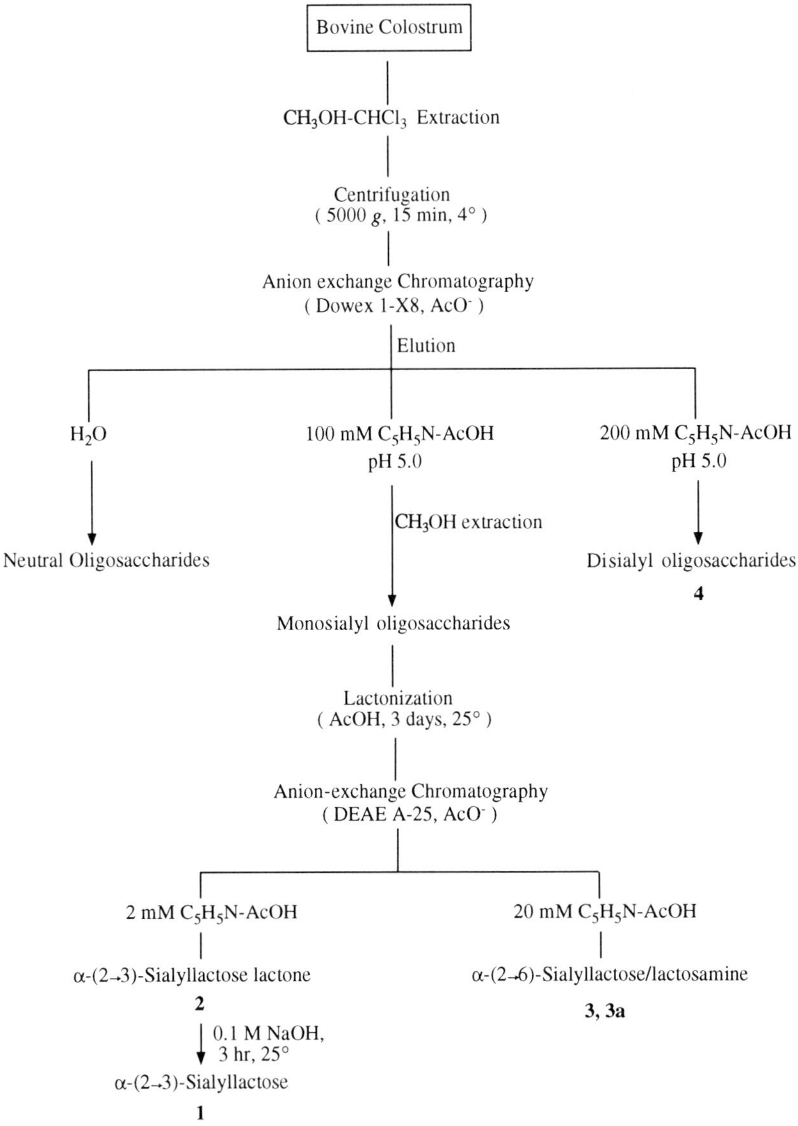

SCHEME 1

The aqueous phase is then concentrated to 500 ml on a rotary evaporator. The solution is passed onto a Dowex 1-X8-400 anion-exchange column (2.5×60 cm, acetate form). The neutral oligosaccharide fraction is eluted with water (5 liters), and the crude monosialyl oligosaccharide fraction [including sialyl α(2→3)-lactose **(1), 3,** and **3a**] is eluted with pyridinium acetate buffer ($C_5H_5N \cdot AcOH$, 100 m*M*, pH 5.0, 1 liter). The disialyl oligosaccharide fraction **(4)** is then eluted with a 200 m*M* concentration of the same buffer (1 liter). The individual fractions are lyophilized. At this stage, they could be processed according to published procedures to obtain the well-separated oligosaccharides.[9,10]

The crude powder containing the monosialyl oligosaccharides (0.8–1.0 g) is extracted three times with methanol (6 ml) to remove residual ninhydrin-positive material. The methanol extracts are evaporated under vacuum. The lactonization of **1** is allowed to proceed in glacial acetic acid (50 ml) for 3 days at room temperature, and then the reaction mixture is lyophilized. The extent of lactonization was followed by TLC in *n*-propanol–water (8 : 2, v/v). The following compounds and the respective R_f values are observed as follows: **1**, R_f 0.21; **2** (lactone), R_f 0.34; and α-D-Neu*p*5Ac-(2→6)-β-D-Gal*p*-(1→4)-D-Glc*p* (**3**; II^6Neu*p*5Ac-Lac), R_f 0.18. The related α-D-Neu*p*5Ac-(2→6)-β-D-Gal*p*-(1→4)-D-Glc*p*NAc **(3a)**, although present, cannot be well differentiated from **1** in this system.

The lyophilized reaction mixture is dissolved in pyridinium acetate buffer (2 m*M*, pH 5.0, 100 ml) and loaded onto a Sephadex-DEAE A-25 column (2.0 × 20 cm, acetate form). The neutral lactone **2** is not retained and is eluted with an additional 100 ml of the same buffer. After lyophilization, **2** is obtained as a white powder (180 mg/liter colostrum); $[\alpha]_D$ + 10.5° (*c* 1.0, 0.1 *M* pyridinium acetate); FAB-MS for $C_{23}H_{37}NO_{18}$ 615.54: (M − 1) 615; IR ν_{max} (KBr): 1750 cm^{-1}; ^1H NMR (D_2O): δ 5.21 (d, 1H, $J_{1,2}$ = 3.2 Hz, αGlc-H1), 4.75 (d, 1H, $J_{1,2}$ = 8.0 Hz, Gal-H1), 4.64 (d, 1H, $J_{1,2}$ = 7.3 Hz, βGlc-H1), 4.30 (m, 1H, NeuAc-H4), 4.21 (dd, 1H, $J_{2,3}$ = 10.6 Hz, $J_{3,4}$ = 2.5 Hz, Gal-H3), 3.89 (NeuAc-H5), 3.68 (NeuAc-H8), 3.59 (NeuAc-H6), 3.56 (NeuAc-H3eq), 2.03 (s, 3H, NAc), 1.77 (t, 1H, $J_{3a,3e}$ = 11.9 Hz, NeuAc-H3ax).

The lactone **2** is hydrolyzed to **1** with 0.1 *M* NaOH (3 hr, 25°), then neutralized with H^+ resin and lyophilized to give a white powder; $[\alpha]_D$ +20.6° (*c* 0.7, 0.1 *M* pyridinium acetate, literature[9] +16.8°); R_f 0.21; FAB-MS for $C_{23}H_{39}NO_{19}$ 633.56: (M − 1) 632, (M + 1) 634; ^1H NMR (D_2O): δ 5.22 (d, 1H, $J_{1,2}$ = 3.8 Hz, αGlc-H1), 4.53 (d, 1H, $J_{1,2}$ = 7.9 Hz, Gal-H1), 4.66 (d, 1H, $J_{1,2}$ = 7.9 Hz, βGlc-H1), 4.12 (dd, 1H, $J_{3,2}$ = 9.9 Hz, $J_{3,4}$ = 3.2 Hz, Gal-H3), 3.75–3.56 (~12H), 3.28 (t, 1H, $J_{2,3}$ = $J_{2,1}$ = 7.9 Hz, βGlc-H2), 2.76 (dd, 1H, $J_{3a,3e}$ = 4.5 Hz, NeuAc-H3eq), 2.03 (s, 3H, NAc), 1.80 (t, 1H, $J_{3a,3e}$ = $J_{3a,4}$ = 12.3 Hz, NeuAc-H3ax). The ^{13}C NMR data agreed with those published.[16] Compounds **3** and **3a** can be obtained by eluting the anion-exchange column described above with 20 m*M* (instead

[16] E. Berman, *Biochemistry* **23,** 3754 (1984).

of 2 m*M*) pyridinium acetate buffer. Compound **3a** elutes after 50 ml buffer has passed through the column; yield: 23 mg/liter colostrum; R_f 0.21; FAB-MS for $C_{25}H_{42}N_2O_{19}$ 674.61: (M − 1) 673; ^{1}H NMR (D_2O: δ 5.22 (d, 1H, $J_{1,2}$ = 3.7 Hz, αGlc-H1), 4.43 (d, 1H, $J_{1,2}$ = 7.7 Hz, Gal-H1), 4.67 (d, 1H, $J_{1,2}$ = 7.9 Hz, βGlc-H1), 3.75–3.56 (~12H), 3.31 (t, 1H, $J_{2,3} = J_{2,1}$ = 7.8 Hz, βGlc-H2), 2.71 (dd, 1H, $J_{3a,3e}$ = 12.3 Hz, $J_{3e,4}$ = 4.9 Hz, NeuAc-H3eq), 2.03 (s, 3H, NAc), 1.76 (t, 1H, $J_{3a,3e} = J_{3a,4}$ = 12.3 Hz, NeuAc-H3ax). Compound **3** elutes after 75 ml buffer has passed through the column; yield: 18.2 mg/liter colostrum; $[\alpha]_D$ 6.95° (*c* 0.5, water); R_f 0.18; FAB-MS for $C_{23}H_{39}NO_{19}$ 633.56: (M − 1) 632; ^{1}H NMR (D_2O): δ 5.20 (d, 1H, $J_{1,2}$ = 1.7 Hz, αGlc-H1), 4.45 (d, 1H, $J_{1,2}$ = 7.6 Hz, Gal-H1), 3.75–3.56 (~12H), 2.67 (dd, 1H, $J_{3a,3e}$ = 11.6 Hz, $J_{3e,4}$ = 4.5 Hz, NeuAc-H3eq), 2.03 (s, 3H, NeuAc-NAc), 2.07 (s, 3H, GlcNAc-NAc), 1.76 (t, 1H, $J_{3a,3e} = J_{3a,4}$ = 12.0 Hz, NeuAc-H3ax).

*Isolation of Sialyl $\alpha(2\rightarrow8)$-sialyl-$\alpha(2\rightarrow3)$-lactose **(4)***. A crude mixture of the disialyl oligosaccharides is collected from three successive 1-liter batches of colostrum processed as described above. The material (~1 g) is dissolved in pyridium acetate (2 m*M*, pH 5.0, 100 ml) and loaded on a 3.0 × 60 cm column of fresh Dowex 1-X8-400 anion-exchange resin (acetate form). The column is rinsed with water (2 liters), eluted with pyridinium acetate (20 m*M*, pH 5.0, 6°, 1 liter) to remove the residual monosialyl oligosaccharides, and then eluted with a higher concentration of the same buffer (200 m*M* pyridinium acetate, pH 5.0, 6°, 2 liters). Three peaks in the eluate of the 200 m*M* buffer are collected and lyophilized separately. The last peak at 710 ml of buffer is identified as **4**. It is lyophilized to an oil, and the quality of the material is improved by desalting on 1.5 × 60 cm Sephadex G-10 column eluted with water. A fluffy white powder is obtained after lyophilization; yield: 69 mg/3 liters colostrum; R_f 0.15; HPLC R_t 3.7 min (0.5 × 25 cm reversed-phase C_{18} column, 1.5 ml/min with 10 m*M* triethylamine/acetic acid, pH 4.8); FAB-MS for $C_{34}H_{56}N_2O_{27}$ 924.82: (M − 1) 923.5, 923.1, 922.6 (three different runs); ^{1}H NMR (D_2O): δ 5.22 (d, 1H, αGlc-H1), 4.52 (d, 1H, Gal-H1), 4.66 (d, 1H, βGlc-H1), 4.11 (dd, 1H, Gal-H3), 3.27 (t, 1H, βGlc-H2), 2.63 (dd, 1H, NeuAc″-H3eq), 2.81 (dd, 1H, NeuAc′-H3eq), 2.01 (s, 3H, NeuAc″-NAc), 2.02 (s, 3H, NeuAc′-NAc), 1.92 (t, 1H, NeuAc″-H3ax), 1.89 (t, 1H, NeuAc′-H3ax).

Direct Conjugation of Sialyl $\alpha(2\rightarrow3)$-Lactose to Bovine Serum Albumin by Reductive Amination

Compound **1** (10.4 mg, 16.5 μmol) and BSA (10.7 mg, 0.16 μmol) are dissolved in borate buffer (0.2 *M*, pH 8.7, 1.5 ml). The solution is stirred at 55° for 1 hr, then $NaCNBH_3$ (26 mg, 470 μmol) is added and reacted for 24 hr. The solution is dialyzed and lyophilized, giving the neoglycoprotein **5** as a white powder; yield: 13.2 mg, 24 NeuAc/BSA, 28% yield based on sialic acid. Note that the increase in weight of the protein is consistent with an increase in molecular weight of 15,000 g/mol.

[10] Coupling of Carbohydrates to Proteins by Diazonium and Phenylisothiocyanate Reactions

By CHERYL M. REICHERT, COLLEEN E. HAYES, and IRWIN J. GOLDSTEIN

Introduction

Haptens are defined as small molecules that are incapable of stimulating antibody formation unless they are chemically linked to a macromolecule. Karl Landsteiner was instrumental in originating procedures such as diazotization whereby low molecular weight haptens may be rendered immunogenic by covalent conjugation to protein.[1] Goebel and Avery and colleagues extended the pioneering work of Landsteiner by coupling the dizaonium salts of *p*-aminobenzyl and *p*-aminophenyl glycosides of a variety of sugars to horse serum globulin.[2] Antibodies to scores of carbohydrates have been prepared by this procedure (see Table I). A modification of this method by Westphal and Feier has become a standard technique in immunochemistry.[3]

The diazotization coupling reaction, however, rarely proceeds with more than 60% efficiency of coupling and is not specific for the side chain of any single specific aminoacyl residue. Diazonium salts attack primarily tyrosyl, histidyl, and lysyl residues of proteins, with diazo linkages to tryptophanyl and arginyl residues also being formed when the diazonium reactant is used in large excess.[4–6] A means whereby the specificity and efficiency of coupling reactions could be increased was suggested by a

[1] K. Landsteiner, "The Specificity of Serological Reactions." Harvard Univ. Press, Cambridge, Massachusetts, 1945.

[2] W. F. Goebel and O. T. Avery, *J. Exp. Med.* **50,** 521 (1929).

[3] O. Westphal and H. Feier, *Chem. Ber.* **89,** 582 (1956).

[4] A. N. Howard and F. Wild, *Biochem. J.* **65,** 651 (1957).

[5] E. W. Gelewitz, W. L. Riedeman, and I. M. Klotz, *Arch. Biochem. Biophys.* **53,** 411 (1954).

[6] M. Tabachnick and H. Sobotka, *J. Biol. Chem.* **235,** 1051 (1960).

Copyright © 1994 by Academic Press, Inc.
All rights of reproduction in any form reserved.

consideration of various commonly used and well-characterized protein reagents, such as phenylisothiocyanate (Edman reagent).[7] Under properly controlled conditions, phenylisothiocyanate reacts almost exclusively and nearly quantitatively with primary amines; under more vigorous conditions, it will also alkylate thiol and hydroxyl groups before denaturing the protein.[8] The principle utilized in this procedure involves the synthesis of carbohydrate derivatives containing the phenylisothiocyanato functional group[9] which serves to alkylate the amino terminal and lysyl ε-amino groups of a protein.

In recent years, numerous additional oligosaccharides have been synthesized as the *p*-isothiocyanatophenyl glycosides and coupled to proteins. Table II lists a number of these saccharides and some of their physical constants. This chapter describes the coupling reactions of carbohydrates to proteins using either the diazotization reaction or the phenylisothiocyanate procedure developed in our laboratory.

Diazonium Coupling of Aminophenyl Glycosides to Protein

Principle

The principle involved in the method is illustrated by the series of reactions below:

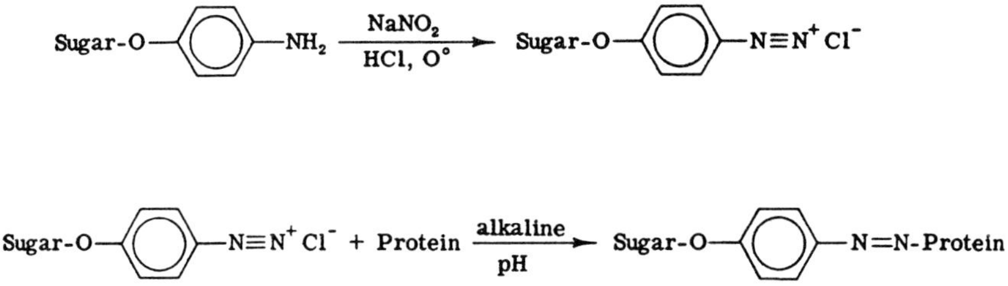

Procedure

Preparation of p-Aminophenyl α-D-Galactopyranoside. To a methanol solution[10] (50 ml) of *p*-nitrophenyl α-D-galactopyranoside (0.473 g, 1.57 mmol) in a round-bottomed flask equipped with a magnetic stirrer is added

[7] P. Edman, Acta Chem. Scand. **4,** 277 (1950).

[8] H. Fraenkel-Conrat, *in* "The Enzymes" (P. D. Boyer, H. Lardy, and K. Myrbäck, eds.), Vol. 1, p. 597. Academic Press, New York, 1959.

[9] D. H. Buss and I. J. Goldstein, *J. Chem. Soc. C,* 1457 (1968).

[10] It is generally preferable to dissolve the sugar glycoside in water (1 ml) before the addition of methanol. Some glycosides may require the addition of more water for solution.

TABLE I

PHYSICAL CONSTANTS OF PHENYLGLYCOSIDES USED IN PROTEIN CONJUGATION

Sugar derivative	Melting point (°C)	$[\alpha]_D$ (°)	$[\alpha]_D$ solvent	Protein used for conjugation
p-Nitrophenyl β-cellobioside	245–246[a]			
	255–256d[b]	−85.1	40% CH_3OH	
p-Aminophenyl β-cellobioside	238–239[a]	+51.3	50% CH_3OH	BSA,[a,f] human γ-globulin[a]
	245d[b]	−52.9	50% CH_3OH	
p-Nitrobenzyl β-cellobioside	199–200[c]	−32.3	H_2O	
p-Aminobenzyl β-cellobioside	188–190d[c]	−35.2	H_2O	Horse serum globulin[c]
p-Nitrobenzyl β-cellobiuronic acid methyl ester	188–189[c]	−48.1	CH_3OH	
p-Aminobenzyl β-cellobiuronide				Horse serum globulin[c]
p-Aminophenyl β-L-colitoside	75[d]	+99	CH_3OH	BSA, ovalbumin[d]
p-Aminophenyl α-L-colitoside	180[d]	−166	CH_3OH	BSA, ovalbumin[d]
p-Nitrophenyl α-L-fucoside	197[e]			
p-Aminophenyl α-L-fucoside	175[e]	−204	CH_3OH	Horse serum globulin,[e] BSA[f]
p-Nitrophenyl β-D-galactoside	180–182[g,h]			
	173–175[i]			
p-Aminophenyl β-D-galactoside	158–159[g,h]	−40.5	CH_3OH	Horse serum globulin, ovalbumin[g]
	153[i]	−40.3	CH_3OH	BSA,[i] porcine and bovine γ-globulin[h]
p-Nitrophenyl α-D-galactoside	169[e]	+225	H_2O	
p-Aminophenyl α-D-galactoside	178[e]	+224	CH_3OH	Horse serum globulin[e]
p-Nitrobenzyl β-D-galactoside	161–162[i]	−32.9	CH_3OH	
p-Aminobenzyl β-D-galactoside	89–90[i]	−50.5	CH_3OH	Horse serum globulin,[i] BSA[f]
p-Aminobenzyl β-D-galacturonic acid methyl ester	108–110[i]	−75.8	H_2O	Horse serum globulin[i]
p-Nitrophenyl β-D-glucoside	165[a,g]	−79.6	CH_3OH	
p-Aminophenyl β-D-glucoside	156–157[a]	+58.6	CH_3OH	BSA,[a,f] human γ-globulin,[a] ovalbumin[g]
	160d[4]			Horse serum globulin,[j] porcine and bovine γ-globulin[h]
p-Nitrophenyl α-D-glucoside	216–217[k]	+227.9	CH_3OH	
p-Aminophenyl α-D-glucoside	185–186[k]	+194.1	CH_3OH	Horse serum globulin,[i] BSA[f]

TABLE I (*continued*)

Sugar derivative	Melting point (°C)	$[\alpha]_D$ (°)	$[\alpha]_D$ solvent	Protein used for conjugation
p-Nitrobenzyl β-D-glucoside	156–157[l]	−47.7	CH_3OH	
p-Aminobenzyl β-D-glucoside	142–143[l]	−61.8	H_2O	Horse serum globulin[l]
p-Aminobenzyl β-D-glucuronide				Horse serum globulin[l]
p-Nitrophenyl *N*-acetyl-β-D-glucosaminide	204[m]	−25.5	Pyridine	
p-Aminophenyl *N*-acetyl-β-D-glucosaminide	228[m]	+12.6		Various albumins[m]
o-Aminophenyl β-D-glucuronide				Hemocyanin[n]
p-Nitrophenyl β-gentiobioside	221–223[b]			
p-Aminophenyl β-gentiobioside	237–238[b]	−79.8	H_2O	BSA[f]
p-Nitrobenzyl β-gentiobioside	120[o]	−46.8	H_2O	
p-Aminobenzyl β-gentiobioside		−49.7	H_2O	Horse serum globulin[o]
p-Nitrophenyl β-lactoside	260d[h]			
p-Aminophenyl β-lactoside	242d[h]			Horse serum globulin[p]
	233[b]	−36.4	H_2O	Porcine and bovine γ-globulin,[h] BSA[f]
p-Nitrophenyl α-D-maltoside	221[b]	+6	H_2O	
p-Aminophenyl α-D-maltoside	91–92[b]	+35.3	50% CH_3OH	Horse serum globulin,[p] BSA[f]
p-Nitrophenyl α-D-mannoside	182[e]	+161	CH_3OH	
p-Aminophenyl α-D-mannoside	164[e]	+128	CH_3OH	Horse serum globulin,[e] BSA[f]
p-Nitrophenyl β-panoside		+40[q]	H_2O	
p-Aminophenyl β-panoside				BSA[r]
p-Nitrophenyl α-L-rhamnoside	179[e]	−144	CH_3OH	
p-Aminophenyl α-L-rhamnoside	166–167[e]			Horse serum globulin[e]
p-Nitrophenyl β-sophoroside	261–262[s]	−67.9	H_2O	
p-Aminophenyl β-sophoroside	211–212[s]			BSA[t]

[a] G. J. Gleich and P. Z. Allen, *Immunochemistry* **2,** 417 (1965).
[b] F. H. Babers and W. F. Goebel, *J. Biol. Chem.* **105,** 473 (1934).
[c] W. F. Goebel, *J. Exp. Med.* **68,** 469 (1938).
[d] O. Lüderitz, O. Westphal, A. M. Staub, and L. LeMinor, *Nature (London)* **188,** 556 (1960).
[e] O. Westphal and H. Feier, *Chem. Ber.* **89,** 582 (1956).
[f] I. J. Goldstein and R. N. Iyer, *Biochim. Biophys. Acta* **121,** 197 (1966).

(*continued*)

TABLE I (*continued*)

[g] W. F. Goebel and O. Avery, *J. Exp. Med.* **50,** 761 (1932).
[h] J. Yariv, M. M. Rapport, and L. Graf, *Biochem. J.* **85,** 383 (1962).
[i] S. M. Beiser, *J. Mol. Biol.* **2,** 125 (1960).
[j] W. F. Goebel, *J. Exp. Med.* **66,** 191 (1937).
[k] W. F. Goebel, F. H. Babers, and O. T. Avery, *J. Exp. Med.* **55,** 761 (1932).
[l] W. F. Goebel, *J. Exp. Med.* **64,** 29 (1936).
[m] O. Westphal and H. Schmidt, *Justus Liebigs Ann. Chem.* **575,** 84 (1951).
[n] I. Corneil and L. Wofsy, *Immunochemistry* **4,** 183 (1967).
[o] W. F. Goebel, *J. Exp. Med.* **72,** 33 (1940).
[p] W. F. Goebel, O. T. Avery, and F. H. Babers, *J. Exp. Med.* **60,** 599 (1934).
[q] R. N. Iyer and I. J. Goldstein, *Immunochemistry* **10,** 313 (1973).
[r] R. S. Martineau, P. Z. Allen, I. J. Goldstein, and R. N. Iyer, *Immunochemistry* **8,** 705 (1971).
[s] R. N. Iyer and I. J. Goldstein, *Carbohydr. Res.* **11,** 241 (1969).
[t] P. Z. Allen, I. J. Goldstein, and R. N. Iyer, *Biochemistry* **6,** 3029 (1967).

platinum oxide (Adams catalyst; 50 mg). The flask is connected to a hydrogenation apparatus which operates at atmospheric pressure. After three alternate evacuations and flushings with H_2, the hydrogenation is allowed to proceed with stirring at room temperature until the calculated volume of H_2 is consumed (~1 hr). The resulting clear solution is filtered to remove catalyst (caution: the catalyst is pyrophoric, and the filter must not be allowed to dry), and, after evaporation of the solvent, *p*-aminophenyl α-D-galactopyranoside is obtained as needles, mp 177°–178°, yield 0.29 g (68%).

Diazotization of p-Aminophenyl α-D-Galactopyranoside and Coupling to Bovine Serum Albumin. The *p*-aminophenyl glycoside (0.2 g, 0.74 mmol) is dissolved in ice-cold 0.1 *M* HCl (10 ml), and ice-cold aqueous 50 m*M*) $NaNO_2$ (15 ml) is added dropwise with stirring. Diazotization of the aromatic amine requires 1 molar equivalent of HNO_2 per mole of amine; $NaNO_2$ is therefore added until a slight excess of HNO_2 is formed as determined by the formation of a blue color with starch–iodide indicator paper. Isolated diazonium salts are unstable and hence are best handled in aqueous solution. The solution of *p*-aminophenyl α-D-galactopyranoside is added with stirring to an ice-cold solution (50 ml) of crystalline bovine serum albumin (BSA, 1.0 g) in 0.15 *M* NaCl, which has been adjusted to pH 9.0 with 0.5 *M* NaOH. The reaction, which causes the mixture to become deep orange in color, is allowed to proceed at 0° for 2 hr, during which time the pH is maintained at 9.0 by the addition of 0.5 *M* NaOH. Following neutralization with 50 m*M* HCl, the orange-red sugar–BSA conjugate is dialyzed exhaustively against distilled water, lyophilized, and the freeze-dried powder stored in the refrigerator.

Conjugation of Phenylisothiocyanato Glycosides to Proteins

Principle

Coupling of phenylisothiocyanato glycosides to proteins is based on the following reaction sequence:

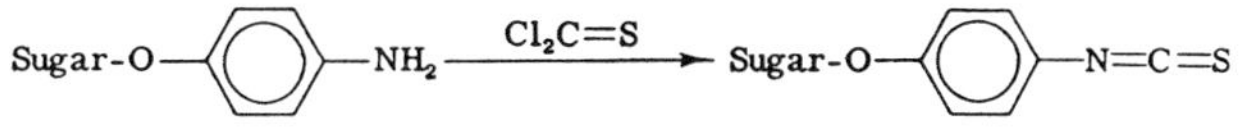

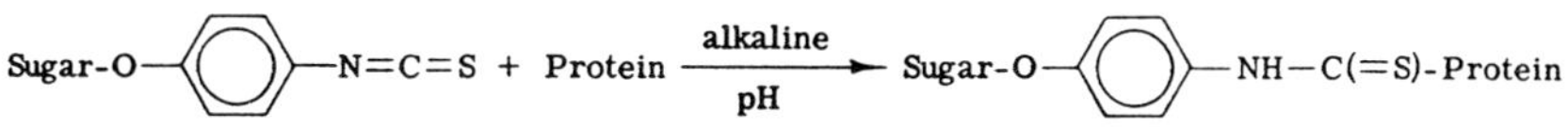

Procedure

Preparation of p-Isothiocyanatophenyl β-Glucopyranoside. Thiophosgene (0.13 ml, 1.69 mmols) is added to a magnetically stirred solution of *p*-aminophenyl β-glucopyranoside (202 mg) in 80% aqueous ethanol (25 ml), and the reaction mixture is allowed to stand at room temperature for 1.5 hr. (This procedure is performed in a well-ventilated hood.) (Within 45 min the pH generally drops to less than 2.0, and the orange solution becomes almost colorless. At this point, thin-layer chromatography (chloroform–methanol, 1 : 1, v/v) shows that all the starting material has reacted and that a single product has formed. Concentration almost to dryness leaves a solid to which water is added. Filtration, and washing the product with water, yields the phenylisothiocyanate derivative (0.200 g, 86%), mp 195°–197°, which has a broad infrared band centered at 2130 cm^{-1}. Recrystallization from ethanol gives needles, mp 195°–197°, $[\alpha]_D^{22}$ −34.7° (*c* 1.09, *N,N*-dimethylformamide).

Coupling of p-Isothiocyanatophenyl β-D-Glucopyranoside to Bovine Serum Albumin. A solution of bovine serum albumin (200 mg) in 0.15 *M* NaCl (20 ml) is brought to pH 9.0 by the addition of 0.1 *N* NaOH. *p*-Isothiocyanatophenyl β-glucoside (150 mg) is added in small portions to the magnetically stirred protein solution. The pH is readjusted to 9.0 with 0.1 *N* NaOH, and the reaction is allowed to continue at room temperature for 6 hr with further additions of 0.1 *N* NaOH solution as required to maintain the pH at 9.0. Although many of the aromatic isothiocyanate derivatives are sparingly soluble in aqueous solution, the reaction appears to proceed smoothly, perhaps because of continued solubilization of the phenylisothiocyanate derivative. The solution is refrigerated overnight, and on the following day the pH is adjusted to 7.0. Unreacted carbohydrate derivative is removed from the protein by dialysis against 0.15 *M* NaCl

TABLE II
PHENYLISOTHIOCYANATO GLYCOSIDES USED IN PROTEIN CONJUGATES[a]

Glycoside	Melting point (°C)	$[\alpha]_D$ (°)	$[\alpha]_D$ solvent	Protein conjugated	Ref.[b]
p-Isothiocyanatophenyl β-D-glucopyranoside	195–197	−34.7		BSA	*1, 2*
p-Isothiocyanatophenyl α-D-glucopyranoside	185–187	+217	DMF	BSA	*1, 2*
p-Isothiocyanatophenyl β-D-galactopyranoside	210–212	−24.1	DMF	BSA	*1, 3*
p-Isothiocyanatophenyl α-D-mannopyranoside				BSA, streptavidin	*2*
p-Isothiocyanatophenyl α-L-rhamnopyranoside				BSA	*3*
p-Isothiocyanatophenyl α-L-fucopyranoside				BSA	*3*
p-Isothiocyanatophenyl β-D-2-acetamido-2-deoxyglucopyranoside	245	−8.8	DMF	BSA	*1, 4*
p-Isothiocyanatophenyl β-lactoside				BSA	*3*
p-Isothiocyanatophenyl 2-*O*-α-D-glucopyranosyl-α-D-glucopyranoside	179–180	+14	DMF	BSA	*5, 6*
p-Isothiocyanatophenyl 2-*O*-α-D-mannopyranosyl-α-D-mannopyranoside		+73	DMF	KLH	*7*
p-Isothiocyanatophenyl 2-*O*-α-D-glucopyranosyl-β-D-galactopyranoside				BSA	*7*
p-Isothiocyanatophenyl 3-*O*-(3,6-dideoxy)-α-D-arabinopyranosyl)-α-D-mannopyranoside		+180		BSA	*8, 9*
p-Isothiocyanatophenyl 3-*O*-(3,6-dideoxy)-α-D-xylopyranosyl-α-D-mannopyranoside		+158		BSA	*8, 10*
p-Isothiocyanatophenyl 3-*O*-(3,6-dideoxy)-α-D-ribopyranosyl-α-D-mannopyranoside		+174		BSA	*11*

TABLE II (*continued*)

Glycoside	Melting point (°C)	$[\alpha]_D$ (°)	$[\alpha]_D$ solvent	Protein conjugated	Ref.[b]
p-Isothiocyanatophenyl 6-*O*-β-D-galactopyranosyl-6-*O*-β-D-galactopyranosyl-β-D-galactopyranoside	210–215	−42		IgA	*12*

[a] BSA, Bovine serum albumin; DMF, dimethylformamide; IgA, immunoglobulin A; KLH, keyhole limpet hemocyanin.

[b] Key to references: (*1*) D. H. Buss and I. J. Goldstein, *J. Chem. Soc. C*, 1460 (1968); (*2*) C. R. McBroom, C. H. Samanen, and I. J. Goldstein, this series, Vol. 28, p. 212; E. Bonfils, C. Mendes, A. Roche, M. Monsigny, and P. Midoux, *Bioconjugate Chem.* **3,** 277 (1992); (*3*) A. C. Roche, M. Barzilay, P. Midoux, S. Junqua, N. Sharon, and M. Monsigny, *J. Cell Biochem.* **22,** 131 (1983); (*4*) A. J. Jones and H. Jobe, *Biochem. J.* **268,** 41 (1990); (*5*) J. Duke, N. Little, and I. J. Goldstein, *Carbohydr. Res.* **27,** 193 (1973); (*6*) S. Ebisu, P. J. Garegg, T. Iversen, and I. J. Goldstein, *J. Immunol.* **121,** 2137 (1978); (*7*) C. M. Reichert and I. J. Goldstein, *J. Immunol.* **122,** 138 (1979); (*8*) G. Ekborg, K. Eklind, P. J. Garegg, B. Gottammar, H. E. Carlsson, A. A. Lindberg, and B. Svenungsson, *Immunochemistry* **14,** 153 (1977); (*9*) G. Ekborg, B. Gotthammar, and P. J. Garegg, *Acta Chem. Scand. Ser B* **29,** 765 (1975); (*10*) K. Eklind, B. Gotthammar, and P. J. Garegg, *Acta Chem. Scand. Ser. B* **30,** 305 (1975); (*11*) B. Gotthammar and P. J. Garegg, *Carbohydr. Res.* **58,** 345 (1977); (*12*) G. Ekborg, B. Vranesic, A. K. Battacharjee, P. Kovac, and P. J. Glaudemans, *Carbohydr. Res.* **142,** 203 (1985).

followed by passage through a Sephadex G-25 column (30 × 2.3 cm), eluting with 0.15 *M* NaCl while monitoring the optical density at 280 nm. Because the aromatic carbohydrate moiety contributes significantly to the molar extinction coefficient of the conjugate, it is necessary either to dilute the more highly concentrated fractions or to monitor the optical density at another wavelength. The subsequent elution of the unreacted *p*-isothiocyanatophenyl β-glucopyranoside may also be followed at 280 nm. Thin-layer chromatography (chloroform–methanol, 1 : 1, v/v) of the pooled fractions of the low molecular weight material shows the free hapten to be altered, and hence not recoverable.

Exhaustive dialysis does not result in the complete removal of the phenylisothiocyanato derivative. Contaminating free hapten may have no deleterious effect if the antigen is to be used for immunization, but it may seriously interfere if the antigen is used to determine the presence of precipitating antibody in the serum of an immunized animal since the soluble hapten molecules may compete with the conjugated protein for the antibody combining sites. In addition, the presence of free hapten may interfere with the determination of the extent of protein modification.

Determination of Number of β-D-Glucopyranosido Residues Incorporated per Mole of Bovine Serum Albumin. The carbohydrate content of the conjugate is quantified by the phenol–sulfuric acid method.[11] The dry weight of the protein–carbohydrate conjugate is estimated from its total nitrogen content, using the ninhydrin method of So and Goldstein.[12] From the nitrogen and carbohydrate analyses, it is estimated that 54 β-D-glucopyranosido residues are incorporated per molecule of bovine serum albumin, which as 57 lysyl groups.[13] Assuming only N-terminal and lysyl modification, the coupling efficiency is about 90%.

Comments

A host of *Salmonella typhimurium* O-antigen oligosaccharides have been generated from their respective polysaccharides, converted to the corresponding *p*-phenylisothiocyanto glycosides, and coupled to various proteins to be used as synthetic vaccines.[14] We have used the phenylisothiocyanate procedure to couple *p*-isothiocyanatophenyl β-D-glucopyranoside, *p*-isothiocyanatophenyl β-L-glucopyranoside, and *p*-isothiocyanatophenyl α-D-mannopyranoside to bovine serum albumin. However, the method may readily be extended to a variety of noncarbohydrate haptens as well as to other natural or synthetic macromolecules. For example, *p*-isothiocyanatophenyl-derivatized haptens may be reacted with solid matrices (i.e., aminoethyl or *p*-aminobenzyl cellulose) to afford solid column absorbents for affinity chromatography.

[11] J. Hodge and B. Hofreiter, *Methods Carbohydr. Chem.* **1,** 388 (1962).
[12] L. L. So and I. J. Goldstein, *J. Biol. Chem.* **242,** 1617 (1967).
[13] T. Peters, Jr., *Clin. Chem.* **14,** 1147 (1968).
[14] S. B. Svenson and A. A. Lindberg, *Prog. Allergy* **33,** 120 (1983).

[11] Coupling of Aldobionic Acids to Proteins Using Water-Soluble Carbodiimide

By JÖRGEN LÖNNGREN and IRWIN J. GOLDSTEIN

Introduction

Synthetic carbohydrate–protein conjugates have proved to be valuable tools in immunochemical investigations. In particular, these substances

Copyright © 1994 by Academic Press, Inc.
All rights of reproduction in any form reserved.

have been used to assess antibody specificities[1,2] and to raise antisera against defined structural moieties in polysaccharides.[1-4] They have also found application in studies of carbohydrate-binding proteins (lectins).[5,6]

This chapter describes a simple method for the preparation of such conjugates.[7] It involves the oxidation of an oligosaccharide to an aldonic acid derivative, which is then coupled to free amino groups in the protein using a water-soluble carbodiimide as the activating agent. The oligosaccharide aldonic acid can be prepared in essentially quantitative yield by oxidation of the oligosaccharide with alkaline iodine solution[8,9] or with bromide–lead carbonate solution.[1] The oligosaccharide aldonic acid reacts with the carbodiimide to form an *o*-acylisourea derivative, which reacts further with the free amino functions (ε-amino groups of lysyl residues) of the protein to form amide linkages. When bovine serum albumin was used as the carrier protein about 35 of the 60 amino groups present in the protein were substituted when the protein/carbohydrate ratio was 1 : 500. By this procedure several di- and trisaccharide aldonic acids (e.g., melibionic, lactobionic, gentiobionic, and maltotrionic acids) have been linked to different proteins such as bovine serum albumin, ovalbumin, and concanavalin A.[7]

The reactivities of several lectins, such as concanavalin A, the *Ricinus communis* lectins, and the α-D-galactopyranosyl-binding lectin of *Griffonia simplicifolia,* against different oligosaccharide aldonate–bovine serum albumin conjugates have been investigated.[7] The conjugates were efficient precipitating agents for the lectins, and the lectin–conjugate interaction involved only the carbohydrate binding site of the lectins. Thus, no nonspecific protein–protein interactions were detected. These properties appear to render the oligosaccharide aldonate–bovine serum albumin conjugates suitable for further studies of lectins.

Oligosaccharide aldonate–bovine serum albumin conjugates have also been used as antigens in immunizing experiments with rabbits.[7,10,11] The

[1] Y. Arakatsu, G. Ashwell, and E. A. Kabat, *J. Immunol.* **97,** 858 (1966).

[2] R. S. Martineau, P. Z. Allen, I. J. Goldstein, and R. N. Iyer, *Immunochemistry* **8,** 705 (1971).

[3] O. Lüderitz, A. M. Staub, and O. Westphal, *Bacteriol. Rev.* **30,** 192 (1966).

[4] R. U. Lemieux, D. R. Bundle, and D. A. Barker, *J. Am. Chem. Soc.* **97,** 4076 (1975).

[5] R. N. Iyer and I. J. Goldstein, *Immunochemistry* **10,** 313 (1973).

[6] I. J. Goldstein, S. Hammarström, and G. Sundblad, *Biochim. Biophys. Acta* **405,** 53 (1975).

[7] J. Lönngren, I. J. Goldstein, and J. E. Niederhuber, *Arch. Biochem. Biophys.* **175,** 661 (1976).

[8] S. Moore and K. P. Link, *J. Biol. Chem.* **133,** 293 (1940).

[9] D. A. Zopf and V. Ginsburg, *Arch. Biochem. Biophys.* **167,** 345 (1975).

[10] S. B. Svenson and A. A. Lindberg, *FEMS Microbiol. Lett.* **1,** 145 (1977).

[11] S. B. Svenson and A. A. Lindberg, *J. Immunol.* 1750 (1978).

investigations showed that the conjugates were potent immunogens, giving rise to high-titer carbohydrate hapten-specific antibodies. Thus, it should be possible to use this technique to raise antibodies to potential immnological determinants isolated from polysaccharides by specific or nonspecific degradations. In the same way, synthetic oligosaccharide derivatives having aglycons containing a carboxyl group could be linked to carrier proteins for immunization experiments.

Procedure

Preparation of Potassium Melibionate. A solution of iodine (2.8 g) in anhydrous methanol (50 ml) is placed in a three-necked, round-bottomed flask (250 ml) equipped with a magnetic stirrer, thermometer, and dropping funnel.[8] The flask is heated to 40°. Melibiose (2.0 g), dissolved in a minimum volume of boiling water, is added all at once to the stirred iodine solution. The resulting reaction mixture is kept at 40°, and a solution of potassium hydroxide (3 g) in methanol (75 ml) is immediately added dropwise over a period of 30 min. Potassium melibionate precipitates as a white solid. The aldonate salt is filtered on a Büchner funnel from the yellow reaction mixture, washed with cold anhydrous methanol (300 ml) and diethyl ether (200 ml), and stored in a vacuum desiccator over potassium hydroxide pellets. The yield is 2.2 g (96%).

Coupling Reaction. Potassium melibionate (1.3 g) is added to a stirred solution of bovine serum albumin (400 mg) in water (4.0 ml), and the pH is adjusted to 4.75 with 1.0 *M* hydrochloric acid.[7] A solution of 1-ethyl-3-(dimethylaminopropyl)carbodiimide hydrochloride (620 mg) in water (1 ml) is added dropwise over a period of 30 min at room temperature. During this time the solution is maintained at pH 4.75 by dropwise addition of 0.5 *M* hydrochloric acid using a pH stat or, if this device is not available, manually. After about 40 min no further acid is consumed, and the solution is allowed to stand at room temperature for 6 hr, during which time the pH drops to about pH 3.5. The reaction, quenched by addition of sodium acetate buffer (5 ml, 1 *M*, pH 5.5), is dialyzed against distilled water (4 times, 4 liters) prior to recovery of the conjugate by freeze-drying. The yield of protein is essentially quantitative, and the degree of substitution (determined by the phenol–sulfuric acid assay[10]) is about 35 (60%).

[12] Coupling of Oligosaccharides to Proteins Using p-Trifluoroacetamidoaniline

By ELISABET KALLIN

Several methods for coupling biologically active oligosaccharides to proteins and other amino-substituted carriers have been described. These methods include, among others, isothiocyanate coupling of p-aminophenyl glycosides[1] and reductive amination of aldehyde functions to amine-containing macromolecules.

Conjunction to isolated complex oligosaccharides requires a mild and highly efficient method. The p-trifluoroacetamidoaniline method described in this chapter is a high-yield, simple, and reproducible method for coupling complex oligosaccharides to proteins. It combines the two mild procedures of reductive amination and isothiocyanate coupling.[2]

The p-trifluoroacetamidoaniline method (Fig. 1) consists of reductive amination of a free oligosaccharide with an aromatic amine (p-trifluoroacetamidoaniline) using sodium cyanoborohydride, and subsequent conjugation of the formed aminoalditols to proteins.[3]

The method has been applied to several oligosaccharides terminating in 4-linked glucose or glucosamine, including fucosylated and sialylated structures,[3] and also to oligosaccharides from bacterial lipopolysaccharides.[4]

The p-trifluoroacetamidoaniline reagent can be prepared in bulk quantities and is a stable crystalline derivative. No large excess of reducing agent or amine is required, although generally a 5-fold excess of reagent is used. The yields are normally around 80–90%. The reaction is carried out at pH 6, at which the aromatic amine is in protonated, and the formation of the intermediate aldimine proceeds smoothly. Sodium cyanoborohydride works excellently at this pH and ensures the formation of aminoalditol derivatives without a significant formation of by-products.

To prevent the p-trifluoroacetamidoaniline derivative from being oxidized, stabilization is performed through a simple N-acetylation step.[5] The N-acetylated derivative is then purified by partitioning between water

[1] D. F. Smith, D. A. Zopf, and V. Ginsburg, this series, Vol. 50, p. 169.
[2] D. H. Buss and I. J. Goldstein, *J. Chem. Soc.* (*C*) 1457 (1968).
[3] E. Kallin, H. Lönn, and T. Norberg, *Glycoconjugate J.* **3**, 311 (1986).
[4] A. Y. Chernyak, A. Weintraug, T. Norberg, and E. Kallin, *Glycoconjugate J.* **7**, 111 (1990).
[5] E. Kallin, H. Lönn, and T. Norberg, *Glycoconjugate J.* **5**, 145 (1988).

Copyright © 1994 by Academic Press, Inc.
All rights of reproduction in any form reserved.

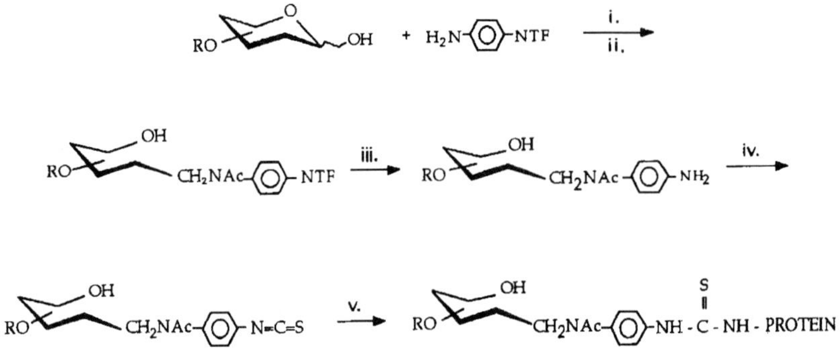

FIG. 1. Synthesis of neoglycoproteins by the *p*-trifluoroacetamidoaniline method. (i) $NaCNBH_3$, (ii) Ac_2O, (iii) aq. NaOH, (iv) $CSCl_2$, (v) protein.

and an organic solvent, and finally by solid-phase extraction using a reversed-phase column.

Conjugation of the N-acetylated *p*-trifluoroacetamidoaniline derivatized oligosaccharides to proteins is achieved by the isothiocyanate coupling method, commonly used for conversion of aromatic amino groups to isothiocyanates.[2] The *N*-trifluoroacetamido protective group is first removed by treatment of the N-acetylated *p*-trifluoroacetamido derivative with aqueous sodium hydroxide. Thiophosgene is then used for preparation of the isothiocyanate derivative.

Subsequent coupling of the isothiocyanate derivative to the desired protein is performed in slightly alkaline buffer solution. By this method an incorporation of 20–25 mol of oligosaccharide per mole of protein was achieved with the use of human serum albumin, and a ratio in the coupling reaction of 30 mol of oligosaccharide per mole of protein.

General Methods

All reactions were performed under nitrogen. Concentrations were performed at <40°C bath temperature. The reaction steps were monitored by thin-layer chromatography (TLC), performed on silica gel 60 F-254 (Merck, Darmstadt, Germany) using EtOAc/HOAc/MeOH/H_2O (6/3/3/2 by vol) as eluant. The spots were visualized with uv light or by charring with sulfuric acid. Bond Elut C-18 cartridges, used for the solid-phase extraction procedure, were from Analytichem International (Harbor City, U.S.A.). Ultrafiltration equipment (Omega cells, 10,000 or 50,000 kDa cutoff) were from Filtron AB (Bjärred, Sweden). Bio-Gel P-2 gel filtration columns (Bio-Rad, Richmond, CA, U.S.A.) were eluted with water.

Gas–liquid chromatography (GLC) was performed on a Hewlett–Packard 5890 instrument with an SE-30 capillary column (0.25 mm × 30 m).

Synthesis of *p*-Trifluoroacetamidoaniline

Trifluoroacetic anhydride (5.0 ml) was added, dropwise at 0°C, to a stirred solution of *p*-nitroaniline (2.76 g) in pyridine (25 ml). After 2 h at room temperature, the mixture was poured into ice-cold water (500 ml). The mixture was stirred for 1 h and then filtered. The solid material was recrystallized from methanol–water to give *p*-trifluoroacetamidonitrobenzene (4.3 g, 92%). Hydrogenation of this product (2.0 g) in ethanol (30 ml) over Pd-C (50 mg) at atmospheric pressure gave, after filtration and concentration, a residue, which was recrystallized from ether–hexane to give *p*-trifluoroacetamidoaniline (1.51 g, 87%), mp 118°C.

Preparation of N-Acetylated *p*-Trifluoroacetamidoaniline Derivatized Oligosaccharides

Concentrated acetic acid (0.2 ml) was added to a solution of oligosaccharide (0.1 mmol) and *p*-trifluoroacetamidoaniline (100 mg) in ethanol/water (2/1, 3 ml). Sodium cyanoborohydride (60 mg) was added, and the mixture was stirred at room temperature for 15–48 h, until TLC indicated that no free oligosaccharide remained. Acetic anhydride (0.5 ml) was added, and the mixture was stirred at room temperature for 2–3 h. The reaction was monitored by TLC, which showed that the acetylated compound corresponded to the slower moving spot.

Water (10 ml) was added to the reaction mixture, and the solution was washed with ethyl acetate (10 ml). The ethyl acetate phase was then washed with water (5 ml). The combined aqueous phases were washed once more with ethyl acetate (10 ml) and then evaporated to 2 ml. The material was then applied to a C-18 solid-phase extraction column (5 g C-18 gel), wetted with methanol, and equilibrated in water. The column was washed with 3 vol of water to wash out salt and unreacted oligosaccharide. Elution of the product was achieved with 3 vol of methanol. The eluate was collected in fractions and analyzed by TLC. Product-containing fractions were combined. Methanol was evaporated and the product lyophilized. The resulting N-acetylated *p*-trifluoroacetamidoaniline derivatized oligosaccharides are stable for years as dried powders and can also be kept in aqueous solution for several weeks.

Conjugation of N-Acetylated *p*-Trifluoroacetamidoaniline Derivatized Oligosaccharides

A solution of an N-acetylated *p*-trifluoroacetamidoaniline derivatized oligosaccharide (0.1 mmol) in aqueous sodium hydroxide (0.5 *M*, 5.0 ml) was left at room temperature for 3 h. Ethanol (10 ml) and acetic acid (0.1 ml) were added, and the solution was neutralized with aqueous sodium hydroxide. The solution was stirred and thiophosgene (0.040 ml) was added. After 10 min, water (5 ml) was added, and the mixture was washed with diethyl ether (10 ml). The diethyl ether phase was washed with water (2 ml), and the combined aqueous phases were evaporated to 2 ml. The solution was immediately added to a previously prepared solution of protein in borate buffer (0.1 *M*, pH 9.4, 25 ml). The mixture was slowly stirred for 15–48 h. When TLC showed no remaining isothiocyanate derivative, the mixture was ultrafiltered, and the remaining solution was washed by repeated additions of water, followed by ultrafiltration. At least four successive ultrafiltrations (4 × 100 ml water) were required to remove all low-molecular-weight material. Alternatively, separation on Bio-Gel can be used. Finally, the product was lyophilized.

The amount of protein used varied depending on the nature of the protein and on the desired degree of substitution. It was found that in order to obtain an incorporation of oligosaccharide of 20–25 mol per mole of protein, 162 mg (0.0025 mmol) of albumin and 93 mg of key hole limpet hemocyanin were ideal. The neoglycoproteins were stable in aqueous solution for several weeks. As lyophilized powders they are stable for years.

Characterization of Neoglycoproteins

The neoglycoproteins were analyzed for oligosaccharide substitution by sugar analysis either by GLC on peracetylated monosaccharide alditol acetates[6] or by a colorimetrical method.

In the sugar analysis glycosidic bonds were hydrolyzed with trifluoroacetic acid (4 *M*, 0.5–1.0 mg glycoprotein/ml) for 4 h at 100°C. Perseitol[6a] or xylose was used as internal standard (5–25 μg/ml). After hydrolysis the solution was evaporated to dryness. Water containing a few drops of concd ammonia was added and the solution was again evaporated. The residue was dissolved in water (0.5 ml), and excess sodium borohydride was added (0.5 mg). After 2 h at room temperature, acetic acid was added until pH 3–4 was reached. After evaporation to dryness, MeOH (0.5

[6] J. S. Sawardeker, J. H. Sloneker, and A. R. Jeanes, *Anal. Chem.* **37,** 1602 (1965).

[6a] D-glycero-D-galacto-heptitol.

ml) was added, and the solution was again evaporated to dryness. This procedure was repeated. The residue was then acetylated for 30 min at 100°C, using pyridine/acetic anhydride (1/2, 0.5 ml). After the mixture had cooled to room temperature, ethanol (0.5 ml) was added, and the mixture was evaporated to dryness. The residue was taken up in water (1.0 ml) and extracted with chloroform (3 × 0.5 ml). The combined chloroform phases were washed with water (3 × 1.0 ml) and evaporated to dryness. Toluene (0.1 ml) was added, and the mixture was again evaporated to dryness. The residue was analyzed by GLC and compared to standard monosaccharide alditol acetates.

For the colorimetric method we used a modification of the anthrone method.[7,8] Anthrone reagent was prepared by dissolving 2.0 g anthrone in 1.0 liter concentrated sulfuric acid. Solutions of glycoprotein in water at different dilutions (1.0 ml/tube, triple samples) were mixed with anthrone reagent (2.0 ml/tube), incubated at 100°C for 15 min, cooled to room temperature, analyzed spectrophotometrically at 620 nm, and compared to standard samples.

[7] R. J. Dimler, W. C. Schaefer, C. S. Wise, and C. E. Rist, *Anal. Chem.* 1411 (1952).
[8] J. H. Roe, *J. Biol. Chem.* **212,** 335 (1955).

Section II

Neoglycolipids

[13] Neoglycolipids of 1-Deoxy-1-phosphatidylethanolaminolactitol Type: Synthesis, Structure Analysis, and Use as Probes for Characterization of Glycosyltransferases

By GOTTFRIED POHLENTZ and HEINZ EGGE

Introduction

Neoglycolipids (NeoGL) of the 1-deoxy-1-phosphatidylethanolaminolactitol type (Lac-PtdEtn; for structures, see Fig. 1), synthesized by coupling an oligosaccharide to phosphatidylethanolamine (PtdEtn) by reductive amination, have been introduced by Tang *et al.*[1] They have been used for studies on antigenicity, receptor function, and lectin and toxin binding by carbohydrates.[1-3] Coupling to phosphatidylethanolamine has also been employed for characterization of N- and O-linked oligosaccharides by liquid secondary ion mass spectrometry.[4,5] We have found that glycosyl-PtdEtns and especially the N-acetylated derivatives can serve as acceptors for glycosphingolipid (GSL) glycosyltransferases. The NeoGL are as good acceptors as the authentic GSL analogs, and they are glycosylated by the same enzymes.[6]

This chapter describes an improved synthesis of Lac-PtdEtn-type neoglycolipids, their N-acetylation, and their use as acceptors for glycosyltransferases and for structure elucidation of oligosaccharides by fast atom bombardment-mass spectrometry (FAB-MS).

Materials

Lactose, dihexadecylglycerophosphoethanolamine (PtdEtn with C_{16} alkyl groups will be referred to as PtdEtn), phosphomolybdic acid, and

[1] P. W. Tang, H. C. Gool, M. Hardy, Y. C. Lee, and T. Feizi, *Biochem. Biophys. Res. Commun.* **132,** 474 (1985).

[2] M. S. Stoll, T. Mizuochi, R. A. Childs, and T. Feizi, *Biochem. J.* **256,** 661 (1988).

[3] T. Pacuszka, R. M. Bradley, and P. H. Fishman, *Biochemistry* **30,** 2563 (1991).

[4] T. Mizuochi, R. W. Loveless, A. M. Lawson, W. Chai, P. J. Lachmann, R. A. Childs, S. Thiel, and T. Feizi, *J. Biol. Chem.* **264,** 13834 (1989).

[5] M. S. Stoll, E. F. Hounsell, A. M. Lawson, W. Chai, and T. Feizi, *Eur. J. Biochem.* **189,** 499 (1990).

[6] G. Pohlentz, S. Schlemm, and H. Egge, *Eur. J. Biochem.* **203,** 387 (1992).

Copyright © 1994 by Academic Press, Inc.
All rights of reproduction in any form reserved.

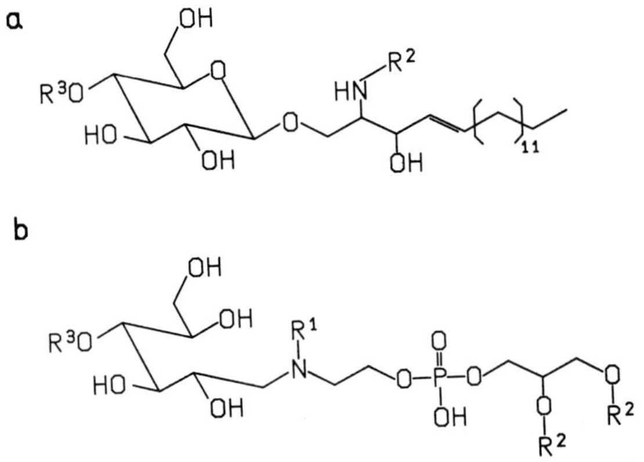

FIG. 1. Structures of (a) glycosphingolipids (glycosylceramides), where R^2 = acyl and R^3 = oligosaccharide, and (b) glycosyl-PtdEtn molecules, where R^1 = H or Ac, R^2 = acyl or alkyl, and R^3 = oligosaccharide.

ceric sulfate are obtained from Fluka (Neu-Ulm, Germany), sialyllactose, diacylglycerophosphoethanolamines (PtdEtns with C_{12}, C_{14}, C_{16}, and C_{18} acyl groups, referred to as PtdEtn-C_{12}, PtdEtn-C_{14}, etc.), and Triton CF-54 are from Sigma (Deisenhofen, Germany).

Triethanolamine (TEA), orcinol, thin-layer chromatography (TLC) plates, silica gel 60, and LiChroprep RP-18 are purchased from Merck (Darmstadt, Germany), and Iatrobeads are obtained from Macherey & Nagel (Düren, Germany). Dialysis tubing (molecular weight cutoff 12,000–14,000) is from Serva (Heidelberg, Germany). Cytidine 5′-monophospho-*N*-acetyl[4,5,6,7,8,9-^{14}C]neuraminic acid (CMP[^{14}C]NeuAc) is from Amersham-Buchler (Braunschweig, Germany) and is used after dilution with the unlabeled sugar nucleotide from Sigma. Scintillation cocktail Picofluor 40 is from Packard (Frankfurt, Germany), and 1-mercapto-2, 3-propanediol (thioglycerol, Tgl) is from Aldrich (Steinheim, Germany). X-Ray film XAR-5 is obtained from Eastman Kodak Company (Rochester, NY). Rats of the Wistar strain (male, 300–350 g) are procured from Hagemann (Extertal, Germany). All reagents and solvents used are of analytical grade quality.

Lacto-*N*-tetraose (LNT) is isolated from human milk as described previously.[7] II^3-Sialylgangliotetraose (II^{3}NeuAc-Gg_4Ose) is kindly provided by Dr. Günter Schwarzmann, Institùt f. Organische und Biochemie, Bonn, Germany.

[7] R. Bruntz, U. Dabrowski, J. Dabrowski, A. Ebersold, J. Peter-Katalinię, and H. Egge, *Biol. Chem. Hoppe-Seyler* **369,** 257 (1988).

Solutions

Reducing reagent: 1% (w/v) sodium cyanoborohydride, 0.1% (by volume) acetic acid in chloroform/methanol (1 : 1, by volume)

Orcinol spray reagent: 0.5% (w/v) orcinol in 15% (by volume) sulfuric acid. This reagent is rather specific for carbohydrates, and the spots on TLC plates are visualized by heating on a hot plate to 120° for 10 min.

Phosphomolybdic acid (PMA) spray reagent: 2.5% (w/v) phosphomolybdic acid, 1% (w/v) ceric sulfate in 6% (by volume) sulfuric acid.[8] The PMA spray reagent is a Stains-all reagent. The spots are visualized by heating the TLC plate to 100° for 10 min.

Solvents

A: Chloroform/methanol/0.2% (w/v) $CaCl_2$, 60 : 35 : 8 (by volume)
B: Chloroform/methanol, 2 : 1 (by volume)
C: Chloroform/methanol, 1 : 1 (by volume)
D: Chloroform/methanol/water, 65 : 25 : 4 (by volume)
E: Chloroform/methanol/water, 24 : 12 : 1.6 (by volume)
F: Chloroform/methanol/water, 60 : 35 : 8 (by volume)
G: Chloroform/methanol, 95 : 5 (by volume)
H: Chloroform/methanol, 90 : 10 (by volume)
I: Chloroform/methanol/water, 75 : 25 : 4 (by volume)
J: Chloroform/methanol/water, 55 : 45 : 10 (by volume)

Methods

Synthesis of 1-Deoxy-1-phosphatidylethanolaminolactitol-Type Neoglycolipids (Method 1)

The NeoGL containing lactose and phosphatidylethanolamines with acyl groups of different chain lengths (lauroyl, myristoyl, palmitoyl, and stearoyl residues) are synthesized following the original method of Tang *et al.*[1] The reactions are carried out in screw-capped vials or flasks. Equimolar amounts of lactose and the respective PtdEtn-C_{12-18} are suspended in solvent C (20 μmol/ml). After adding freshly prepared reducing reagent (1 ml per 10 mg lactose) and heating the mixture to 60° for 16 hr, the solvent is evaporated under a stream of nitrogen. The residue is suspended

[8] The solution is prepared by first adding water and then the appropriate amount of sulfuric acid, otherwise a clear solution is not obtained.

in the same volume of water, dialyzed (twice for 4 hr against 5 liters of water), and lyophilized. For final purification the raw product is applied to a silica gel column, and the Lac-PtdEtn components are eluted with solvent D. The products are characterized by FAB-MS (data not shown). The final yields are listed in Table I. The relatively low yields can be explained by two side reactions: (1) direct reduction of the sugar owing to the excess of reducing reagent at the beginning of the reaction and (2) formation of a disubstituted PtdEtn (which obviously occurs only with mono- and disaccharides).

Therefore, method 1 has been improved as follows. First, only dihexadecyl-PtdEtn was used because of its greater chemical stability and the superior glycosyltransferase acceptor qualities of the resulting neoglycolipids (see section on glycosyl-PtdEtn molecules as acceptors for glycosyltransferases). Second, because more complex oligosaccharides are almost insoluble in the organic solvent mixtures used for the coupling reaction, dimethyl sulfoxide (DMSO) is employed as mediator solvent that strongly increases the solubility of the sugars and thereby leads to higher yields. DMSO is converted to some extent to a water-soluble product (not characterized; see Fig. 2) that can be removed by dialysis. Third, by the succes-

TABLE I
YIELDS OBTAINED FROM SYNTHESIS OF Lac-PtdEtn SPECIES[a]

PtdEtn	Oligosaccharide	Method	Yield (%)
C_{12}-acyl	Lactose	1	38.1
C_{14}-acyl	Lactose	1	24.6
C_{16}-acyl	Lactose	1	31.3
C_{18}-acyl	Lactose	1	34.7
C_{16}-alkyl	Glucose	1	33.2
C_{16}-alkyl	LNT	1	27.4
C_{16}-alkyl	Lactose	2	72.6
C_{16}-alkyl	NeuAc-Lac	2	72.8
C_{16}-alkyl	LNT	2	44.3
C_{16}-alkyl	II^3NeuAc-Gg_4Ose	2	67.6

[a] Oligosaccharides and various PtdEtn types as indicated were coupled by reductive amination following the procedures described in the text as Methods 1 and 2. The yields were calculated in relation to the amount of oligosaccharide.

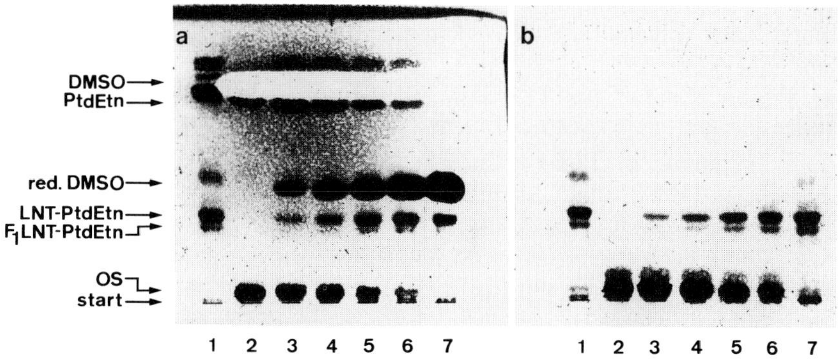

FIG. 2. Monitoring by TLC of the reaction course during the synthesis of lacto-*N*-tetraosyl-PtdEtn. Lacto-*N*-tetraose (LNT) and PtdEtn were coupled using Method 2 as described in the text, and aliquots of the reaction mixture were applied to a TLC plate at the times indicated. The chromatogram was developed in solvent A, and the spots were visualized by PMA (a) and orcinol staining (b). Lane 1, Reaction mixture after dialysis; lane 2, reaction mixture before adding the reducing reagent (0 min); lane 3, reaction mixture after adding reducing reagent (45 min reaction time); lanes 4–7, 1.5, 3.5, 6, and 24 hr reaction times, respectively. DMSO, Dimethyl sulfoxide; red. DMSO, reduced DMSO; F_1LNT, fucosyllacto-*N*-tetraose; OS, oligosaccharide.

sive addition of smaller portions of the reducing reagent, the side reaction involving direct reduction of the oligosaccharide is diminished in favor of the reductive amination. Finally, to avoid the disubstituted by-products and to achieve quantitative reaction of the oligosaccharides, a 2-fold molar excess of PtdEtn is used.

Method 2

The oligosaccharide (up to 100 mg) is suspended in dimethylsulfoxide (100 mg/ml) and then dissolved in methanol (20 mg/ml). Two equivalents of PtdEtn is dissolved in solvent B (5 mg/ml). After combining both solutions the mixture is allowed to stand for 2 hr at 60°. Then the reducing reagent (freshly prepared; 1 ml/20 mg oligosaccharide) is added in portions of 0.5 to 2 ml during the reaction time (up to 24 hr at 60°), and the course of the reaction is controlled by thin-layer chromatography (solvent A or D, visualized with orcinol or PMA spray reagent). When residual oligosaccharide is no longer detectable, the solvent is evaporated, and the residue is suspended in 10 to 20 ml water, dialyzed (twice for 4 hr against 5 liters water), and lyophilized. For final purification the raw product is applied to a silica gel column (20 to 200 ml bed volume), and the neoglycolipid is eluted with solvent D, F, I, or J, depending on the type of oligosaccharide moiety.

Figure 2 shows the TLC-monitored time course of the coupling of PtdEtn to a lacto-*N*-tetraose (LNT) preparation contaminated with about 10% fucopentose (Fuc-LNT). The time-dependent formation of the NeoGL and the product originating from DMSO (Fig. 2, red DMSO) can be observed as well as the complete disappearance of free oligosaccharides after 24 hr. The "reduced" DMSO is removed during the dialysis step (see lane 1, Fig. 2), and the Fuc-LNT-PtdEtn is separated by chromatography on silica gel. Method 2 has been successfully used with glucose, lactose, sialyllactose, LNT, II^3NeuAc-Gg_4Ose, and blood group A trisaccharide [GalNAc(Fuc)Gal].[9] The yields shown in Table I are significantly higher than those obtained with Method 1.

N-Acetylation of Glycosylphosphatidylethanolamines

The purified glycosyl-PtdEtn is dissolved in a chloroform–methanol(–water) mixture (solvent C or D, 2 mg/ml). After addition of powdered $NaHCO_3$ (2 mg/mg glycosyl-PtdEtn), acetic anhydride (20 μl/mg glycosyl-PtdEtn) is added in five portions every 15 min. The course of the reaction is monitored by TLC (solvent A or D, staining with orcinol or PMA).[10] After completion of the reaction the solvent is evaporated under a stream of nitrogen.

For desalting of the residues, two alternative procedures are performed depending on the amount of glycosyl-PtdEtn. Larger amounts (>2 mg) of NeoGL are suspended in 10 to 20 ml water, dialyzed, and lyophilized as described above. Smaller amounts of glycosyl-PtdEtn are separated from the salts using reversed-phase chromatography. A small column (Pasteur Pipette, bed volume 1 to 1.5 ml) is filled with RP-18 gel suspended in solvent C. The gel is washed with 10 ml solvent C, acetonitrile, and water. After application of the aqueous suspension of the above residue (1 to 2 ml) the column is successively eluted with 10 ml each of water, acetonitrile, and solvent C. The NeoGL is eluted with solvent C. Final purification of

[9] B. Klima, G. Pohlentz, D. Schindler, and H. Egge, *Biol. Chem. Hoppe-Seyler* **373**, 989 (1992).

[10] The N-acetylated glycosyl-PtdEtns [glycosyl-PtdEtn(NAc)] migrated slightly slower than the parent compounds. Loss of the positive charge of the amino group induced a(n additional) negative charge. This effect obviously overcame the decrease of polarity caused by the *N*-acetyl group.

the raw products is carried out by silica gel chromatography as described above.

With this N-acetylation procedure O-acetylation is observed in some cases (i.e., 1 to 3 hydroxyl groups are additionally acetylated; 10 to 20%). The *O*-acetyl groups can be easily removed with sodium methoxide. The lyophilizate or the RP-18 eluate obtained after desalting is dissolved in the smallest possible volume of methanol, and sodium methoxide is added from a stock solution to a final concentration of 10 m*M*. The mixture is stirred for 2 hr at room temperature, and the course of reaction is monitored by TLC. After the reaction is completed, the solvent is evaporated under a stream of nitrogen. The residue can be directly used for final purification. The N-acetylated glycosyl-PtdEtn compounds are usually obtained in high yields (60 to 80%; Table II).

N-Acetylation of Lacto-N-tetraose-phosphatidylethanolamine with [^{14}C]Acetic Anhydride

Radiolabeled NeoGL can be useful tools for investigations on glycolipid metabolism in cell culture. Glycosyl-PtdEtns can be labeled by N-acetylation with [^{3}H]- or [^{14}C]acetic anhydride. In a screw-capped vial 5 mg LNT-PtdEtn (3.7 μmol) is dissolved in 2 ml methanol, and 10 mg powdered $NaHCO_3$ is added. The ampoule containing [^{14}C]acetic anhy-

TABLE II
YIELDS OBTAINED FROM N-ACETYLATION OF GLYCOSYL-PtdEtn SPECIES[a]

Glycosyl-PtdEtn	Yield (%)
Lac-PtdEtn-C_{12}	59.7
Lac-PtdEtn-C_{14}	44.5
Lac-PtdEtn-C_{16}	39.2
Lac-PtdEtn-C_{18}	81.8
Lac-PtdEtn	65.0
NeuAc-Lac-PtdEtn	82.5
LNT-PtdEtn	86.5

[a] Oligosaccharide-PtdEtn molecules as indicated were N-acetylated by reaction with acetic anhydride as described in the text. The yields were calculated in relation to the amount of neoglycolipid.

dride (250 μCi) is cooled with liquid nitrogen, opened, and 50 μl toluene added. The toluene solution is immediately injected into the reaction mixture using a glass–Teflon syringe. The ampoule is washed with an additional 50 μl of toluene, and the wash solution is combined with the reaction mixture. After stirring for 30 min at room temperature, four 10-μl portions of nonlabeled acetic anhydride are added at 15-min intervals. A TLC control (solvent A, localization of the product by scanning for radioactivity and subsequent staining with orcinol) shows that the reaction is completed.

The solvent is evaporated under a stream of nitrogen. The residue is suspended in 5 ml water, dialyzed (one against 500 ml saturated $NaHCO_3$ solution and twice against 500 ml water), and lyophilized. The lyophilizate is applied to an Iatrobeads column (10 ml bed volume), and elution is performed with solvent D. Fractions (2 ml each) are collected, and the fractionation is monitored by TLC (solvent A, localization by scanning for radioactivity). Fractions containing radioactive material comigrating with nonlabeled LNT-PtdEtn(NAc) are combined. The LNT-PtdEtn[^{14}C](NAc) is obtained in a chemical yield of 86.5% with a specific radioactivity of 9.4 Ci/mol (radioactivity yield 24.0%).

Glycosyl-phosphatidylethanolamines and N-Acetylated Glycosyl-phosphatidylethanolamines as Acceptors for Sialyltransferases

In order for the glycosyl-PtdEtn compounds to serve as substitutes for natural GSL in glycosyltransferase assays, it must be ascertained that neoglycolipids are converted to the analogous products by the same enzymes which act on natural GSL and that they exhibit similar apparent kinetic constants. To this end we have used glycosyl-PtdEtns and the N-acetylated derivatives as acceptors for sialyltransferases from rat liver Golgi, identified the transferase products by FAB-MS, and demonstrated by competition experiments that the NeoGL and the authentic GSL analog are sialylated by the same respective enzymes.

Golgi Preparation. The preparation of Golgi vesicles follows the method of Sandberg *et al.*[11] In short, male rats of the Wistar strain (300–350 g) are decapitated, and the livers are quickly excised, minced in ice-cold 0.3 *M* sucrose, and homogenized in a Potter–Elvehjem glass–Teflon homogenizer with 10 strokes (pestle speed 500 rpm, 25 ml of 0.3 *M* sucrose per 10 g liver). All further steps are performed at 4°. The homogenate is

[11] P.-O. Sandberg, L. Marzella, and H. Glaumann, *Exp. Cell Res.* **130,** 393 (1980).

centrifuged at 7500 *g* for 20 min. The supernatant is decanted, and the pellets are rehomogenized and centrifuged as described above. The combined supernatants are adjusted to 1.15 *M* sucrose using a 2 *M* sucrose stock solution. The resulting suspension is transfered to centrifuge tubes and overlaid with a sucrose gradient consisting of one layer of 1.1 *M* sucrose and one layer of 0.3 *M* sucrose. The subsequent centrifugation is run for 2 hr at 150,000 *g*. The layers of Golgi fraction (at the 1.1/0.3 *M* sucrose interface) are collected, diluted with 0.3 *M* sucrose, and centrifuged down on a cushion of 1.1 *M* sucrose (0.25 to 0.5 ml). Again the Golgi layers are collected, combined, and stored at −70° in portions of 100 to 500 μl.

The protein content of the membrane fractions is determined by the method of Lowry *et al.*[12] using bovine serum albumin as standard. Golgi preparations obtained with the above procedure are usually enriched 40- to 60-fold in Golgi-specific glycosyltransferase activities, and contamination with other membranes is below 5%.[13]

Sialyltransferase Assays. In a total volume of 50 μl, assay solutions contain up to 200 μ*M* glycolipid acceptor (LacCer, G_{M3}, G_{M1}, or Lac-PtdEtn-type NeoGL), 0.3% (w/v) Triton CF-54, 150 m*M* sodium cacodylate–hydrochloric acid buffer (pH 6.6), 10 m*M* MgCl2, 10 m*M* 2-mercaptoethanol, 1 m*M* CMP[^{14}C]NeuAc [5000 to 10,000 counts/min (cpm)/nmol], and 50 μg of Golgi protein. The assay mixtures are incubated for 30 min at 37°. Assays and product separation by Sephadex G-25 gel chromatography are executed as described before.[13] The radioactivity of the products is determined in a liquid scintillation counter. Appropriate blanks without exogenous glycolipid acceptor are run for each determination. All experiments are performed at least in duplicate, and mean values are presented. Unless otherwise stated standard deviations (SD) are below 5%. For product identification, Sephadex G-25 eluates are dried under a stream of nitrogen, redissolved in 250 μl chloroform/methanol/water, 16 : 8 : 1 (by volume), and applied to silica gel 60 TLC plates. Chromatograms are developed in solvent A. Radiolabeled material is visualized by autoradiography.

For FAB-MS analysis, the Sephadex G-25 eluates of 10 assays are combined, and the sialylation products are separated on a column of Iatrobeads (bed volume 10 ml, solvent D, 2-ml fractions). The reaction

[12] O. H. Lowry, N. J. Rosebrough, A. L. Farr, and R. J. Randall, *J. Biol. Chem.* **193,** 265 (1951).

[13] H. K. M. Yusuf, G. Pohlentz, G. Schwarzmann, and K. Sandhoff, *Eur. J. Biochem.* **134,** 47 (1983).

products are detected by examining aliquots of the fractions for radioactivity. The respective fractions are combined and, after evaporation of the solvent, are analyzed by FAB-MS.

Results. When Lac-PtdEtn molecules and the N-acetylated derivatives were used as acceptors for sialyltransferases from rat liver Golgi, the following results were obtained (Fig. 3a): (1) all NeoGLs tested were sialylated; (2) chain lengths of the acyl or alkyl residues in the PtdEtn moieties seemed to have minor effects on the reaction rates; (3) N-acetylation led to a strong increase in the reaction rates [up to 220% of that obtained with the authentic acceptor lactosylceramide (LacCer); this phenomenon was most probably due to the loss of the positive charge of the PtdEtn amino group, and a similar effect has been observed for ganglioside

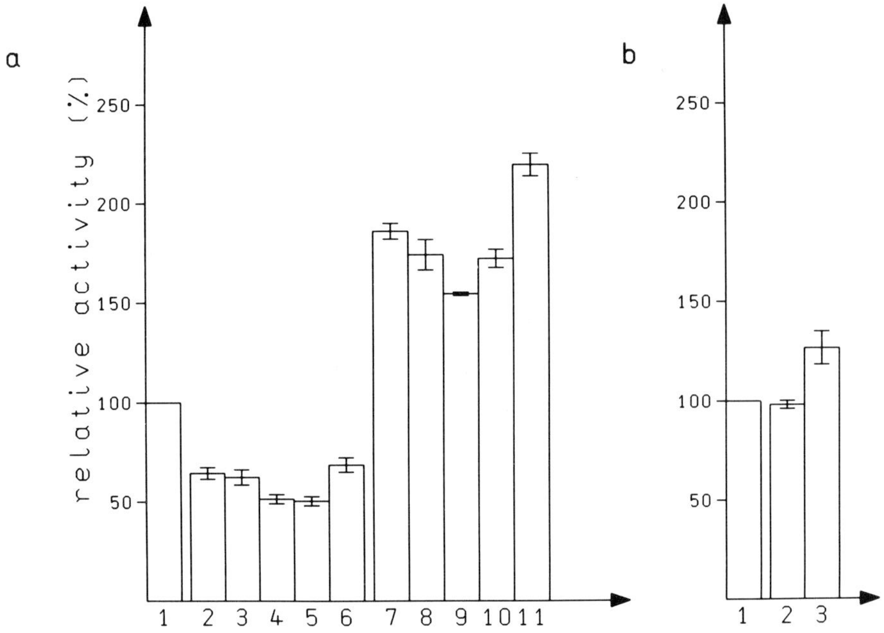

FIG. 3. Relative sialyltransferase activities obtained with (a) LacCer and PtdEtn analogs and (b) G_{M3} and NeuAc-Lac-PtdEtns as acceptors. [Adapted from G. Pohlentz, S. Schlemm, and H. Egge, *Eur. J. Biochem.* **203,** 238 (1992).] The acceptors were as follows: (a) 1, LacCer; 2, Lac-PtdEtn-C_{12}; 3, Lac-PtdEtn-C_{14}; 4, Lac-PtdEtn-C_{16}; 5, Lac-PtdEtn-C_{18}; 6, Lac-PtdEtn; 7, Lac-PtdEtn(NAc)-C_{12}; 8, Lac-PtdEtn(NAc)-C_{14}; 9, Lac-PtdEtn(NAc)-C_{16}; 10, Lac-PtdEtn(NAc)-C_{18}; 11, Lac-PtdEtn(NAc); (b) 1, G_{M3}; 2, NeuAc-Lac-PtdEtn; 3, NeuAc-Lac-PtdEtn(NAc) (PtdEtn without addition stands for the C_{16} ether compound).

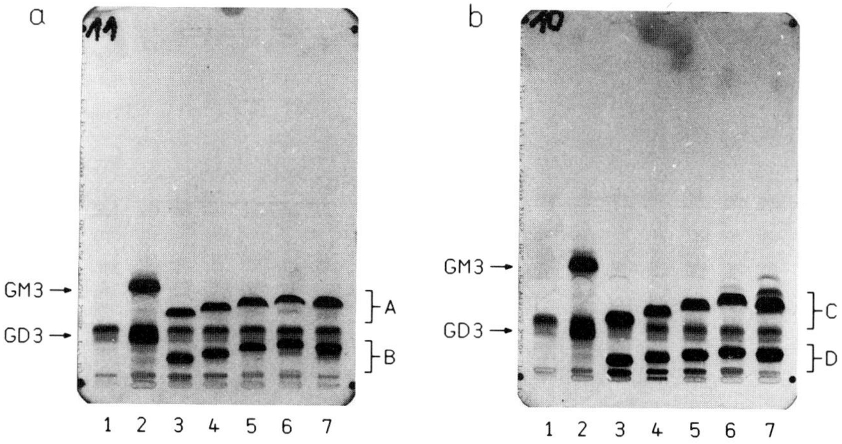

FIG. 4. Separation by TLC of the sialyltransferase products obtained from LacCer and Lac-PtdEtn species. Golgi vesicles were incubated with the acceptors as indicated below and CMP[^{14}C]NeuAc as described in the text. The radioactive products were separated with solvent A and visualized by autoradiography. [Adapted from G. Pohlentz, S. Schlemm, and H. Egge, *Eur. J. Biochem.* **203,** 238 (1992).] The acceptors were as follows: (a) lane 1, none; lane 2, LacCer; lane 3, Lac-PtdEtn-C_{12}; lane 4, Lac-PtdEtn-C_{14}; lane 5, Lac-PtdEtn-C_{16}; lane 6, Lac-PtdEtn-C_{18}; lane 7, Lac-PtdEtn; (b) lane 1, none; lane 2, LacCer; lane 3, Lac-PtdEtn(NAc)-C_{12}; lane 4, Lac-PtdEtn(NAc)-C_{14}; lane 5, Lac-PtdEtn(NAc)-C_{16}; lane 6, Lac-PtdEtn(NAc)-C_{18}; lane 7, Lac-PtdEtn(NAc). The presumed products were as follows: A, NeuAc-Lac-PtdEtn species; B, $NeuAc_2$-Lac-PtdEtn species; C, NeuAc-Lac-PtdEtn(NAc) species; D, $NeuAc_2$-Lac-PtdEtn(NAc) species.

G_{M3} derivatives[14]]; and (4) in both series of neoglycolipids (nonacetylated and acetylated) the highest activities were found with the C_{16} ether derivatives.

Separation by TLC of the various sialylated Lac-PtdEtn species is shown in Fig. 4. By analogy with the authentic acceptor LacCer that was sialylated to G_{M3} and subsequently to G_{D3}, all Lac-PtdEtn species were converted to the corresponding mono- and disialylated products. This result was corroborated by FAB-MS analysis of the sialylation products.[6]

Ganglioside G_{M3} was sialylated to G_{D3} by rat liver sialyltransferase II. The NeoGL analogs of G_{M3}, NeuAc-Lac-PtdEtn and NeuAc-Lac-PtdEtn(NAc), were converted to the corresponding disialylated derivatives by rat liver Golgi sialyltransferases. The relative reaction rates were found to be about 100 and 130%, respectively, compared to the rate

[14] D. Klein, G. Pohlentz, G. Schwarzmann, and K. Sandhoff, *Eur. J. Biochem.* **167,** 417 (1987).

obtained with G_{M3} (Fig. 3b). The sialylation products obtained from the neoglycolipids could be unambiguously identified as $NeuAc_2$-Lac-PtdEtn and $NeuAc_2$-Lac-PtdEtn(NAc) by FAB-MS analysis.[6]

Finally the ganglioside G_{M1} analog II^3NeuAc-Gg_4Ose-PtdEtn(NAc) was synthesized and tested as an acceptor for sialyltransferases. Like G_{M1} which was sialylated to G_{D1a}, the NeoGL was utilized to yield IV^3, II^3NeuAc-Gg_4Ose-PtdEtn(NAc). The relative reaction rate compared to that obtained with G_{M1} was 157 ± 20%. The TLC analysis of the sialylation product is shown in Fig. 5, and the structure of the presumed product was corroborated by FAB-MS (see below).

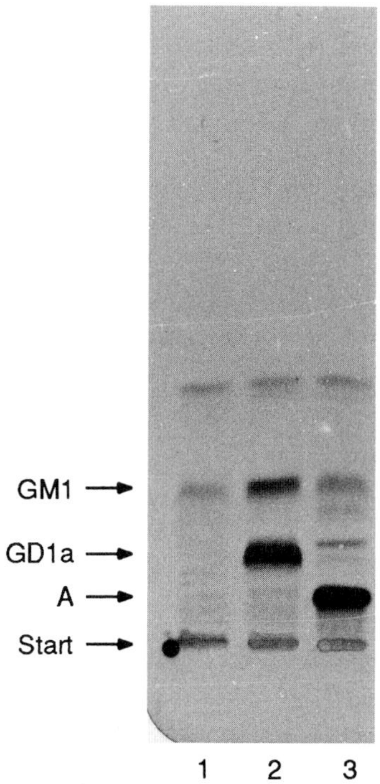

FIG. 5. Separation by TLC of the sialyltransferase products obtained from G_{M1} and II^3NeuAc-Gg_4Ose-PtdEtn(NAc). Golgi vesicles were incubated with the acceptors as indicated below and CMP[^{14}C]NeuAc as described in the text. The radioactive products were separated with solvent A and visualized by autoradiography. The acceptors were as follows: lane 1, none; lane 2, G_{M1}; lane 3, II^3NeuAc-Gg_4Ose-PtdEtn(NAc). The presumed product (A) was IV^3,II^3NeuAc-Gg_4Ose-PtdEtn(NAc).

Given that NeoGL serve as substitutes for GSL in studies on glycolipid metabolism, both kinds of lipids should be glycosylated by the same enzymes. We were able to demonstrate, by competition experiments, that G_{M3} and NeuAc-Lac-PtdEtn(NAc) and G_{M1} and II^3NeuAc-Gg_4Ose-PtdEtn(NAc) were sialylated by the same sialyltransferases, namely, sialyltransferase II and sialyltransferase IV, respectively. This kind of competition experiment has been used before to prove the identity of enzymes converting two or more substrates.[15,16] Briefly, both substrates G_{M3} and NeuAc-Lac-PtdEtn(NAc) [or G_{M1} and II^3NeuAc-Gg_4Ose-PtdEtn(NAc)] were used in the sialyltransferase assay at the same time at various partial concentrations, keeping the total substrate concentration constant. For two independent enzymes, each recognizing only one of two substrates a and b, the total reaction velocity v_t can be calculated from the partial velocities v_a and v_b given by the respective Michaelis equations:

$$v_t = v_a + v_b = \frac{V_a}{1 + K_a/[a]} + \frac{V_b}{1 + K_b/[b]} \tag{1}$$

If both substrates are utilized by the same enzyme and each substrate acts as a competitive inhibitor of the other ($K_m = K_i$), the total velocity is given by[17]

$$v'_t = v'_a + v'_b = \frac{V_a}{1 + (K_a/[a])(1 + [b]/K_b)} + \frac{V_b}{1 + (K_b/[b])(1 + [a]/K_a)} \tag{2}$$

The total reaction velocities can be calculated from Eqs. (1) and (2) with K_m and V_{max} values determined simultaneously using the same Golgi preparation and can be compared with the experimentally obtained v_t values.

The results of competition experiments using G_{M3} and NeuAc-Lac-PtdEtn(NAc) and G_{M1} and II^3NeuAc-Gg_4Ose-PtdEtn(NAc), respectively, as substrates and mutual inhibitors for sialyltransferases are shown in the Figs. 6 and 7. In both cases the experimentally determined reaction velocities follow the curves calculated from Eq. (2) for one enzyme. From these results we concluded that the neoglycolipids of the Lac-PtdEtn

[15] G. Pohlentz, D. Klein, G. Schwarzmann, D. Schmitz, and K. Sandhoff, *Proc. Natl. Acad. Sci. U.S.A.* **85,** 7044 (1988).

[16] H. Iber, R. Kaufmann, G. Pohlentz, G. Schwarzmann, and K. Sandhoff, *FEBS Lett.* **248,** 18 (1989).

[17] M. Dixon and E. C. Webb, "Enzymes." Longmans, London, 1979.

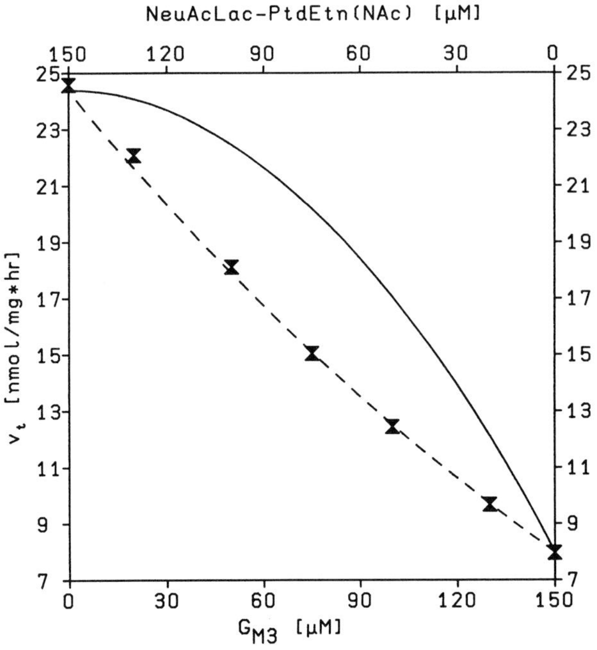

FIG. 6. Competition between ganglioside G_{M3} and the neoglycolipid NeuAc-Lac-PtdEtn (NAc) in the sialyltransferase assay. G_{M3} and NeuAc-Lac-PtdEtn(NAc) were used as acceptors for sialyltransferase in various partial concentrations, keeping the total substrate concentration at 150 μM. The kinetic constants were determined in a parallel experiment with the same Golgi preparation [G_{M3}: K_m = 145.2 μM, V_{max} = 15.7 nmol mg^{-1} hr^{-1}; NeuAc-Lac-PtdEtn(NAc): K_m = 271.4 μM, V_{max} = 68.5 nmol mg^{-1} hr^{-1}]. Total reaction velocities either experimentally determined (⧖) or calculated for one enzyme [Eq. (2); – – –] or two different enzymes [Eq. (1); —] are plotted versus the partial substrate concentrations. [Adapted from G. Pohlentz, S. Schlemm, and H. Egge, *Eur. J. Biochem.* **203,** 238 (1992).]

type and the corresponding GSL analogs were sialylated by the same sialyltransferases.

Fast Atom Bombardment-Mass Spectrometry Analysis of N-Acetylated Neoglycolipids of 1-Deoxy-1-phosphatidylethanolaminolactitol Type

The FAB mass spectra are recorded on a VG analytical ZAB-HF reverse geometry mass spectrometer (V.G. Analytical, Manchester, UK) as described previously.[18,19] Xenon is used for atom bombardment, and

[18] H. Egge, J. Peter-Katalinic, G. Reuter, R. Schauer, R. Ghidoni, S. Sonnino, and G. Tettamanti, *Chem. Phys. Lipids* **37,** 127 (1985).

[19] J. Müthing, H. Egge, B. Kniep, and P. F. Mühlradt, *Eur. J. Biochem.* **163,** 407 (1987).

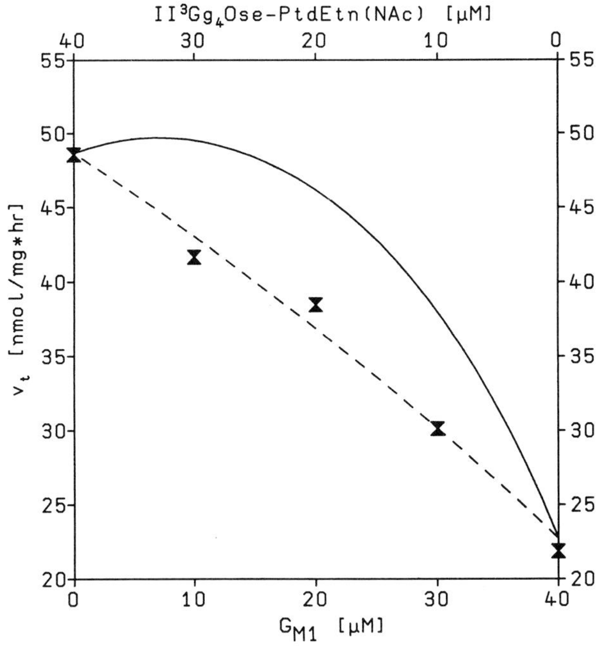

FIG. 7. Competition between ganglioside G_{M1} and the neoglycolipid II^3NeuAc-Gg_4Ose-PtdEtn(NAc) in the sialyltransferase assay. G_{M1} and II^3NeuAc-Gg^4Ose-PtdEtn(NAc) were used as acceptors for sialyltransferase in various partial concentrations, keeping the total substrate concentration at 40 μM. The kinetic constants were determined in a parallel experiment with the same Golgi preparation [G_{M1}: K_m = 64.5 μM, V_{max} = 59.3 nmol mg^{-1} h^{-1}; II^3NeuAc-Gg_4Ose-PtdEtn(NAc): K_m = 42.2 μM, V_{max} = 100.0 nmol mg^{-1} hr^{-1}]. Total reaction velocities either experimentally determined (⌛) or calculated for one enzyme [Eq. (2); – – –] or two different enzymes [Eq. (1); —] are plotted versus the partial substrate concentrations.

either positive ions [FAB(+)] or negative ions [FAB(−)] are detected. 1-Mercapto-2,3-propanediol (thioglycerol, Tgl) and triethanolamine (TEA) are used as matrices in the FAB(+) and FAB(−) mode, respectively.

The spectra are recorded and evaluated on a SAM II/68 K computer (KWS, Ettlingen, Germany) using the DP10 program of AMD (Harpstedt, Germany). Plotting and processing of data are performed on the central computer system of the University of Bonn. Samples of GSL from natural sources often consist of up to 10 (or more) different molecular species, owing to the heterogeneity in the ceramide residue (different fatty acids: C_{16} to C_{26}, unsaturated, hydroxylated; C_{18}- or C_{20}-sphingosine, sphinganine, phytosphingosine). This diversity generally leads to low relative intensities of molecular ions in FAB spectra of GSL as compared to those

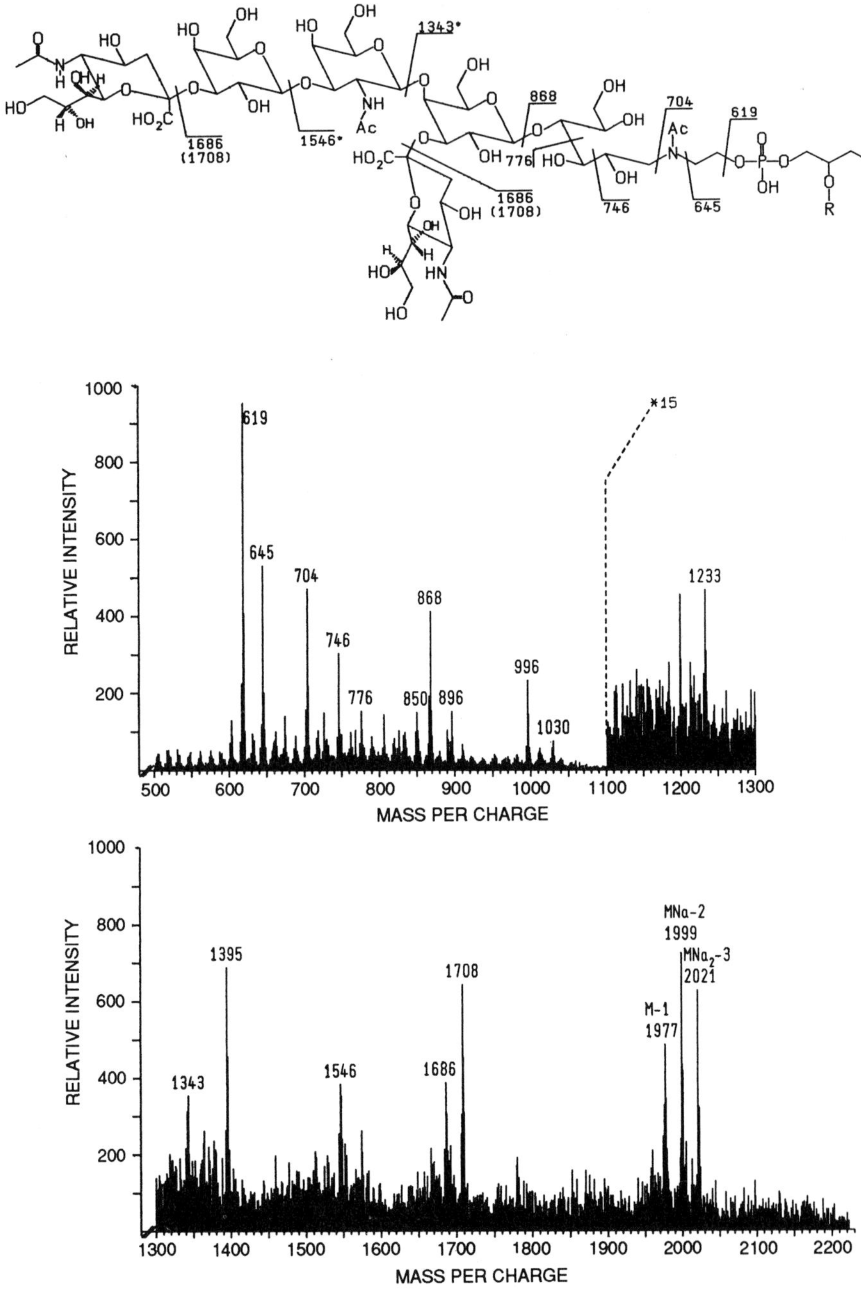

FIG. 8. Fragmentation scheme and FAB(−) mass spectrum of IV[3], II[3]NeuAc-Gg_4Ose-PtdEtn(NAc).

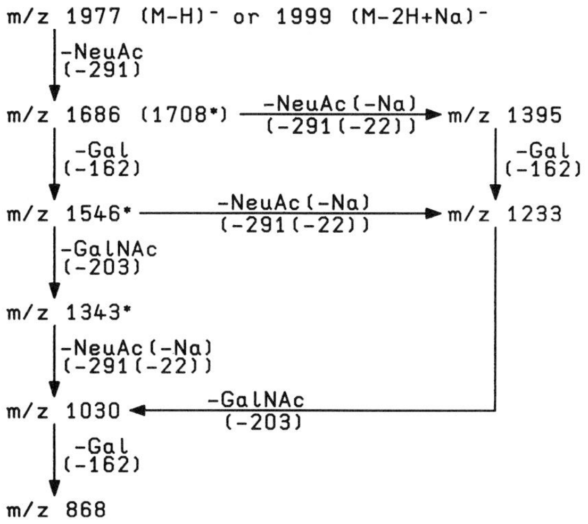

FIG. 9. Fragmentation sequence revealed from the FAB(−) mass spectrum of IV^3, II^{3}NeuAc-Gg$_4$Ose-PtdEtn(NAc). The *m/z* values marked by an asterisk (*) belong to sodium-containing fragments.

obtained from a sample containing one defined lipid moiety. Consequently, FAB mass spectra of the neoglycolipids showed intense pseudomolecular ion and fragment peaks.

Two additional improvements are made. First, use of C_{16} ether PtdEtn enhances the lucidity of the spectra since no peaks resulting from fatty acid cleavage and no diacylglycerol peaks are obtained. Second, N-acetylation has two effects: (i) owing to loss of the positive charge of the Etn amino group the spectrum intensity in the negative ion mode strongly increases [see FAB(−) section], and (ii) N-acetylation enabled the formation of a very stable oxazolidinium ion detectable in positive ion mode [see FAB(+) section].

Fast Atom Bombardment Mass Spectrometry in Negative Ion Mode. A comparison between the FAB(−) spectra of Lac-PtdEtn and Lac-PtdEtn(NAc) shows that N-acetylation leads to a significant increase in the signal-to-noise ratio.[20] In general, intense molecular ions and characteristic fragment ions are observed. As an example, the FAB(−) spectrum of IV^3,II^{3}NeuAc-Gg$_4$Ose-PtdEtn(NAc) obtained by enzymatic sialylation of II^{3}NeuAc-Gg$_4$Ose-PtdEtn(NAc) (see above) and the corresponding fragmentation scheme are shown in Fig. 8. Besides $(M - H)^-$ (*m/z* 1977), two additional molecular ion peaks at *m/z* 1999 [$(M - 2H + Na)^-$] and *m/z*

[20] G. Pohlentz, S. Schlemm, B. Klima, and H. Egge, *Chem. Phys. Lipids* **70,** 83 (1994).

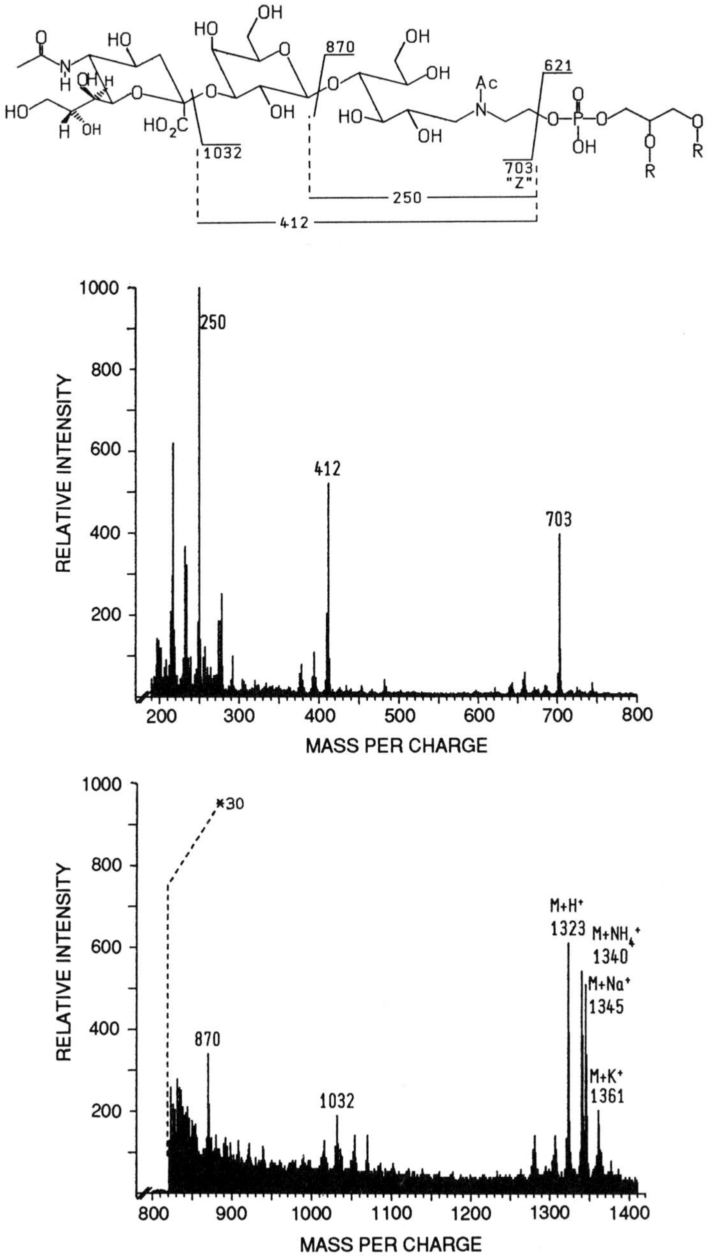

FIG. 10. Fragmentation scheme and FAB(+) mass spectrum of NeuAc-Lac-PtdEtn(NAc).

2021 [$(M - 3H + 2Na)^-$] are detected. The most dominant fragmentation leads to the phosphatidic acid fragment (*m/z* 619). The fragment ions at *m/z* 645, 704, 746, and 776 resulting from cleavage of the Etn and the reduced glucose, respectively, are observed in the FAB(−) spectra of all N-acetylated neoglycolipids analyzed.[20] Fragmentations resulting from elimination of the respective terminal monosaccharide units from the former nonreducing end of the NeoGL are highly diagnostic for the oligosaccharide sequence. Moreover, branching points of the carbohydrate chain can be localized by alternative elimination of the terminal sugar. Thus, the FAB(−) spectrum of IV^3,II^3NeuAc-Gg_4Ose-PtdEtn(NAc) (Fig. 8) can be interpreted by the fragmentation sequence shown in Fig. 9, which is in complete agreement with the structure of the NeoGL.

Fast Atom Bombardment Mass Spectrometry in Positive Ion Mode. The interpretation of FAB(+) spectra of glycosyl-PtdEtns (NAc) is demonstrated exemplarily with the spectrum of chemically synthesized NeuAc-Lac-PtdEtn(NAc). The spectrum and the corresponding fragmentation scheme are shown in Fig. 10. Besides $M + H^+$ (*m/z* 1323), in most cases additional pseudomolecular ion peaks were found [like $M + NH_4^+$ (*m/z* 1340), $M + Na^+$ (*m/z* 1345), or $M + K^+$ (*m/z* 1361)]. Starting with the molecular ions, fragmentation from the former nonreducing end of the oligosaccharide chain seems to play a minor role in FAB(+) spectra of the NeoGL. If at all detectable, the corresponding peaks are of low intensity {for NeuAc-Lac-PtdEtn(NAc): $M + H^+$ − NeuAc [−291 atomic mass units (amu)] → *m/z* 1032 −Gal (−162 amu) → *m/z* 870}. By cleavage of the phosphatidic acid–ethanolamine bond an N-acetylated glycosyl-Etn fragment, the so-called fragment Z, is formed.[20] Presumably fragment Z is a mesomery-stabilized oxazolidinium ion. Starting with Z cleavage of the glycosidic bonds from the "nonreducing" terminal leads to a series of intense secondary fragment ions from which the oligosaccharide sequence can be deduced [for NeuAc-Lac-PtdEtn(NAc): "fragment Z" *m/z* 777 − NeuAc (−291 amu) → *m/z* 412 − Gal (−162 amu) → *m/z* 250].

[14] Ceramide Glycanase from the Leech *Macrobdella decora* and Oligosaccharide-Transferring Activity

By Yu-Teh Li and Su-Chen Li

Introduction

Ceramide glycanase (CGase) is an endoglycosidase that releases the intact glycan chain from various glycosphingolipids by cleaving the linkage between the ceramide and the glycan chain. This enzyme is found in the leech *Macrobdella*[1] and in other annelids such as the European leech, *Hirudo medicinalis*,[2] and the earthworm, *Lumbricus terrestris*.[3] A similar enzyme (endoglycoceramidase) has also been isolated from *Rhodococcus*[4] and *Corynebacterium*[5] species.

Many exoglycosidases can catalyze transglycosylation of a monosaccharide in addition to hydrolysis. In contrast, the transfer of an intact complex oligosaccharide by endoglycosidases is less common. The CGase isolated from the leech can transfer the intact oligosaccharide from various glycosphingolipids (GSLs) to suitable acceptors. This oligosaccharide-transferring reaction can be used to synthesize neoglycoconjugates for studying the biological functions expressed by glycan chains in GSLs.

Assay Methods

Principle

The following equations illustrate the hydrolytic activity of CGase:

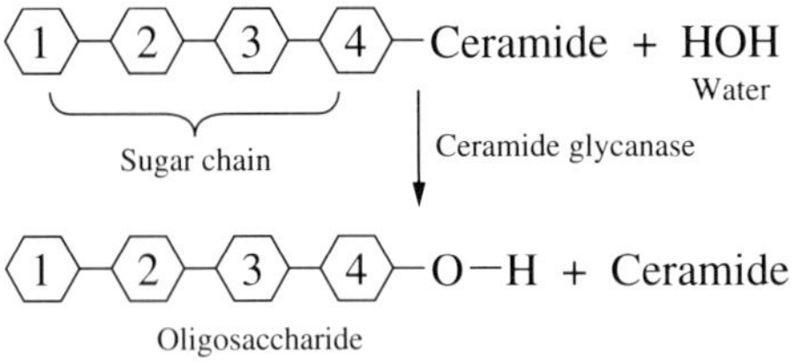

[1] B. Zhou, S.-C. Li, R. A. Laine, R. T. C. Huang, and Y.-T. Li, *J. Biol. Chem.* **264,** 12272 (1989).

[2] S.-C. Li, R. DeGasperi, J. E. Muldrey, and Y.-T. Li, *Biochem. Biophys. Res. Commun.* **141,** 346 (1986).

[3] Y.-T. Li, Y. Ishikawa, and S.-C. Li, *Biochem. Biophys. Res. Commun.* **149,** 167 (1987).

[4] M. Ito and T. Yamagata, *J. Biol. Chem.* **261,** 14278 (1986).

[5] H. Ashida, K. Yamamoto, H. Kumagai, and T. Tochikura, *Eur. J. Biochem.* **205,** 729 (1993).

Copyright © 1994 by Academic Press, Inc.
All rights of reproduction in any form reserved.

The oligosaccharide-transfer activity of the enzyme can be represented as

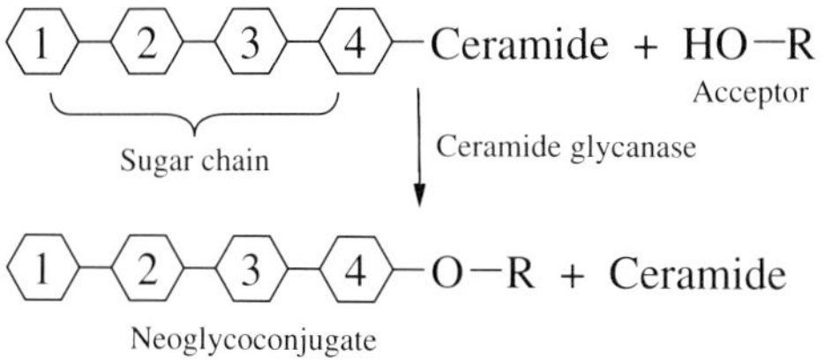

Reagents for Hydrolysis of Glycosphingolipids

[^{3}H]G_{M1} (II$^3\alpha$NeuAc-GgOse$_4$Cer) or nonradioactive GSLs
Sodium cholate
Sodium acetate buffer, 50 m*M*, pH 5.0
CGase

Reagents for Oligosaccharide Transfer

GSL
Sodium cholate
Acceptor (such as 1-octanol)
Sodium acetate buffer, 50 m*M*, pH 5.0
CGase

Procedures

Hydrolysis. The standard assay mixture contains [^{3}H]G_{M1}, 30 nmol, 10,000 counts/min (cpm); sodium cholate, 200 μg; and an appropriate amount of CGase in 200 μl of 50 m*M* sodium acetate buffer, pH 5.0. After incubation at 37° for a preset time, the liberated radioactive oligosaccharide is analyzed as described.[1] When a nonradioactive GSL is used as the substrate, the reaction is terminated by adding 5 volumes of chloroform–methanol (2 : 1, v/v). The mixture is vortexed and centrifuged at 7000 *g* for 5 min at room temperature to separate the organic phase (lower) and the aqueous phase (upper). The two phases are individually evaporated to dryness. For the detection of the released oligosaccharide, the aqueous phase is analyzed by thin-layer chromatography (TLC) using *n*-butanol–acetic acid–water (2 : 1 : 1, v/v/v) as the developing solvent.[6] Oligosaccharides containing sialic acid are revealed by resorcinol,[7] whereas those containing neutral sugars are visualized by diphenylamine reagent.[8] For the analysis of the ceramides released by the enzyme, the organic phase is

[6] M. Kitamikado, M. Ito, and Y.-T. Li, *J. Biol. Chem.* **256,** 3906 (1981).
[7] L. Svennerholm, *Biochim. Biophys. Acta* **24,** 604 (1957).
[8] G. Harris and I. C. MacWilliams, *Chem. Ind.* (*London*), 249 (1954).

analyzed by TLC using chloroform–methanol (9 : 1, v/v) as the developing solvent.[1] The ceramides are revealed by staining the plate with Coomassie Brilliant blue.[9] Because sphingosine and alkyl alcohols are not stained by Coomassie Brilliant blue, the release of oligosaccharides is used to determine the hydrolysis of lyso-G_{M1}, lyso-LacCer, and alkyl-β-lactosides. Quantitative estimation of the stained bands on TLC plates is accomplished by scanning the plate using a Shimadzu CS-930 TLC scanner (Tokyo, Japan) as described by Ando *et al.*[10]

Oligosaccharide Transfer. A GSL substrate, 30 nmol in 50 μl chloroform–methanol (2 : 1, v/v), and sodium cholate, 50 μg in 2.5 μl of water, are pipetted into a 10 × 75 mm test tube and evaporated to dryness. Twenty microliters of the acceptor in liquid form is added to the tube followed by the addition of 25 μl of 50 m*M* sodium acetate buffer, pH 6.0. The mixture is sonicated for 1 min in an ultrasonic water bath. The reaction is initiated by the addition of 0.1 unit of CGase in 5 μl of 50 m*M* sodium acetate buffer, pH 6.0. When the acceptor is in a solid form, 20 μl containing 1 mg of the acceptor dissolved in chloroform–methanol (2 : 1, v/v) and sodium cholate, in 2.5 μl of water, are pipetted into the tube and evaporated to dryness. Then, 20 μl of water is added before the addition of the buffer and the enzyme as described above. After incubating at 37° in a shaking water bath for 17 hr, 10 μl of the incubation mixture is directly analyzed by thin-layer chromatography (TLC) using 1-butanol–acetic acid–water (2 : 1 : 1, v/v/v)[6] as the developing solvent. To reveal the glycoconjugates, the plate is sprayed with diphenylamine reagent[8] and heated at 110° for 15 to 20 min.

Definition of Unit and Specific Activity. One unit of CGase activity is defined as the amount of enzyme which hydrolyzes 1 μmol of G_{M1} per minute at 37°. The specific activity of the enzyme is expressed as units per milligram protein. Protein concentrations are determined by the method of Lowry *et al.*[11] with bovine serum albumin (BSA) as the standard.

Procedure for Ceramide Glycanase Purification

All operations are performed at 0°–5° except that chromatography on octyl-Sepharose and Matrex gel blue A is run at room temperature. Centrifugations are carried out at 20,000–30,000 *g* for 30 min in a Sorvall

[9] K. Nakamura and S. Handa, *Anal. Biochem.* **142,** 406 (1984).

[10] S. Ando, N.-C. Chang, and R. K. Yu, *Anal. Biochem.* **89,** 437 (1978).

[11] O. H. Lowry, N. J. Rosebrough, A. L. Farr, and R. T. Randall, *J. Biol. Chem.* **193,** 265 (1951).

RC5C refrigerated centrifuge. Ultrafiltrations are carried out using Amicon (Danvers, MA) stirred cells and PM10 membranes.

Step 1: Crude Extract. Leeches (1 kg) are homogenized with 2.5 liters of distilled water in a Waring blendor at 30-sec intervals for a total of 2 min. The homogenate is centrifuged to obtain a water extract. After adjusting the pH of the extract to 4.8 with saturated citric acid, the precipitate is removed by centrifugation, and the supernatant is adjusted to pH 6.5 with saturated Na_2HPO_4.

Step 2: Protamine Sulfate and Ammonium Sulfate Fractionation. To the solution obtained in Step 1, protamine sulfate (2 g/100 ml water) is added dropwise until the precipitate ceases to appear. At this point, the final concentration of protamine sulfate is about 0.5 mg/ml. The precipitate is removed by centrifugation, and the supernatant is brought to 30% saturation with solid ammonium sulfate (176 g/liter). After standing for 2 hr, the precipitate is removed by centrifugation, and the supernatant is further brought to 55% saturation with solid ammonium sulfate (162 g/liter) and left overnight.

Step 3: Octyl-Sepharose Column Chromatography. After standing overnight, the precipitate is collected by centrifugation and dissolved in 100 ml of 50 m*M* sodium phosphate buffer, pH 7.0, to obtain a crude CGase preparation. An equal volume of 2% sodium cholate solution is mixed with the preparation and the mixture is applied to an octyl-Sepharose column (2.5 × 20 cm; Pharmacia, Piscataway, NJ) equilibrated with 1% sodium cholate solution. The column is subsequently washed extensively with the same solution to remove the unadsorbed proteins. Under these conditions, all exoglycosidases are washed off the column. After the absorbance at 280 nm of the effluent drops below 0.01, the column is washed further with approximately 4 liters of water. The CGase retained by the column is then eluted with 500 ml of 1% octyl β-glucoside. Fractions of 10 ml are collected. The fractions containing CGase are pooled and concentrated to about 20 ml by ultrafiltration.

Step 4: Matrex Gel Blue A Column Chromatography. The preparation from Step 3 is dialyzed against 50 m*M* sodium acetate buffer, pH 6.0, and applied to a Matrex gel blue A column (2.5 × 20 cm; Amicon) equilibrated with 50 m*M* sodium acetate buffer, pH 6.0. The column is washed with the same buffer and then with water to remove the unadsorbed proteins, and the CGase is eluted with 1% sodium cholate. The active fractions are pooled and concentrated to 4.8 ml by ultrafiltration.

Step 5: BioGel A-0.5m Column Chromatography. One-half of the preparation (2.5 ml) from Step 4 is applied to a BioGel A-0.5m column (1.5 × 80 cm; Bio-Rad, Richmond, CA) equilibrated with 50 m*M* sodium phosphate buffer, pH 7.0. The column is eluted with the same buffer, and the first protein peak which contains CGase activity is pooled and concentrated to 1.6 ml by ultrafiltration. Table I summarizes a typical purification scheme and the recovery of CGase from 1 kg of leeches.

Procedure for Synthesis of Octyl-II^{3}NeuAc-GgOse$_4$ Using Ceramide Glycanase

Step 1. The G_{M1} (3 mg) is mixed with 3 mg sodium cholate in 0.15 ml of water, 1.2 ml of 1-octanol, and 1.65 ml of 50 m*M* sodium acetate buffer, pH 6.0. The mixture is briefly sonicated in an ultrasonic water bath, followed by the addition of 5 milliunits (mU) of CGase in 0.3 ml of 50 m*M* sodium acetate buffer, pH 6.0, and incubated at 37° in a shaking water bath for 17 hr.

Step 2. The incubation mixture is subjected to Folch partitioning by vortexing with 15 ml of chloroform–methanol (2 : 1, v/v). The enzymatically synthesized octylglycan, octyl-II^{3}NeuAc-GgOse$_4$, is recovered in the aqueous phase. The organic phase is evaporated to dryness, dissolved in 15 ml of chloroform–methanol (2 : 1, v/v), and subjected to a second Folch partitioning by vortexing with 3 ml of water. The two aqueous phases are combined and evaporated to dryness.

Step 3. The dried material from Step 2 is dissolved in 1 ml of water and applied to a BioGel P-2 column (0.5 × 92 cm) previously equilibrated with water. The column is eluted with water at 7 ml/hr, and 1.2-ml fractions are collected. One hundred microliters from each fraction is evaporated to dryness using a Savant (Farmingdale, NY) Speed-Vac concentrator and analyzed by TLC using the solvent system and the spray reagent

TABLE I
PURIFICATION OF CERAMIDE GLYCANASE FROM LEECHES[a]

Step	Total protein (mg)	Total units (nmol/min)	Specific activity (mU/mg)	Purification (-fold)	Recovery (%)
Crude extract	18,860	4160	0.22	1.00	100
Protamine sulfate treatment	11,450	3320	0.29	1.32	80
$(NH_4)_2SO_4$ (30–55%)	4750	2580	0.54	2.45	62
Octyl-Sepharose	84	1760	20.91	95.23	42
Matrex gel blue A	28.4	1370	48.27	219.27	33
BioGel A-0.5 m	3.6	710	197.22	896.45	17

[a] Purification started with 1 kg of *Macrobdella decora*.

described in the section on oligosaccharide transfer. Under these conditions, II^3NeuAc-GgOse$_4$ and octyl-II^3NeuAc-GgOse$_4$ appear in the same fractions. The fractions containing the two compounds are pooled and lyophilized.

Step 4. The dried material from Step 3, dissolved in 1 ml of water, is subsequently applied to a Sep-Pak Vac/3 cm^3 C_{18} cartridge. After washing with 50 ml of water to remove the free oligosaccharidc (II^3NeuAc-GgOse$_4$) and other unadsorbed materials, the cartridge is eluted with 70 ml of methanol–water (15 : 85, v/v) to elute octyl-II^3NeuAc-GgOse$_4$. The eluate is evaporated to dryness to obtain approximately 700 μg of pure octyl-II^3NeuAc-GgOse$_4$.

Properties of Ceramide Glycanase

Molecular Weight. The final enzyme preparation shows one major band stained by Coomassie Brilliant blue on sodium dodecyl sulfate (SDS)–polyacrylamide gel electrophoresis with a molecular mass of 54 kDa. The enzyme, however, shows a molecular mass of 330 kDa on BioGel A-0.5m gel filtration. For some unknown reason, the enzyme cannot enter the gel on native polyacrylamide gel electrophoresis even with 5% gels.

pH Optimum and Stability. The CGase is stable between pH 4.5 and 8.5 and shows the highest activity at pH 5.0 using G_{M1} as substrate. The p*I* of the enzyme is between pH 4.7 and 4.9. At a concentration of 5 mg/ml, the enzyme does not lose any appreciable activity at −20° for 3 months or at 37° for 1 week.

K_m. The rate of hydrolysis increases with the increment of G_{M1} concentration and reaches the maximum at a concentration of 150 μM. The K_m for G_{M1} is estimated to be 1.54×10^{-5} *M* from the double-reciprocal plot.[12]

Effect of Metal Ions. The effects of metal ions on CGase show that Zn^{2+} at 5 m*M* inhibits 60% of the activity, whereas Cu^{2+}, Ag^{+}, and Hg^{2+} at 1 m*M* inhibit 64%, 80%, and 95% of the activity, respectively. The ions Ba^{2+}, Ca^{2+}, Co^{2+}, Mg^{2+}, and Mn^{2+}, at 5 m*M*, do not have appreciable effect.

Effect of Detergents. The hydrolysis of GSLs by CGase in the absence of a detergent is very slow. The reaction is stimulated by bile salts. Among various bile salts tested, sodium cholate is most effective, except for the hydrolysis of LacCer (Table II). It should be noted that the maximal solubility of sodium cholate in 50 m*M* sodium acetate buffer, pH 5.0, is close to 1 μg/μl. Above this concentration, CGase activity is severely inhibited. For the hydrolysis of LacCer, sodium taurodeoxycholate at 2.5

[12] M. Dixon and E. C. Webb, "Enzymes," p. 55. Longmans, London, 1979.

TABLE II
EFFECTS OF DETERGENTS ON HYDROLYSIS OF GLYCOSPHINGOLIPIDS BY CERAMIDE GLYCANASE[a]

Detergents	Hydrolysis (%)						
	G_{M1}	G_{M2}	G_{M3}	G_{D1a}	$GbOse_5Cer$	$GbOse_4Cer$	LacCer
Sodium taurodeoxycholate	100	100	100	100	100	100	100
Sodium cholate	232	100	204	120	209	423	65
Sodium deoxycholate	16	0	47	0	4	0	3
Sodium taurocholate	24	2	10	0	44	0	12
Sodium taurochenodeoxycholate	64	49	61	37	112	286	36
Sodium taurodehydrocholate	8	15	21	0	0	0	2
Sodium taurolithocholate	0	0	0	0	0	—	—
Tween 20	0	—[b]	—	—	—	—	—
Triton X-100	0	—	—	—	—	—	—
Nonidet P-40 (NP-40)	0	—	—	—	—	—	—

[a] In all cases the reaction mixture contained 30 nmol of the GSL substrate. For the hydrolysis of G_{M1}, G_{M2}, G_{M3}, G_{D1a}, $GbOse_5Cer$, and $GbOse_4Cer$, each GSL was incubated with 0.2 mU of the enzyme at 37° for 1 hr, whereas LacCer was incubated with 0.4 mU of the enzyme at 37° for 2 hr. Each incubation mixture contained one of the detergents at a final concentration of 1 $\mu g/\mu l$. The CGase activity was assayed by quantitative TLC according to the conditions described in the text.

[b] —, Not determined.

$\mu g/\mu l$ is most effective. Under the same conditions, the hydrolysis of G_{M1} requires a much lower concentration (1.5 $\mu g/\mu l$) of the detergent to reach the maximal activity. These results may reflect the difference in the solubility of the two GSLs.

Substrate Specificity. Among various GSLs studied, $GgOse_3Cer$ and $GgOse_4Cer$ are the two best substrates for CGase, followed by G_{M2} and G_{M1}. As is evident from Table III, the rate of hydrolysis of G_{M1} is considerably slower than that of $GgOse_4Cer$. Ganglioside G_{D1a}, in turn, is hydrolyzed much slower than G_{M1}. Also the rate of hydrolysis of G_{T1b} is slower than that of G_{D1a}, suggesting that there is an inverse relationship between the rate of hydrolysis and the number of sialic acids attached to the glycolipid substrates. This indicates that sialic acids may interfere with binding of the substrate and enzyme. With the exception of GlcCer, the rate of hydrolysis of LacCer, which is about 20% the rate of G_{M1}, is the slowest among various GSLs. The deacyl GSLs such as lyso-G_{M1} and lyso-LacCer are hydrolyzed by CGase only at 10% of the rate of intact GSLs. Apparently fatty acyl residues may be important for binding between the substrate and the enzyme.

The shortest carbohydrate chain on a GSL which can be efficiently hydrolyzed by CGase is LacCer. We studied the hydrolysis of five disac-

TABLE III
HYDROLYSIS OF GLYCOSPHINGOLIPIDS BY CERAMIDE GLYCANASE[a]

Glycosphingolipid	Relative activity (%)
G_{M1}	100
$GgOse_3Cer$	197
$GgOse_4Cer$	185
G_{M2}	115
$GbOse_3Cer$	67
$GbOse_4Cer$	50
$GbOse_5Cer$	46
G_{D1a}	41
nLcOse$_4$Cer	38
G_{M3}	35
LacCer^{3}II-SO_4	34
nLcOse$_3$Cer	31
G_{T1b}	22
LacCer	19
GlcCer	5

[a] The CGase (0.2 mU) was incubated with 30 nmol each of different GSLs in 200 μl according to the standard assay conditions. The incubation time varied depending on the substrate: for G_{M1}, $GgOse_3Cer$, $GgOse_4Cer$, G_{M2}, and $GbOse_3Cer$, 30 min; for $GbOse_4Cer$, $Gbose_5$ Cer, G_{D1a}, nLcOse$_4$Cer, G_{M3}, LacCer^{3}II-SO_4, nLcOse$_3$Cer, and G_{T1b}, 1 hr; for LacCer and GlcCer, 2 hr. The activity was determined by quantitative TLC as described in the text.

charides, namely, Galβ1→4GlcNAcβ→, Galα1→4Galβ→, Manβ1→4-Glcβ→, Glcα1→4Glcβ→, and Galα1→6Glcβ→, attached to ceramide or to an alkyl chain. As shown in Table IV, Galα1→4GalβCer and Galβ1→4-GlcNAcβ-CETE are not hydrolyzed by the enzyme. In these two glycolipids, the first sugar attached to the ceramide or 2-(2-carbomethoxythylthio)ethyl (CETE) is either Gal or GlcNAc instead of Glc. These results indicate that the identity of the first sugar attached to the hydrophobic portion of the glycolipid substrate is recognized by the enzyme. In addition, the second sugar and its linkage to the Glc moiety also affect the rate of hydrolysis. Manβ1→4GlcβCer is hydrolyzed at a rate 40% of that for LacCer, and the rate of hydrolysis of *Al*-LacCer (conversion of C-6 of Gal in LacCer to an aldehyde group) is only 5% in comparison to LacCer. The substrates Glcα1→4Glcβ-dodecanol and Galα1→6GlcβCer,

TABLE IV
HYDROLYSIS OF GLYCOLIPIDS WITH DIFFERENT DISACCHARIDES BY CERAMIDE GLYCANASE[a]

Substrate[b]	Enzyme (mU)	Incubation time (hr)	Hydrolysis (%)
Galβ1 → 4GlcβCer	0.5	2	100
Manβ1 → 4GlcβCer	0.5	2	40
Al-Galβ1 → 4GlcβCer	0.5	2	5
Glcα1 → 4Glcβ-dodecanol	0.5	4	0
Galα1 → 6GlcβCer	0.5	2	0
Galα1 → 4GalβCer	0.5	2	0
Galβ1 → 4GlcNAcβ-CETE	0.9	4	0

[a] Each substrate (30 nmol) was incubated with the indicated amount of CGase at 37°. The assay conditions are described in the text.
[b] *Al*-Gal, C-6 of Gal is an aldehyde group; CETE, 2-(2-carbomethoxyethylthio)ethyl.

on the other hand, are completely refractory to CGase. These results clearly indicate that the nature and linkage of the sugar unit next to the Glc moiety can also affect the rate of hydrolysis. The fact that the conversion of LacCer to *Al*-LacCer renders the substrate resistant to CGase may suggest the possible interaction of the enzyme with the C-6 region of the Gal residue in LacCer.

To understand the effect of the structure and hydrophobicity of the aglycon portion of glycolipids on the rate of hydrolysis, we examine the hydrolysis of several synthetic β-lactosides by CGase. As shown in Table V, short-chain alkyl β-lactosides such as methyl, ethyl, and *n*-propyl β-lactoside are refractory to CGase. The increase in chain length of the alkyl group starting from *n*-butyl β-lactoside results in the concomitant increase in the rate of hydrolysis, which reaches a maximum at 14 carbons. Based on this result, if dodecyl β-lactoside is available, it should also be hydrolyzed by this enzyme. The rate of hydrolysis of stearyl β-lactoside, on the other hand, is much slower than that of myristyl β-lactoside. Other synthetic lactosides such as 2-(octadecylthio)ethyl (OTE)- and CETE-β-lactosides are also hydrolyzed. *p*-Nitrophenyl, benzyl, and phytyl β-lactosides are resistant to hydrolysis. The above results indicate that the chain length, structure, and hydrophobicity of the aglycon portion of the glycolipids have a profound effect on the rate of hydrolysis.

In addition to the GSLs listed in Table III, the leech CGase also hydrolyzed unusual GSLs such as 3-OMeGalβ1→3GalNAcα1→3[6′-*O*-(2-aminoethylphosphonyl)Galα1→2]-(2-aminoethylphosphonyl1→6)Gal β1→4Glcβ1→1Cer, IV^3-(3′-SO_3-GlcA)-*n*$LcOse_4$Cer, GalNAcα1→3Gal

TABLE V
HYDROLYSIS OF SYNTHETIC LACTOSIDES BY CERAMIDE GLYCANASE[a,b]

Synthetic β-lactoside	Hydrolysis (%)	
	−TDC	+TDC
PNP-lactoside	0	0
Benzyl lactoside	0	0
Phytyl lactoside	0	0
Methyl lactoside	0	0
Ethyl lactoside	0	0
n-Propyl lactoside	0	0
n-Butyl lactoside	0.3	0.1
n-Octyl lactoside	2.5	6.7
Myristyl lactoside	13.3	34.5
Stearyl lactoside	0.5	4.6
OTE-lactoside	1.7	3.5
CETE-lactoside	2.8	13.8

[a] Synthetic β-lactosides (30 nmol) were incubated with 0.9 mU of CGase in 200 μl of 50 m*M* sodium acetate buffer, pH 5.0, at 37° for 3 hr. The TDC concentration was 1 μg/μl.

[b] TDC, Sodium taurodeoxycholate; PNP, *p*-nitrophenyl; OTE, 2-(octadecylthio)ethyl; and CETE, 2-(2-carbomethoxyethylthio) ethyl.

(2→1αFuc)β1→3GlcNAcβ1→3Galβ1→4GlcCer, and polyglycosylceramides, although at a much slower rate. This indicates that the CGase has a broad specificity, which makes this enzyme useful for structural analysis of GSLs.

Acceptor Specificity of Oligosaccharide Transfer. Figure 1 shows that CGase transfers the oligosaccharide II^3NeuAc-$GgOse_4$ from G_{M1} to eight different 1-alkanols as well as 1,8-octanediol and 4-phenyl-1-butanol. The short-chain 1-alkanols such as methanol, ethanol, 1-propanol, 1-butanol, and 1-pentanol inhibit the hydrolysis of G_{M1} by CGase, and no transfer of the oligosaccharide is observed. Among 1-hexanol, 1-octanol, 1-dodecanol, 1-tetradecanol, 1-hexadecanol, and 1-octadecanol, the first three are good acceptors, and 1-octanol is the best acceptor. Under the standard assay conditions, the formation of octyl-II^3NeuAc-$GgOse_4$ is approximately one-third of that of the released free II^3NeuAc-$GgOse_4$. The higher 1-alkanols with chain lengths longer than 12 carbons are poor acceptors because of low solubility in water. The CGase also effectively transfers

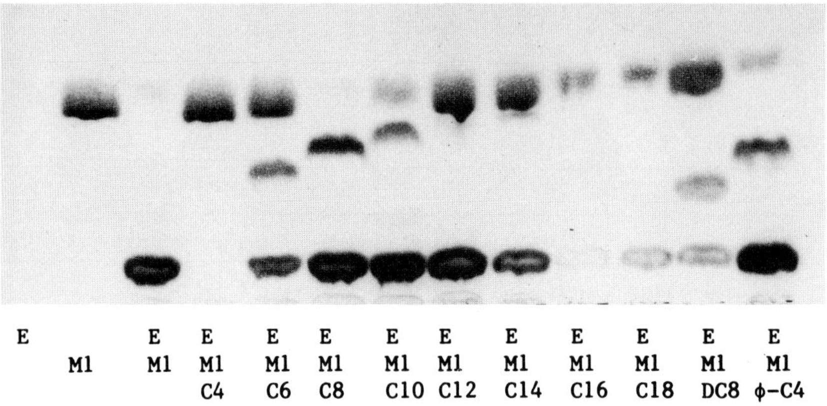

FIG. 1. Transfer of II³NeuAc-GgOse$_4$ from G_{M1} to various 1-alkanols. E, CGase; M1, G_{M1}; C4, 1-butanol; C6, 1-hexanol; C8, 1-octanol; C10, 1-decanol; C12, 1-dodecanol; C14, 1-tetradecanol; C16, 1-hexadecanol; C18, 1-octadecanol; DC8, 1,8-octanediol; ϕ-C4, 4-phenyl-1-butanol. Incubations were carried out at 37° for 17 hr according to the conditions described in the text.

II^{3}NeuAc-GgOse$_4$ to 1,8-octanediol and 4-phenyl-1-butanol. From the fact that only one product is detected in the transfer of the oligosaccharide to 1,8-octanediol and from the TLC mobility of the product (see Fig. 1, second lane from the right), we tentatively conclude that the oligosaccharide is transferred to only one of the two primary hydroxyl groups in the diol. Although 1-butanol cannot serve as an acceptor, phenyl-1-butanol is a surprisingly good acceptor. This indicates that hydrophobicity may be one of the important factors which influences the ability of an alkanol to accept the oligosaccharide. In contrast to primary alkanols, secondary alkanols such as 2-hexanol and 2-octanol are not able to serve as acceptors. The CGase is also able to transfer the oligosaccharide from other GSLs such as G_{M2}, G_{M3}, G_{D1a}, GbOse$_5$Cer, GbOse$_4$Cer, GbOse$_3$Cer, and LacCer to 1-hexanol and 1-octanol.

Acceptors Useful for Synthesis of Neoglycoconjugates. As simple alkylglycans without a functional group on the alkyl chain cannot be chemically linked to proteins and other matrices to form neoglycoconjugates, we investigated the transfer of the oligosaccharide from G_{M1} to several acceptors which contain a primary hydroxyl group at one end and an additional functional group at the other end of the alkyl chain. As shown in Fig. 2, CGase is able to transfer the oligosaccharide from G_{M1} to $CF_3CO—NH(CH_2)_5CH_2OH$ (lane 5), $(CH_3)_3CO—CO—NH(CH_2)_5CH_2OH$ (lane 7), and $CH_2{=}CH(CH_2)_7CH_2OH$ (lane 9). The functional group in these alkylglycans can be covalently linked to suitable matrices to form

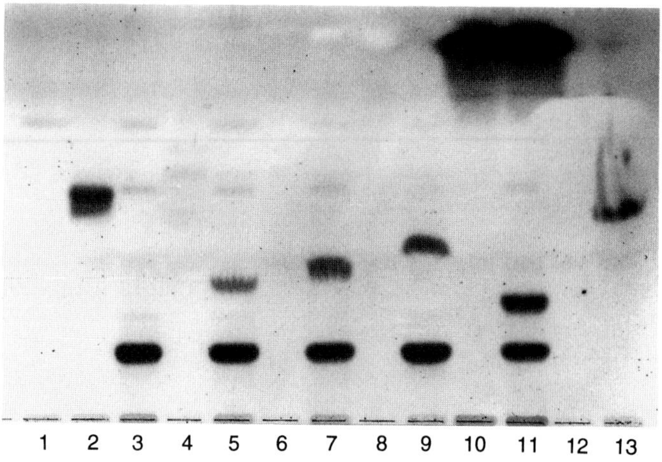

FIG. 2. Transfer of II^3NeuAc-$GgOse_4$ from G_{M1} to three 1-alkanol derivatives useful for the synthesis of neoglycoconjugates and to 1,2 : 3,4-diisopropylidene-D-galactopyranose (DIG). Lane 1, CGase only; lane 2, G_{M1} only; lane 3, G_{M1} + CGase; lane 4, CF_3CO—$NH(CH_2)_5CH_2OH$ only; lane 5, G_{M1} + CF_3CO—NH—$(CH_2)_5CH_2OH$ + CGase; lane 6, $(CH_3)_3CO$—$CONH(CH_2)_5CH_2OH$ only; lane 7, G_{M1} + $(CH_3)_3CO$—$CONH(CH_2)_5CH_2OH$ + CGase; lane 8, CH_2=$CH(CH_2)_7CH_2OH$ only; lane 9, G_{M1} + CH_2=$CH(CH_2)_7CH_2OH$ + CGase; lane 10, DIG only; lane 11, G_{M1} + DIG + CGase; lane 12, $NH_2(CH_2)_5CH_2OH$ only; lane 13, G_{M1} + $NH_2(CH_2)_5CH_2OH$ + CGase. Incubations were carried out at 37° for 17 hr according to the conditions described in the text.

neoglycoconjugates. Although CF_3CO—$NH(CH_2)_5CH_2OH$ is a good acceptor, $NH_2(CH_2)_5CH_2OH$ not only failed to serve as an acceptor, but also inhibits the hydrolysis of G_{M1} by CGase (Fig. 2, lane 13). The CGase also effectively transfers the oligosaccharide to $(HOCH_2)_3C$—NH—CO—$(CH_2)_4$—COOMe. The number and position of the oligosaccharide attached to this acceptor remain to be elucidated. In addition to alkanol derivatives, CGase also transfers the oligosaccharide from G_{M1} to the primary hydroxyl group of 1,2 : 3,4-di-*O*-isopropylidene-D-galactopyranose (Fig. 2, lane 11). Thus, this transglycosylation reaction can also be used to elongate a sugar chain. 1,2 : 5,6-Di-*O*-isopropylidene-D-glucofuranose, however, is not able to serve as an acceptor. This again suggests that the hydroxyl group on the acceptor has to be a primary alcohol.

No effective method is currently available for linking an oligosaccharide to an aglycon without modifying the sugar unit of the reducing terminus of the sugar chain. Ceramide glycanase not only transfers the intact oligosaccharide from GSLs to suitable acceptors but also retains the original structure of the sugar chain, including the anomeric configuration of the reducing sugar unit being transferred. Thus, the transglycosylation reaction catalyzed by CGase should become useful for synthesizing neoglycoconjugates to study the biological functions expressed by glycan chains in GSLs.

[15] Synthesis of Sialyl Lewis X Ganglioside and Analogs

By AKIRA HASEGAWA and MAKOTO KISO

Introduction

Sialyl Lewis X (sLe^x) ganglioside (**A**, see Scheme 1) was first isolated[1] from human kidney and found[2] to be widespread as a tumor-associated ganglioside antigen. It has been demonstrated[3–10] that a family of receptors, the selectins [lectin–epithelial growth factor (EGF)–complement binding–cell adhesion molecules], such as L-selectin (leukocyte adhesion molecule-1), E-selectin (endothelial leukocyte adhesion molecule-1), and P-selectin (granule membrane protein), recognize the sLe^x determinant, α-Neu5Ac-(2→3)-β-D-Gal-(1→4)-[α-L-Fuc-(1→3)]-β-GlcNAc, which is found as the terminal carbohydrate structure in both cell membrane glyco-

[1] H. Pauvala, *J. Biol. Chem.* **251,** 7517 (1976).

[2] K. Fukushima, H. Hirota, P. I. Terasaki, A. Watanabe, H. Togashi, D. Chia, N. Sayama, Y. Fukushi, S. Nudelman, and S. Hakomori, *Cancer Res.* **44,** 5279 (1984).

[3] D. Tyrell, P. James, N. Rao, C. Foxall, S. Abbas, F. Dasgupta, M. Nashed, A. Hasegawa, M. Kiso, D. Asa, J. Kidd, and B. K. Brandley, *Proc. Natl. Acad. Sci. U.S.A.* **88,** 10372 (1991).

[4] M. J. Polley, M. L. Phillips, E. Wayner, E. Nudelman, A. K. Singhal, S. Hakomori, and J. C. Paulson, *Proc. Natl. Acad. Sci. U.S.A.* **88,** 6224 (1991).

[5] A. Tanaka, K. Ohmori, N. Takahashi, K. Tsuyuoka, A. Yago, K. Zenita, A. Hasegawa, and R. Kannagi, *Biochem. Biophys. Res. Commun.* **179,** 713 (1991).

[6] M. Larkin, T. L. Ahern, M. S. Stoll, M. Shaffer, D. Sako, J. O'Brien, C.-T. Yuen, A. M. Lawson, R. A. Child, K. M. Barone, P. R. Langer-Safer, A. Hasegawa, M. Kiso, G. R. Larson, and T. Feizi, *J. Biol. Chem.* **267,** 13601 (1992).

[7] C. Foxall, S. R. Matson, D. Dowbenko, C. Fennie, L. A. Lasky, M. Kiso, A. Hasegawa, A. Asa, and B. K. Brandley, *J. Cell Biol.* **117,** 895 (1992).

[8] P. J. Green, T. Tamatani, T. Watanabe, M. Miyasaki, A. Hasegawa, M. Kiso, C.-T. Yuen, M. S. Stoll, and T. Feizi, *Biochem. Biophys. Res. Commun.* **188,** 241 (1992).

[9] K. Ohmori, A. Takada, T. Yoneda, Y. Buma, K. Hirashima, K. Tsuyuoka, A. Hasegawa, and R. Kannagi, *Blood* **81,** 101 (1993).

[10] P. Kotovuori, E. Tontti, R. Pigott, M. Shepherd, M. Kiso, A. Hasegawa, R. Renkonen, P. Nortamo, D. C. Altieri, and C. Gahmberg, *Glycoconjugate J.* **3,** 131 (1993).

Copyright © 1994 by Academic Press, Inc.
All rights of reproduction in any form reserved.

lipids and glycoproteins, by use of chemically synthesized sLe^x, sLe^x ganglioside, and analogs.

In the first part of this chapter we describe an efficient method[11a,b] for the synthesis of sLe^x ganglioside and a position isomer, sialyl α(2→6)-Le^x ganglioside. To investigate the role of sialic acid and L-fucose moieties for the functions of sLe^x, syntheses of the chemically modified sialic acid- and L-fucose-containing sLe^x ganglioside analogs are then described. These gangliosides and analogs could be used as effective probes for elucidation of cell–cell adhesion mechanisms.

Synthesis of Sialyl Lewis X Ganglioside and Sialyl α(2→6)-Lewis X Ganglioside Analog

Syntheses of sLe^x ganglioside (**A**) and its sialyl α(2→6) positional isomer[12] (**B**) by sialylation of the Gal residue are performed according to Scheme 1. Dibutyltin oxide-mediated, selective etherification of 2-(trimethylsilyl)ethyl *O*-β-D-galactopyranosyl-(1→4)-β-D-glucopyranoside[13] (**1**), using 4-methoxybenzyl chloride and tetrabutylammonium bromide, gives the 3′-*O*-(4-methoxybenzyl)derivative[14] (**2**, 74%). Treatment of **2** with benzyl bromide in *N,N*-dimethylformamide in the presence of sodium hydride affords the benzyl derivative **3** (79%) which, after treatment with 2,3-dichloro-5,6-dicyanobenzoquinone[15] in dichloromethane–water, gives the 3′-hydroxy compound (**4**, 70%). Glycosylation of **4** with 2,4,6-tri-*O*-acetyl-2-deoxy-2-phthalimido-D-glucopyranosyl bromide[16] (**5**) in the presence of silver carbonate and silver perchlorate gives the desired β-glycoside **6** (94%). O-Deacetylation of **6**, followed by heating with hydrazine hydrate in aqueous 95% ethanol and subsequent N-acetylation, affords 2-(trimethylsilyl)ethyl *O*-(2-acetamido-2-deoxy-β-D-glucopyranosyl)-(1→3)-*O*-(2,4,6-tri-*O*-benzyl-β-D-galactopyranosyl)-(1→4)-2,3,6-tri-*O*-benzyl-β-D-glucopyranoside (**7**, 88%), which is converted to the trisaccharide glycosyl acceptor **8** (84%) by 4,6-O-benzylidenation.

[11a] A. Kameyama, H. Ishida, M. Kiso, and A. Hasegawa, *Carbohydr. Res.* **209,** c1 (1991).
[11b] A. Kameyama, H. Ishida, M. Kiso, and A. Hasegawa, *J. Carbohydr. Chem.* **10,** 549 (1991).
[12] A. Kameyama, H. Ishida, M. Kiso, and A. Hasegawa, *J. Carbohydr. Chem.* **10,** 729 (1991).
[13] K. P. R. Kartha, A. Kameyama, M. Kiso, and A. Hasegawa, *J. Carbohydr. Chem.* **8,** 145 (1989).
[14] A. Kameyama, H. Ishida, M. Kiso, and A. Hasegawa, *Carbohydr. Res.* **200,** 269 (1990).
[15] Y. Oikawa, T. Tanaka, K. Horita, T. Yoshida, and O. Yonemitsu, *Tetrahedron Lett.* **25,** 5393 (1984).
[16] R. U. Lemieux, T. Takeda, and B. Y. Chung, *ACS Symp. Ser.* **39,** 90 (1976).

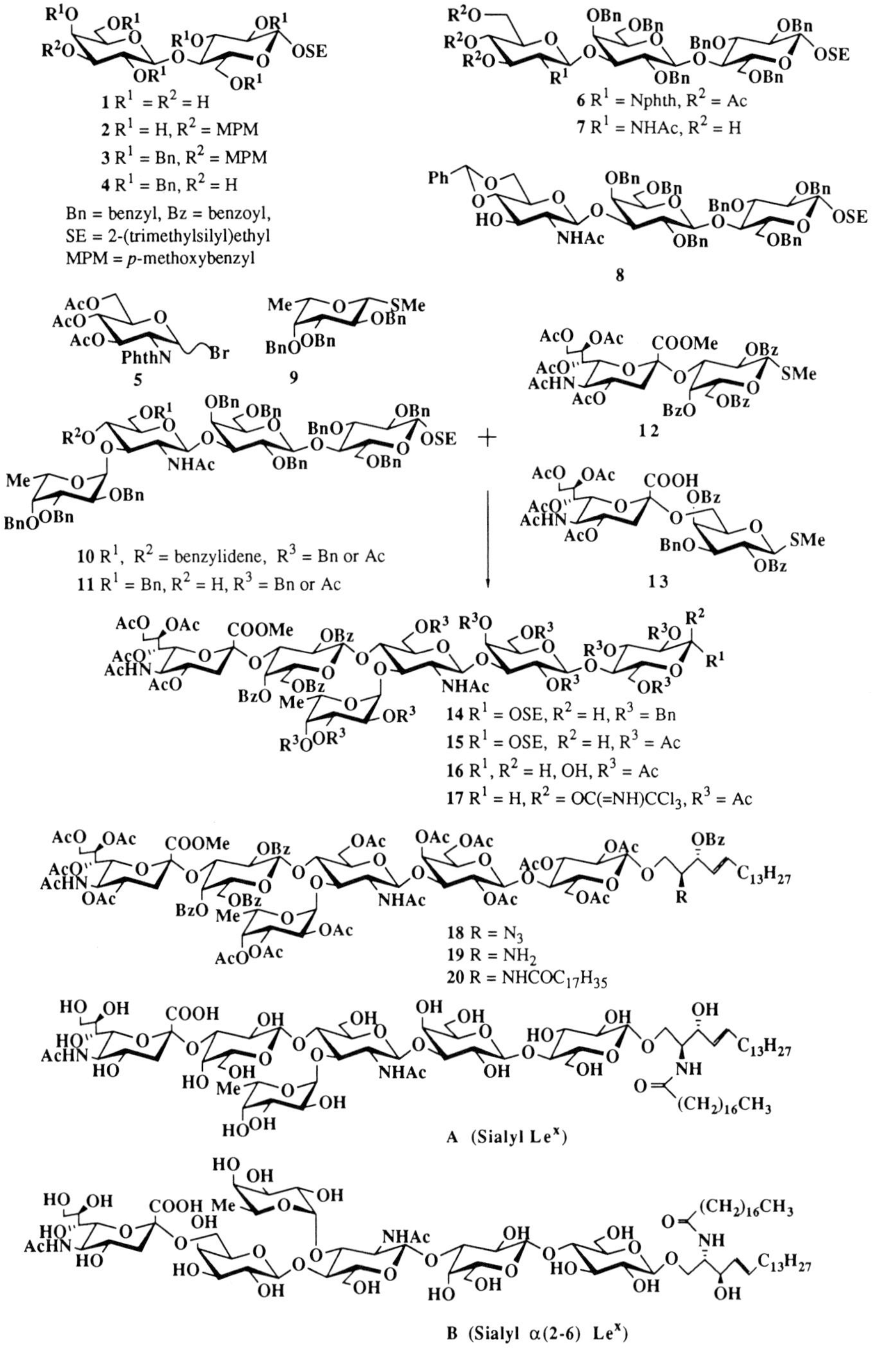

SCHEME 1. Synthesis of sLe^x ganglioside and its position isomer.

Glycosylation of **8** with methyl 2,3,4-tri-*O*-benzyl-1-thio-β-L-fucopyranoside[11b,17] (**9**) in the presence of dimethyl(methylthio)sulfonium triflate[18,19] (DMTST) in benzene gives an 86% yield of the derived tetrasaccharide **10**. Reductive ring opening[20] of the benzylidene group in **10** with sodium cyanoborohydride–hydrogen chloride affords **11** (75%). The glycosylation of **11** is effected with **12**, synthesized[14] from the coupling product of methyl (methyl 5-acetamido-4,7,8,9-tetra-*O*-acetyl-3,5-dideoxy-2-thio-D-*glycero*-D-*galacto*-2-nonulopyranosid)onate[21] and 2-(trimethylsilyl)ethyl 6-*O*-benzoyl-β-D-galactopyranoside[22] via O-benzoylation, replacement of the 2-(trimethylsilyl)ethyl group by acetyl, and introduction of the methylthio group with (methylthio)trimethylsilane. Reaction in dichloromethane for 20 hr at room temperature in the presence of DMTST gives the hexasaccharide **14** (41%). Removal of the benzyl groups from **14** by catalytic hydrogenolysis over 10% Pd–C in 3 : 1 (v/v) ethanol–acetic acid and subsequent acetylation give **15** (81%). Selective removal of the 2-(trimethylsilyl)ethyl group from **15** by treatment[23,24] with trifluoroacetic acid in dichloromethane for 1 hr at room temperature gives the 1-hydroxy compound **16** (94%).

Treatment of **16** with trichloroacetonitrile in the presence of 1,8-diazabicyclo[5.4.0]undec-7-ene (DBU) for 3 hr at 0° gives[25,26] the αtrichloroacetimidate **17** (91%). Final glycosylation of (2*S*,3*R*,4*E*)-2-azido-3-*O*-benzoyl-4-octadecene-1,3-diol[27,28] with **17** by use of boron trifluoride etherate[24,29] for 3 hr at 0° affords only the expected β-glycoside **18** (56%). Selective reduction[27,30] of the azido group in **18** with hydrogen sulfide in aqueous 83% pyridine gives the amine **19**, and this on condensation with octadecanoic acid in the presence of 1-(3-dimethylaminopropyl)-3-ethylcarbodi-

[17] F. Yamazaki, S. Sato, T. Nukada, Y. Ito, and T. Ogawa, *Carbohydr. Res.* **201,** 31 (1990).
[18] P. Fügedi and P. J. Garegg, *Carbohydr. Res.* **149,** c9 (1986).
[19] O. Kanie, M. Kiso, and A. Hasegawa, *J. Carbohydr. Chem.* **7,** 501 (1988); T. Murase, H. Ishida, M. Kiso, and A. Hasegawa, *Carbohydr. Res.* **184,** c1 (1988).
[20] P. J. Garegg, H. Hultberg, and S. Wallin, *Carbohydr. Res.* **108,** 97 (1982).
[21] A. Hasegawa, H. Ohki, T. Nagahama, H. Ishida, and M. Kiso, *Carbohydr. Res.* **212,** 277 (1991).
[22] T. Murase, A. Kameyama, K. P. R. Kartha, H. Ishida, M. Kiso, and A. Hasegawa, *J. Carbohydr. Chem.* **8,** 265 (1989).
[23] K. Jansson, S. Ahlfors, T. Frejd, J. Kihlberg, G. Magnusson, J. Dahmén, G. Noori, and K. Stenvall, *J. Org. Chem.* **53,** 5629 (1988).
[24] T. Murase, H. Ishida, M. Kiso, and A. Hasegawa, *Carbohydr. Res.* **188,** 71 (1989).
[25] M. Numata, M. Sugimoto, K. Koike, and T. Oguma, *Carbohydr. Res.* **163,** 209 (1987).
[26] R. R. Schmidt and J. Michel, *Angew. Chem., Int. Ed. Engl.* **19,** 731 (1980).
[27] Y. Ito, M. Kiso, and A. Hasegawa, *J. Carbohydr. Chem.* **8,** 285 (1989).
[28] R. R. Schmidt and P. Zimmermann, *Angew. Chem., Int. Ed. Engl.* **25,** 725 (1986).
[29] R. R. Schmidt and G. Grundler, *Synthesis,* 885 (1981).
[30] T. Adachi, Y. Yamada, I. Inoue, and M. Saneyoshi, *Synthesis,* 45 (1977).

imide (DAC) in dichloromethane for 16 hr at room temperature affords the sialyl Le^x derivative **20** (81%). Finally, O-deacylation of **20** with sodium methoxide in methanol and saponification of the methyl ester group yield sLe^x ganglioside (**A**, quantitative). When reacted with sialyl $\alpha(2\rightarrow6)$-Gal derivative **13**,[31] as described for the synthesis of **14**, compound **11** gives the sialyl $\alpha(2\rightarrow6)$-Gal residue-containing sLe^x hexasaccharide, which is converted to the sialyl $\alpha(2\rightarrow6)$ position isomer (**B**) of sLe^x ganglioside (**A**) according to the method described for the synthesis of **A**. A number of methods[32–36] for the synthesis of sLe^x oligosaccharide and analogs,[37–41] as well as conformational studies,[3,42–44] have been reported.

Detailed Procedures

2-(Trimethylsilyl)ethyl O-[3-O-(4-Methoxybenzyl)-β-D-Galactopyranosyl]-(1→4)-β-D-Glucopyranoside (**2**). A suspension of **1**[13] (2.8 g, 6.33 mmol) and dibutyltin oxide (2.38 g) in methanol (28 ml) is stirred and heated for 4 hr at 45°, then concentrated. To a solution of the residue in benzene (28 ml) are added 4-methoxybenzyl chloride (2.58 ml), tetrabutylammonium bromide (1.05 g), and Molecular Sieve 4Å (MS-4Å; 2.8 g), and the mixture is stirred and boiled under reflux for 3 hr, then concentrated. Column chromatography [30 : 1 (v/v) dichloromethane–methanol] of the residue

[31] A. Hasegawa, K. Hotta, A. Kameyama, H. Ishida, and M. Kiso, *J. Carbohydr. Chem.* **10,** 439 (1991).

[32] K. C. Nicolaou, T. J. Caulfield, H. Kataoka, and N. H. Stylianides, *J. Am. Chem. Soc.* **112,** 3693 (1990).

[33] K. C. Nicolaou, C. W. Hummel, N. J. Bockovich, and C.-H. Wong, *J. Chem. Soc. Chem. Commun.*, 870 (1991).

[34] G. E. Ball, R. S. O'Neill, J. E. Schltz, J. B. Lowe, B. W. Weston, J. O. Nagy, E. G. Brown, C. J. Hobbs, and M. D. Bednarski, *J. Am. Chem. Soc.* **114,** 5449 (1992).

[35] M. M. Sim, H. Kondo, and C.-H. Wong, *J. Am. Chem. Soc.* **115,** 2260 (1993).

[36] S. J. Danishefsky, K. Koseki, D. A. Griffith, J. Gervay, J. M. Peterson, F. E. McDonald, and T. Oriyama, *J. Am. Chem. Soc.* **114,** 8331 (1992).

[37] A. Hasegawa, T. Ando, A. Kameyama, and M. Kiso, *J. Carbohydr. Chem.* **11,** 645 (1992).

[38] H. Furui, M. Kiso, and A. Hasegawa, *Carbohydr. Res.* **229,** c1 (1992).

[39] R. M. Nelson, S. Dolich, A. Aruffo, O. Cecconi, and M. P. Bevilacqua, *J. Clin. Invest.* **91,** 1157 (1993).

[40] M. Kiso, H. Furui, K. Ando, and A. Hasegawa, *J. Carbohydr. Chem.* **12,** 673 (1993).

[41] A. Hasegawa, K. Fushimi, H. Ishida, and M. Kiso, *J. Carbohydr. Chem.* **12,** 1203 (1993).

[42] Y.-C. Lin, C. W. Hummel, D.-H. Huang, Y. Ichikawa, K. C. Nicolaou, and C.-H. Wong, *J. Am. Chem. Soc.* **114,** 5452 (1992).

[43] E. L. Berg, M. K. Robinson, O. Mansson, E. C. Butcher, and J. L. Magnani, *J. Biol. Chem.* **266,** 14869 (1991).

[44] D. V. Erbe, B. A. Wolitzky, L. G. Presta, C. R. Norton, R. J. Ramos, D. K. Burns, J. M. Rumberger, B. N. N. Rao, C. Foxall, B. K. Brandley, and L. A. Lasky, *J. Cell Biol.* **119,** 215 (1992).

on silica gel (200 g) gives **2** (2.65 g, 74%). Recrystallization from diethyl ether gives needles with mp 183.5°–184.5°, $[\alpha]_D$ −3.2° [*c* 1, 1 : 1 (v/v) dichloromethane–methanol].

*2-(Trimethylsilyl)ethyl O-[2,4,6-tri-O-benzyl-3-O-(4-methoxybenzyl)-β-D-galactopyranosyl]-(1→4)-2,3,6-tri-O-benzyl-β-D-glucopyranoside (**3**).* To a solution of **2** (3.05 g, 5.4 mmol) in *N,N*-dimethylformamide (20 ml) is added a suspension of sodium hydride in oil (1.95 g, 60% sodium hydride by weight). The mixture is stirred for 30 min at 0°, benzyl bromide (5.8 ml, 8.1 mmol) is added dropwise, and stirring is continued for 16 hr at room temperature. When the reaction is complete, methanol (1 ml) is added, and the mixture is concentrated and extracted with dichloromethane. The extract is washed with water, dried (Na_2SO_4), and concentrated. Column chromatography [1 : 6 (v/v) ethyl acetate–hexane] of the residue on silica gel (300 g) gives **3** (4.75 g, 80%), isolated as a syrup, $[\alpha]_D$ −7.5° (*c* 0.4, dichloromethane).

*2-(Trimethylsilyl)ethyl O-[2,4,6-tri-O-benzyl-β-D-galactopyranosyl)-(1→4)-2,3,6-tri-O-benzyl-β-D-glucopyranoside (**4**).* To a stirred solution of **3** (4.74 g, 4.3 mmol) in dichloromethane (54 ml) are added 2,3-dichloro-5,6-dicyano-benzoquinone (1.46 g, 64 mmol) and water (3 ml), and stirring is continued for 1 hr at room temperature. The precipitate is collected and washed with dichloromethane, and the combined filtrate and washings are washed with water, dried (Na_2SO_4), and concentrated. Column chromatography [1 : 5 (v/v) ethyl acetate–hexane] of the residue on silica gel (200 g) gives **4** (2.94 g, 70%), isolated as a syrup, $[\alpha]_D$ +0.2° (*c* 0.8, dichloromethane).

*2-(Trimethylsilyl)ethyl O-(3,4,6-tri-O-acetyl-2-deoxy-2-phthalimido-β-D-glucopyranosyl)-(1→3)-O-(2,4,6-tri-O-benzyl-β-D-galactopyranosyl)-(1→4)-2,3,6-tri-O-benzyl-β-D-glucopyranoside (**6**).* To a solution of **4** (2.9 g, 2.9 mmol) in dichloromethane (7 ml) are added silver carbonate (1.7 g, 6.2 mmol), silver perchlorate (1.3 g, 6.3 mmol), and powdered MS-4Å (3.0 g), and the mixture is stirred for 20 hr at room temperature in the dark (mixture A). A solution of **5** (3.2 g, 6.4 mmol) in dichloromethane (7 ml) is treated with powdered MS-4Å (3 g) as above and then added to mixture A at room temperature. After vigorous stirring for 16 hr, the precipitate is collected and washed with dichloromethane, and the combined filtrate and washings are concentrated. Column chromatography [1 : 2 (v/v) ethyl acetate–hexane] of the residue on silica gel (200 g) affords **6** (3.9 g, 94%), isolated as a syrup, $[\alpha]_D$ −3.7° (*c* 1.1, chloroform).

*2-(Trimethylsilyl)ethyl O-(2-acetamido-2-deoxy-β-D-glucopyranosyl)-(1→3)-O-(2,4,6-tri-O-benzyl-β-D-galactopyranosyl)-(1→4)-2,3,6-tri-O-benzyl-β-D-glucopyranoside (**7**).* A solution of **6** (3.9 g, 2.8 mmol) in methanol (15 ml) is stirred with sodium methoxide (100 mg) for 2 hr at room temperature. The mixture is treated with Amberlite IR-120 (H^+) resin and concentrated, and a solution of the residue in aqueous 95% ethanol (40 ml) is treated with hydrazine hydrate (1 ml) for 2 hr under reflux. The precipitate is collected and washed with ethanol, and the combined filtrate and washings are concentrated. The residue is treated with acetic anhydride (1 ml) in methanol (50 ml) for 2 hr at room temperature, pyridine (1.5 ml) is added, the mixture is concentrated, and a solution of the residue in dichloromethane is washed with 2 *M* HCl, water, and 1 *M* Na_2CO_3, dried (Na_2SO_4), and concentrated. Column chromatography [4 : 1 (v/v) ethyl acetate–hexane] of the residue on silica gel (200 g) gives **7** (2.9 g, 88%), isolated as a syrup, $[\alpha]_D$ −7.3° (*c* 0.4, chloroform).

*2-(Trimethylsilyl)ethyl O-(2-acetamido-4,6-O-benzylidene-2-deoxy-β-D-glucopyranosyl)-(1→3)-O-(2,4,6-tri-O-benzyl-β-D-galactopyranosyl)-(1→4)-2,3,6-tri-O-benzyl-β-D-glucopyranoside (**8**).* To a solution of **7** (2.9 g, 2.5 mmol) in *N,N*-dimethylformamide (15 ml) are added benzaldehyde dimethyl acetal (0.73 ml, 4.88 mmol), *p*-toluenesulfonic acid monohydrate (30 mg), and Drierite (3 g). The mixture is stirred for 16 hr at room temperature, then neutralized with Amberlite IR-410 (HO^-) resin and concentrated. Column chromatography [4 : 1 (v/v) ethyl acetate–hexane] of the residue on silica gel (150 g) gives **8** (2.6 g, 83.5%), isolated as a syrup, $[\alpha]_D$ −23° (*c* 1, dichloromethane).

*2-(Trimethylsilyl)ethyl O-(2,3,4-tri-O-benzyl-α-L-fucopyranosyl)-(1→3)-O-(2-acetamido-4,6-O-benzylidene-2-deoxy-β-D-glucopyranosyl)-(1→3)-O-(2,4,6-tri-O-benzyl-β-D-galactopyranosyl)-(1→4)-2,3,6-tri-O-benzyl-β-D-glucopyranoside (**10**).* To a solution of **8** (1.35 g, 1.06 mmol) and **9**[11] (560 mg, 1.27 mmol) in dry benzene (20 ml) is added powdered MS-4Å (4 g), and the mixture is stirred for 4 hr at room temperature. DMTST (980 mg, 3.8 mmol) and MS-4Å (920 mg) are added to the stirred mixture at 6°, and stirring is continued for 4 hr at 6°. Methanol (3 ml) and triethylamine (1 ml) are added to the mixture and stirred for 30 min. The precipitate is filtered off and washed with dichloromethane. The filtrate and washings are combined, and the solution is washed with water, dried (Na_2SO_4), and concentrated. Column chromatography [1 : 2 (v/v) ethyl acetate–hexane] of the residue on silica gel (60 g) gives **10** (1.53 g, 86%) as an amorphous mass, $[\alpha]_D$ −37.5° (*c* 0.9, chloroform).

*2-(Trimethylsilyl)ethyl O-(2,3,4-tri-O-benzyl-α-L-fucopyranosyl)-(1→3)-O-(2-acetamido-6-O-benzyl-2-deoxy-β-D-glucopyranosyl)-(1→3)-O-(2,4,6-tri-O-benzyl-β-D-galactopyranosyl)-(1→4)-2,3,6-tri-O-benzyl-β-D-glucopyranoside (**11**).* To a solution of **10** (610 mg, 0.36 mmol) in dry tetrahydrofuran (10 ml) is added MS-4Å (2 g), and the mixture is stirred for 1 hr at room temperature. Sodium cyanoborohydride (340 mg, 5.4 mmol) is gradually added under N_2. After the reagent has dissolved, hydrogen chloride in ether is added in small portions at room temperature until the evolution of gas ceases. Thin-layer chromatography (TLC) indicates that the reaction is complete after 5 min. The mixture is diluted with dichloromethane (50 ml) and water (10 ml), then filtered. The organic layer is washed with 2 *M* hydrochloric acid and water, dried (Na_2SO_4), and concentrated. Column chromatography [1 : 1 (v/v) ethyl acetate–hexane] of the residue on silica gel (40 g) gives **11** (460 mg, 75%) as an amorphous mass, $[\alpha]_D$ −19.5° (*c* 1.1, chloroform).

*Methyl O-(methyl 5-acetamido-4,7,8,9-tetra-O-acetyl-3,5-dideoxy-*D*-glycero-α-*D*-galacto-2-nonulopyranosylonate)-(2→3)-2,4,6-tri-O-benzoyl-1-thio-β-*D*-galactopyranoside (**12**).* To a solution of 2-(trimethylsilyl)ethyl *O*-(methyl 5-acetamido-4,7,8,9-tetra-*O*-acetyl-3,5-dideoxy-D-*glycero*-α-D-*galacto*-2-nonulopyranosylonate)-(2→3)-6-*O*-benzoyl-β-D-galactopyranoside[21,24,45,46] (1.2 g, 1.4 mmol) in pyridine (6 ml) are added benzoic anhydride (1.27 g, 5.6 mmol) and 4-dimethylaminopyridine (200 mg, 1.4 mmol), and the mixture is stirred for 10 hr at room temperature. Methanol (1 ml) is added to the mixture and concentrated, then the residue is extracted with dichloromethane. The extract is washed with 2 *M* hydrochloric acid and water, dried (Na_2SO_4), and concentrated. Column chromatography [3 : 1 (v/v) ethyl acetate–hexane] of the residue on silica gel (60 g) gives the 2,4,6-tri-*O*-benzoyl derivative (1.35 g, 90%) as an amorphous mass, $[\alpha]_D$ +24° (*c* 0.8, chloroform). To a solution of the tri-*O*-benzoyl derivative (1.84 g, 1.72 mmol) in dry toluene (10 ml) and acetic anhydride (2.6 ml) is added boron trifluoride etherate (0.45 ml), and the mixture is stirred at room temperature. After 2 hr, dichloromethane (50 ml) is added, and the solution is washed with 1 *M* sodium hydrogen carbonate, dried (Na_2SO_4), and concentrated. Column chromatography [50 : 1 (v/v) dichloromethane–methanol] of the residue on silica gel (100 g) affords the 1-*O*-acetyl derivative (1.6 g, 93%), isolated as a syrup, $[\alpha]_D$ +41° (*c* 0.9, chloroform). To a solution of the 1-*O*-acetyl derivative (1.6 g, 1.6 mmol) in dry dichloromethane (10 ml) are added, with stirring, methylthio(trimethyl)silane (480 mg, 4.0 mmol) and boron trifluoride etherate (0.4 ml), and the mixture is stirred for 2 hr at room temperature. Dichloromethane (50 ml) is added, and the solution is washed with 1 *M* sodium hydrogen carbonate, dried (Na_2SO_4), and concentrated. Column chromatography [80 : 1 (v/v) dichloromethane–methanol] of the residue on silica gel (100 g) gives **12** (1.3 g, 82%), isolated as a syrup, $[\alpha]_D$ +34° (*c* 0.7, chloroform).

[45] T. Murasa, H. Ishida, M. Kiso, and A. Hasegawa, *Carbohydr. Res.* **184,** c1 (1988).
[46] A. Hassegawa, T. Nagahama, H. Ohki, K. Hotta, H. Ishida, and M. Kiso, *J. Carbohydr. Chem.* **10,** 493 (1991).

*O-(Methyl 5-acetamido-4,7,8,9-tetra-O-acetyl-3,5-dideoxy-D-glycero-α-D-galacto-2-nonulopyranosylonate)-(2→3)-O-(2,4,6-tri-O-benzoyl-β-D-galactopyranosyl)-(1→4)-O-[(2,3,4-tri-O-acetyl-α-L-fucopyranosyl)-(1→3)]-O-(2-acetamido-6-O-acetyl-2-deoxy-β-D-glucopyranosyl)-(1→3)-O-(2,4,6-tri-O-acetyl-β-D-galactopyranosyl)-(1→4)-2,3,6-tri-O-acetyl-α-D-glucopyranosyl trichloroacetimidate (**17**).* To a solution of **11** (473 mg, 0.28 mmol) and **12** (417 mg, 0.42 mmol) in dry dichloromethane (8 ml) is added MS-4Å (1 g), and the mixture is stirred for 4 hr at room temperature. Then DMTST (325 mg, 1.26 mmol) and MS-4Å (220 mg) are added to the stirred mixture, and the stirring is continued for 20 hr at room temperature; the progress of the reaction is monitored by TLC. Methanol (1 ml) and triethylamine (0.5 ml) are added to the mixture, and the precipitate is filtered off and washed with dichloromethane. The combined filtrate and washings are washed with water, dried (Na_2SO_4), and concentrated. Column chromatography [4 : 1 (v/v) ethyl acetate–hexane] of the residue on silica gel (30 g) gives **14** (300 mg, 41%) as an amorphous mass, $[\alpha]_D$ −14.5° (*c* 0.8, chloroform). A solution of **14** (460 mg, 0.17 mmol) in ethanol (60 ml) and acetic acid (22 ml) is hydrogenolyzed in the presence of 10% Pd–C (400 mg) for 4 days at 45°, then filtered and concentrated. The residue is acetylated with acetic anhydride (3 ml)–pyridine (5 ml) for 16 hr at room temperature. The product is purified by chromatography on a column of silica gel (40 g) with 6 : 1 (v/v) ethyl acetate–hexane, to give **15** (305 mg, 81%) as an amorphous mass, $[\alpha]_D$ −20.2° (*c* 0.7, chloroform). To a solution of **15** (305 mg, 0.14 mmol) in dry dichloromethane (1 ml) is added trifluoroacetic acid (0.1 ml), and the mixture is stirred for 1 hr at room temperature. Ethyl acetate (1 ml) is added to the mixture, which is then concentrated. Column chromatography [6 : 1 (v/v) ethyl acetate–hexane] of the residue on silica gel (30 g) gives **16** (274 mg, 94%) as an amorphous mass, $[\alpha]_D$ −8.4° (*c* 0.9, chloroform). To a stirred solution of **16** (146 mg, 0.07 mmol) in dry dichloromethane (1 ml), cooled to −5°, are added trichloroacetonitrile (0.3 ml) and DBU (11 mg). The mixture is stirred for 3 hr at 0°, then concentrated. Column chromatography [30 : 1 (v/v) dichloromethane–methanol] of the residue on silica gel (20 g) gives **17** (142 mg, 91%) as an amorphous mass, $[\alpha]_D$ +1.5° (*c* 1.4, chloroform).

*Sialyl Lewis X Ganglioside (**A**).* To a solution of **17** (142 mg, 64 μmol) and (2*S*,3*R*,4*E*)-2-azido-3-*O*-benzoyl-4-octadecene-1,3-diol[27,28] (56 mg, 130 μmol) in dry dichloromethane (2 ml) is added MS-4Å (AW-300, 1.4 g), and the mixture is stirred for 30 min at room temperature and then cooled to 0°. Boron trifluoride etherate (3.5 μl) is added to the mixture, and this is stirred for 3 hr at 0°. The precipitate is filtered off and washed with dichloromethane. The solution is successively washed with 1 *M* sodium carbonate and water, dried (Na_2SO_4), and concentrated. Column chromatography [40 : 1 (v/v) dichloromethane–methanol] of the residue on silica gel (20 g) gives **18** (89 mg, 56%) as an amorphous mass, $[\alpha]_D$ −23° (*c* 0.9, chloroform). Hydrogen sulfide is bubbled through a stirred solution of **18** (116 mg, 47 μmol) in aqueous 83% pyridine (12 ml) for 2 days at 0°. The reaction is monitored by TLC. The mixture is concentrated, and the residue (**19**) is stirred with octadecanoic acid (27 mg, 92 μmol) and 1-(3-dimethylaminopropyl)-3-ethylcarbodiimide (DAC, 27 mg, 95 μmol) in dichloromethane (2 ml) for 16 hr at room temperature. Dichloromethane (30 ml) is added, and the mixture is washed with water, dried (Na_2SO_4), and concentrated. Column chromatography [30 : 1 (v/v) dichloromethane–methanol] of the residue on silica gel (25 g) gives **20** (103 mg, 81%) as an amorphous mass, $[\alpha]_D$ −11.8° (*c* 0.78, chloroform). To a solution of **20** (103 mg, 34 μmol) in methanol (5 ml) is added sodium methoxide (30 mg); the mixture is stirred for 24 hr at 40°, and water (0.5 ml) is added. The solution is stirred for 8 hr at room temperature, neutralized with Amberlite IR-120 (H^+) resin, and filtered. The resin is washed with 1 : 1 (v/v) water–methanol, and the combined filtrate and washings are concentrated. Column chromatography [5 : 4 : 0.7 (v/v/v) chloroform–methanol–water] of the residue on Sephadex LH-20 (50 g) gives sialyl Le^x ganglioside (**A**, 65 mg, quantitative) as an amorphous mass, $[\alpha]_D$ −17.5° [*c* 0.6, 5 : 4 : 0.7 (v/v/v) chloroform–methanol–water].

Synthesis of Sialyl Lewis X Ganglioside (Pentasaccharide)

We describe here the stereocontrolled synthesis[37] of sLe^x ganglioside, α-Neu5Ac-(2→3)-β-D-Gal-(1→4)-[α-L-Fuc-(1→3)]-β-D-GlcNAc-(1→3)-β-D-Gal-(1→1)-ceramide, which contains a pentasaccharide. The glycosylation (Scheme 2) of 2-(trimethylsilyl)ethyl 2,4,6-tri-*O*-benzyl-β-D-galactopyranoside (**24**), prepared from 2-(trimethylsilyl)ethyl β-D-galactopyranoside[47] (**21**) via selective 3-O-methoxybenzylation, O-benzylation, and oxidative removal of the 4-methoxybenzyl group as described above (see detailed procedure), is effected with **5**. Reaction in the presence of silver carbonate, silver perchlorate, and powdered MS-4Å in dichloromethane gives the desired β-glycoside **25** (76%). O-Deacetylation of **25**, followed by de-N-phthaloylation, N-acetylation of the product, and subsequent 4,6-O-benzylidenation, affords the disaccharide glycosyl acceptor **27** (72%). Glycosylation of **27** with the fucose donor **9** in benzene for 10 hr at 5°–10°, using DMTST, gives the α-trisaccharide **28** (97%). Reductive ring opening of the benzylidene group in **28**, according to the method of Garegg *et al.*,[20] affords compound **29** (77%). Glycosylation of **29** with the sialylgalactose

[47] K. Jansson, T. Frejd, J. Kihlberg, and G. Mugnusson, *Tetrahedron Lett.* **27,** 753 (1986).

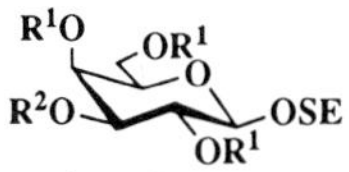

21 $R^1 = R^2 = H$
22 $R^1 = H, R^2 = MPM$
23 $R^1 = Bn, R^2 = MPM$
24 $R^1 = Bn, R^2 = H$

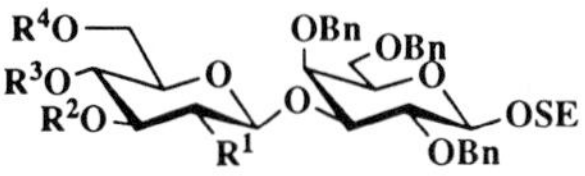

	R^1	R^2	R^3	R^4
25	Nphth	Ac	Ac	Ac
26	NHAc	H	H	H
27	NHAc	H	benzylidene	

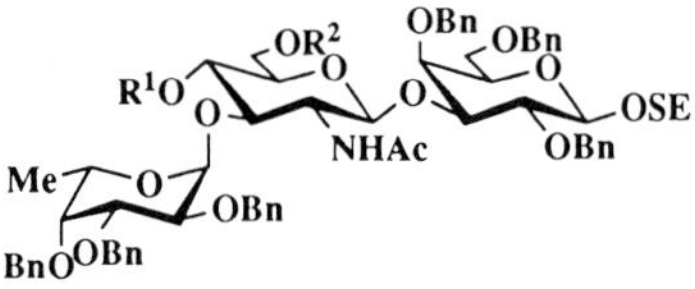

28 $R^1 = R^2$ = benzylidene
29 $R^1 = H, R^2 = Bn$

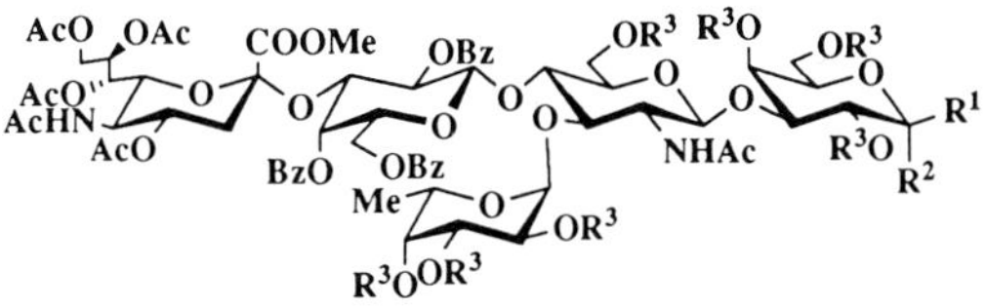

	R^1	R^2	R^3
30	OSE	H	Bn
31	OSE	H	Ac
32	H, OH		Ac
33	H	$OC(=NH)CCl_3$	H

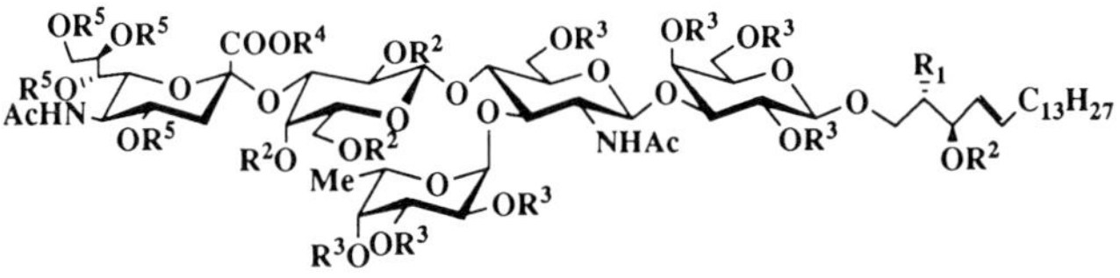

	R^1	R^2	R^3	R^4	R^5
34	N_3	Bz	Ac	Me	Ac
35	$NHCOC_{17}H_{35}$	Bz	Ac	Me	Ac
36	$NHCOC_{17}H_{35}$	H	H	H	H

SCHEME 2. Synthesis of sLe^x pentasaccharide ganglioside.

donor **12** in dichloromethane by use of DMTST gives the pentasaccharide **30** (57%). Compound **30** is converted via removal of the benzyl groups by catalytic hydrogenolysis over Pd–C, subsequent acetylation, selective removal of the 2-(trimethylsilyl)ethyl group, and trichloroacetimidation, to *O*-(methyl 5-acetamido-4,7,8,9-tetra-*O*-acetyl-3,5-dideoxy-D-*glycero*-α-D-galacto-2-nonulopyranosylonate)-(2→3)-*O*-(2,4,6-tri-*O*-benzoyl-β-D-galactopyranosyl)-(1→4)-*O*-[(2,3,4-tri-*O*-acetyl-α-L-fucopyranosyl)-(1→3)]-*O*-(2-acetamido-6-*O*-acetyl-2-deoxy-β-D-glucopyranosyl)-(1→3)-2,4,6-tri-*O*-acetyl-α-D-galactopyranosyl trichloroacetimidate (**33**) in good yield. Condensation of **33** with (2*S*,3*R*,4*E*)-2-azido-3-*O*-benzoyl-4-octadecene-1,3-diol[27,28] in dichloromethane for 8 hr at 0°, using boron trifluoride etherate and MS-4Å, affords only the expected β-glycoside **34** (71%). According to the method described for the total synthesis of sLe^x ganglioside, compound **34** is transformed into the title ganglioside **36** in good yields, via reduction of the azido group, coupling with octadecanoic acid, O-deacylation, and hydrolysis of the methyl ester group. On the other hand, O-deacylation of **31** and subsequent hydrolysis of the methyl ester group gave sLe^x pentasaccharide epitope, quantitatively.

Synthesis of Deoxyfucose-Containing Sialyl Lewis X Ganglioside Analogs

We describe here the synthesis[48] of the 2-, 3-, and 4-deoxyfucose-containing sLe^x oligosaccharides and their ceramide analogs, in order to clarify the structural features of the fucose moiety for the recognition of selectins, according to the method described in the previous sections.

For the synthesis of the desired sLe^x analogs, we employ the methyl 1-thioglycosides[48] (**37–39**) of deoxy-L-fucose as the glycosyl donors and compound **27** as a suitably protected glycosyl acceptor (Scheme 3). Glycosylation of **27** with the methyl 1-thioglycoside derivatives (**37–39**) of the respective deoxyfucoses in benzene for 10 hr at 5°–10°, using DMTST as a promoter, gives exclusively the α-glycosides **40** (86%), **42** (82%), and **44** (57%). These are transformed by reductive ring opening of the benzylidene acetal with sodium cyanoborohydride into the glycosyl acceptors **41** (97%), **43** (80%), and **45** (87%). The DMTST-promoted glycosylation of **41**, **43**, and **45** with the sialyl galactose donor **12** affords the desired pentasaccharides **46**, **50**, and **54**, which are converted via reductive removal of the benzyl groups, O-acetylation, selective removal of the 2-(trimethylsilyl) ethyl group, and subsequent imidate formation to the corresponding α-trichloroacetimidates **48**, **52**, and **56** in good yields. Glycosylation of (2*S*,3*R*,4*E*)-2-azido-3-*O*-benzoyl-4-octadecene-1,3-diol[27,28] with **48**, **52**, or

[48] A. Hasegawa, T. Ando, M. Kato, H. Ishida, and M. Kiso, *Carbohydr. Res.* **257,** 67 (1994).

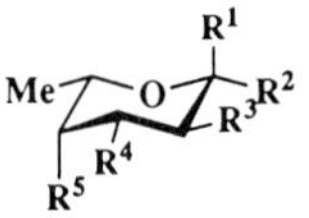

	R^1	R^2	R^3	R^4	R^5
37	H , SMe		H	OBz	OBz
38	H	SMe	OBn	H	OBn
39	H	SMe	OBn	OBn	H

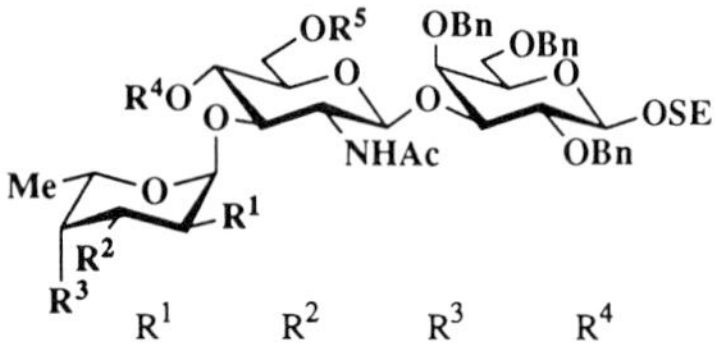

	R^1	R^2	R^3	R^4	R^5
40	H	OBz	OBz	benzylidene	
41	H	OBz	OBz	H	Bn
42	OBn	H	OBn	benzylidene	
43	OBn	H	OBn	H	Bn
44	OBn	OBn	H	benzylidene	
45	OBn	OBn	H	H	Bn

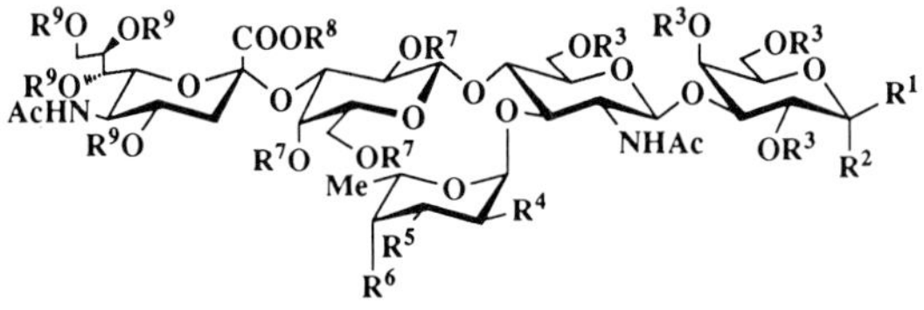

	R^1	R^2	R^3	R^4	R^5	R^6	R^7	R^8	R^9
46	OSE	H	Bn	H	OBz	OBz	Bz	Me	Ac
47	OSE	H	Ac	H	OBz	OBz	Bz	Me	Ac
48	H	$OC(=NH)CCl_3$	Ac	H	OBz	OBz	Bz	Me	Ac
49	OSE	H	H	H	OH	OH	H	H	H
50	OSE	H	Bn	OBn	H	OBn	Bz	Me	Ac
51	OSE	H	Ac	OAc	H	OAc	Bz	Me	Ac
52	H	$OC(=NH)CCl_3$	Ac	OAc	H	OAc	Bz	Me	Ac
53	OSE	H	H	OH	H	OH	H	H	H
54	OSE	H	Bn	OBn	OBn	H	Bz	Me	Ac
55	OSE	H	Ac	OAc	OAc	H	Bz	Me	Ac
56	H	$OC(=NH)CCl_3$	Ac	OAc	OAc	H	Bz	Me	Ac
57	OSE	H	H	OH	OH	H	H	H	H

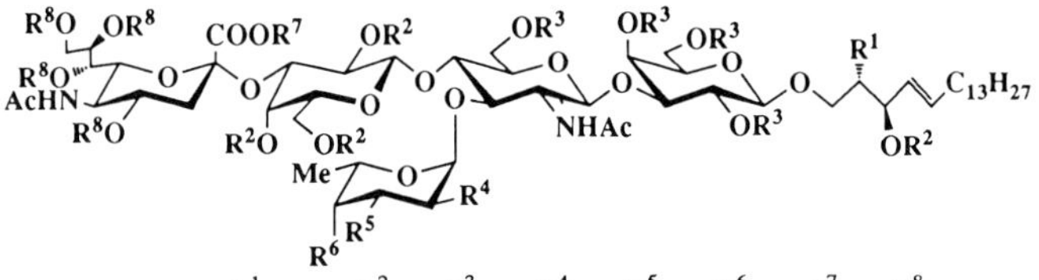

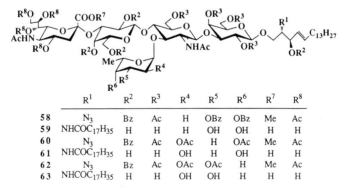

	R^1	R^2	R^3	R^4	R^5	R^6	R^7	R^8
58	N_3	Bz	Ac	H	OBz	OBz	Me	Ac
59	$NHCOC_{17}H_{35}$	H	H	H	OH	OH	H	H
60	N_3	Bz	Ac	OAc	H	OAc	Me	Ac
61	$NHCOC_{17}H_{35}$	H	H	OH	H	OH	H	H
62	N_3	Bz	Ac	OAc	OAc	H	Me	Ac
63	$NHCOC_{17}H_{35}$	H	H	OH	OH	H	H	H

SCHEME 3. Synthesis of deoxyfucose-containing sLe^x gangliosides.

56 in the presence of boron trifluoride etherate affords the β-glycosides **58** (85%), **60** (64%), and **62** (60%), respectively, which are converted in good yields, via selective reduction of the azido group, coupling with octadecanoic acid by use of DBU, O-deacylation, and deesterification, to the target gangliosides **59** (55%), **61** (97%), and **62** (82%), according to a similar procedure as described in the previous sections. Sialyl Le^x oligosaccharide analogs **49**, **53**, and **57** are obtained in good yields from **47**, **51**, or **55** via de-O-acetylation and subsequent deesterification of the methyl ester group, respectively.

Synthesis of Chemically Modified Sialic Acid-Containing Sialyl Lewis X Ganglioside Analogs

We describe here the stereocontrolled synthesis[49] of sLe^x ganglioside analogs containing C_7-Neu5Ac, C_8-Neu5Ac, and 8-epi-Neu5Ac, clarifying the structural requirements of the sialic acid moiety for selectin recognition (Scheme 4).

Methyl *O*-(methyl 5-acetamido-4,7-di-*O*-acetyl-3,5-dideoxy-β-L-*arabino*-2-heptulopyranosylonate)-(2→3)-, methyl *O*-(methyl 5-acetamido-4,7,8-tri-*O*-acetyl-3,5-dideoxy-α-D-*galacto*-2-octulopyranosylonate)-(2→3)-, and methyl *O*-(methyl 5-acetamido-4,7,8,9-tetra-*O*-acetyl-3,5-dideoxy-L-*glycero*-β-D-*galacto*-2-nonulopyranosylonate)-(2→3)-2,4,6-tri-*O*-benzoyl-1-thio-β-D-galactopyranosides (**75**, **79**, and **83**) are selected as the glycosyl donors, and compound **29** as the trisaccharide glycosyl acceptor in the synthesis of sLe^x ganglioside analogs. Glycosylation of **29** with **75**, **79**, and **83**, yields intermediates that can then, by introduction of the ceramide moiety, be transformed to the title products, **98**, **101**, and **104**. Glycosylation of 2-(trimethylsilyl)ethyl 6-*O*-benzoyl-β-D-galactopyranoside (**7**) with the phenyl or methyl 2-thioglycoside derivatives (**67**,[49] **69**,[49] and **70**[50]) of the respective sialic acids, using *N*-iodosuccinimide (NIS)–trifluoromethanesulfonic acid as a promoter[46,51,52] in acetonitrile for 2 hr at −35°, gives the three required 2-(trimethylsilyl)ethyl (2*S*)-sialyl-(2→3)-β-D-galactopyranosides, **72** (45%), **76** (45%), and **80** (41%). These are converted via O-benzoylation (**73**, **77**, and **81**), selective transformation of the 2-(trimethylsilyl)ethyl group into an acetyl by treatment[23] with boron trifluoride etherate in toluene–acetic anhydride, and introduction of the methylthio group with methylthio(trimethyl)silane to the corresponding

[49] M. Yoshida, A. Uchimura, M. Kiso, and A. Hasegawa, *Glycoconjugate J.* **10,** 3 (1993).

[50] A. Hasegawa, K. Adachi, M. Yoshida, and M. Kiso, *J. Carbohydr. Chem.* **11,** 95 (1992).

[51] P. Konradsson, U. E. Udodong, and B. Fraser-Reid, *Tetrahedron Lett.* **31,** 4313 (1990).

[52] G. H. Veeneman, S. H. van Leevwen, and J. H. van Boom, *Tetrahedron Lett.* **31,** 1331 (1990).

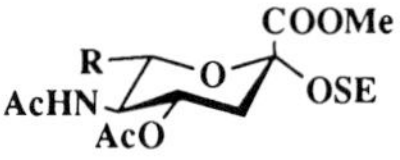

64 R = X^1
65 R = Y^1

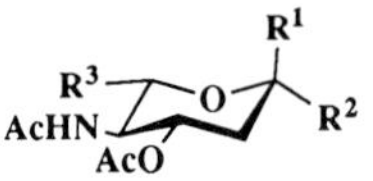

	R^1, R^2	R^3
66	OAc, COOMe	X^1
67	SPh, COOMe	X^1
68	OAc, COOMe	Y^1
69	SPh, COOMe	Y^1

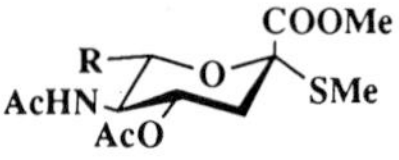

R = Z^1
70

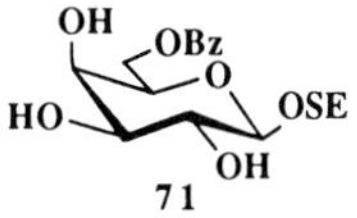

71

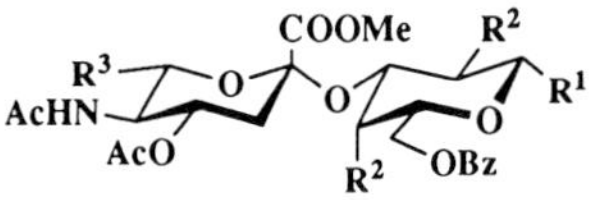

	R^1	R^2	R^3
72	OSE	OH	X^1
73	OSE	OBz	X^1
74	OAc	OBz	X^1
75	SMe	OBz	X^1
76	OSE	OH	Y^1
77	OSE	OBz	Y^1
78	OAc	OBz	Y^1
79	SMe	OBz	Y^1
80	OSE	OH	Z^1
81	OSE	OBz	Z^1
82	OAc	OBz	Z^1
83	SMe	OBz	Z^1

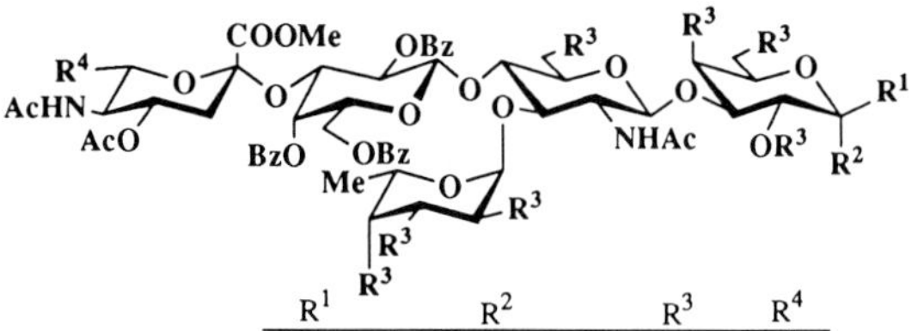

	R^1	R^2	R^3	R^4
84	OSE	H	OBn	X^1
85	OSE	H	OAc	X^1
86	H, OH		OAc	X^1
87	H	OC(=NH)CCl_3	OAc	X^1
88	OSE	H	OBn	Y^1
89	OSE	H	OAc	Y^1
90	H, OH		OAc	Y^1
91	H	OC(=NH)CCl_3	OAc	Y^1
92	OSE	H	OBn	Z^1
93	OSE	H	OAc	Z^1
94	H, OH		OAc	Z^1
95	H	OC(=NH)CCl_3	OAc	Z^1

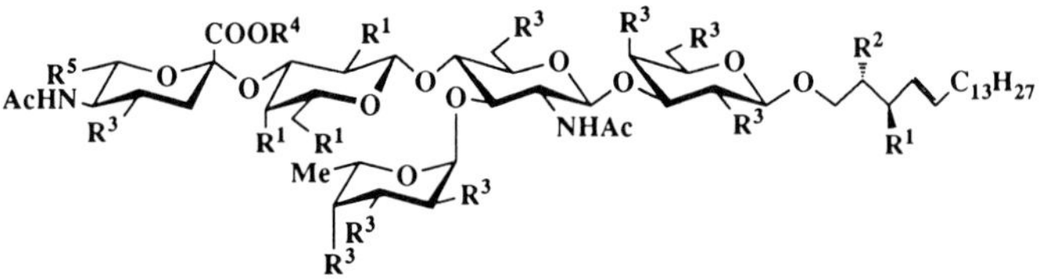

	R^1	R^2	R^3	R^4	R^5
96	OBz	N_3	OAc	Me	X^1
97	OBz	$NHCOC_{17}H_{35}$	OAc	Me	X^1
98	OH	$NHCOC_{17}H_{35}$	OH	H	X^2
99	OBz	N_3	OAc	Me	Y^1
100	OBz	$NHCOC_{17}H_{35}$	OAc	Me	Y^1
101	OH	$NHCOC_{17}H_{35}$	OH	H	Y^2
102	OBz	N_3	OAc	Me	Z^1
103	OBz	$NHCOC_{17}H_{35}$	OAc	Me	Z^1
104	OH	$NHCOC_{17}H_{35}$	OH	H	Z^2

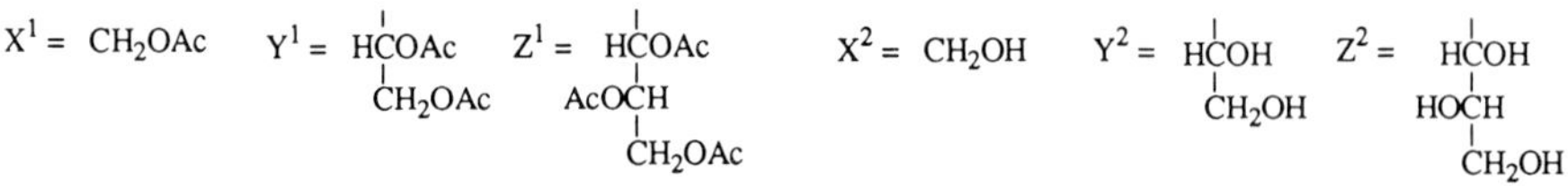

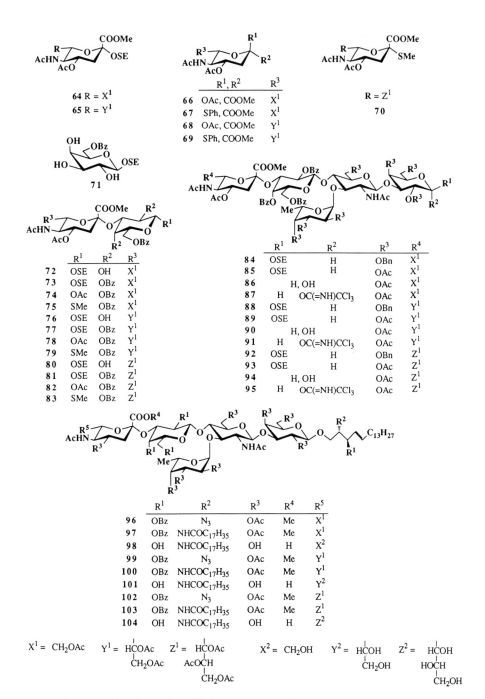

SCHEME 4. Synthesis of modified Neu5Ac-containing sLe^x ganglioside analogs.

glycosyl donors (**75**, **79**, and **83**) in good yields. Glycosylation of the trisaccharide acceptor **29** with these donors (1.45 equivalents with respect to the acceptor), in dichloromethane for 48 hr at 5° in the presence of DMTST and MS-4Å, gives the expected β-glycosides, **84** (53%), **88** (53%), and **92** (49%), respectively. Catalytic hydrogenolysis (10% Pd–C) in ethanol–acetic acid for 3 days at 45° of the benzyl groups in **84**, **88**, or **92**, and subsequent O-acetylation gave the per-*O*-acyl derivatives, **85** (85%), **89** (87%), and **93** (83%), after column chromatography. Compounds **85**, **89**, and **93** are converted to the corresponding α-trichloroacetimidates, **87**, **91**, and **95**, according to the method described above, and these, on coupling with (2*S*,3*R*,4*E*)-2-azido-3-*O*-benzoyl-4-octadecene-1,3-diol, give the required β-glycosides, **96** (68%), **99** (43%), and **102** (46%). Finally, these are transformed via selective reduction of the azido group, condensation with octadecanoic acid, O-deacylation, and deesterification of the methyl ester group into the target sLe^x ganglioside analogs, **98**, **101**, and **104**, in good yields, as described for the synthesis of sLe^x ganglioside.

Acknowledgments

This work was supported in part by Grants-in-Aid (No. 04250102 and No. 03660132) for Scientific Research on Priority Areas from the Ministry of Education, Science, and Culture of Japan.

[16] Synthesis of Ganglioside G_{M3} and Analogs Containing Modified Sialic Acids and Ceramides

By MAKOTO KISO and AKIRA HASEGAWA

Introduction

Ganglioside G_{M3}, first isolated from horse erythrocytes by Yamakawa and Suzuki[1] in 1952, is the major ganglioside component in erythrocytes of many animal species. In addition to the known biological functions of

[1] T. Yamakawa and S. Suzuki, *J. Biochem.* (*Tokyo*) **39,** 383 (1952).

Copyright © 1994 by Academic Press, Inc.
All rights of reproduction in any form reserved.

G_{M3},[2–6] much attention has been focused[7,8] on ganglioside G_{M3} and its degradation products because of their involvement in cell proliferation, differentiation, oncogenesis, modulation of transmembrane signaling, and so on. Ganglioside G_{M3} has also been found to be a potential immunosuppressor[9] and a substrate of the *trans*-sialidase of *Trypanosoma cruzi*.[10] In view of these facts, it is of interest to investigate the relationship between the structures of sialic acid and ceramide moieties and the biological functions of G_{M3} at the molecular level. This chapter describes a facile, preparative synthesis not only of natural ganglioside G_{M3} but also of various types of analogs containing a variety of modified sialic acids and ceramides.

Synthesis of Ganglioside G_{M3} Using 2-Thioglycoside of *N*-Acetylneuraminic Acid

The first synthesis of ganglioside G_{M3} was achieved[11] by the use of methyl 5-acetamido-4,7,8,9-tetra-*O*-acetyl-2-chloro-2,3,5-trideoxy-D-*glycero*-β-D-*galacto*-2-nonulopyranosonate but proceeded in very low yield. The problematic step in ganglioside synthesis is the α-stereoselective glycosylation of sialic acid, and a variety of approaches[12] have been attempted. Particularly annoying is the formation of the 2,3-dehydro derivative of sialic acid (compound **7** in Scheme 1) and the thermodynamically favored β-glycosides. More efficient methods for the synthesis of ganglioside G_{M3} by overcoming these difficulties have been reported.[13,14]

A highly efficient, regio- and α-stereoselective glycosylation of sialic acid to the 3-OH of the galactose moiety has been achieved[15–18] by using the

[2] G. J. M. Hooghwinkel, P. F. Barri, and G. W. Bruyn, *Neurology* **16,** 934 (1966).
[3] N. Handa and S. Handa, *J. Exp. Med.* **35,** 331 (1965).
[4] K. Uemura, M. Yuzawa, and T. Takemori, *J. Biochem.* (*Tokyo*) **83,** 463 (1978).
[5] A. Gorio, G. Carmignoto, F. Facci, and M. Finesso, *Brain Res.* **197,** 236 (1980).
[6] Y. Suzuki, M. Matsunaga, and M. Matsumoto, *J. Biol. Chem.* **260,** 1362 (1985).
[7] S. Hakomori, *J. Biol. Chem.* **265,** 18713 (1990).
[8] W. Song, M. F. Vacca, R. Welti, and D. A. Rintoul, *J. Biol. Chem.* **266,** 10174 (1991).
[9] S. Ladisch, H. Becker, and L. Ulsh, *Biochim. Biophys. Acta* **1125,** 180 (1992).
[10] S. Schenkman, J. Man-Shiow, G. W. Hart, and V. Nussenzweig, *Cell* (*Cambridge, Mass.*) **65,** 1117 (1991).
[11] M. Sugimoto and T. Ogawa, *Glycoconjugate J.* **2,** 5 (1985).
[12] K. Okamoto and T. Goto, *Tetrahedron* **46,** 5835 (1990).
[13] T. Murase, H. Ishida, M. Kiso, and A. Hasegawa, *Carbohydr. Res.* **188,** 71 (1989).
[14] M. Numata, M. Sugimoto, Y. Ito, and T. Ogawa, *Carbohydr. Res.* **203,** 205 (1990).
[15] T. Murase, H. Ishida, M. Kiso, and A. Hasegawa, *Carbohydr. Res.* **184,** c1 (1988).
[16] A. Hasegawa, H. Ohki, T. Nagahama, H. Ishida, and M. Kiso, *Carbohydr. Res.* **212,** 277 (1991).
[17] A. Hasegawa, T. Nagahama, H. Ohki, K. Hotta, H. Ishida, and M. Kiso, *J. Carbohydr. Chem.* **10,** 493 (1991).
[18] A. Hasegawa, H.-K. Ishida, and M. Kiso, *J. Carbohydr. Chem.* **12,** 371 (1993).

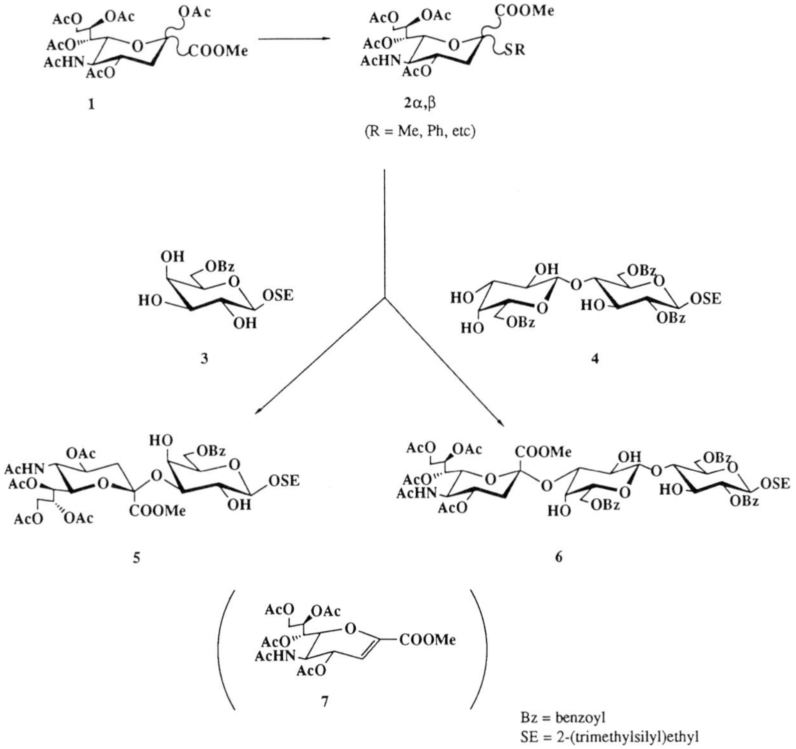

SCHEME 1. Regio- and α-stereoselective glycoside synthesis of sialic acids.

2-thioglycosides (**2α,β**) of *N*-acetylneuraminic acid (Neu5Ac) as glycosyl donors and the 2-(trimethylsilyl)ethyl (SE)[19] glycosides (**3** and **4**) of "lightly protected" galactose or lactose as glycosyl acceptors in the presence of thiophilic promoters such as dimethyl(methylthio)sulfonium triflate[20] (DMTST) or *N*-iodosuccinimide–trifluoromethanesulfonic acid[21,22] (NIS–TfOH) in acetonitrile (Scheme 1).

In the initial studies,[13,15,23] the methyl α-2-thioglycoside of Neu5Ac (**2α**, R = Me) was employed as a glycosyl donor. Glycosylation of lightly protected 2-(trimethylsilyl)ethyl 6-*O*-benzoyl-β-D-galactopyranoside[15,23]

[19] G. Magnusson, *Trends Glycosci. Glycotechnol.* **4,** 358 (1992).

[20] P. Fügedi, P. J. Garegg, H. Lönn, and T. Norberg, *Glycoconjugate J.* **4,** 97 (1987).

[21] G. H. Veeneman, S. H. van Leeuwen, and J. H. van Boom, *Tetrahedron Lett.* **31,** 1331 (1990).

[22] P. Konradsson, U. E. Udodong, and B. Fraser-Reid, *Tetrahedron Lett.* **31,** 4313 (1990).

[23] T. Murase, A. Kameyama, K. P. R. Kartha, H. Ishida, M. Kiso, and A. Hasegawa, *J. Carbohydr. Chem.* **8,** 265 (1989).

(**3**) or 2-(trimethylsilyl)ethyl *O*-(6-*O*-benzoyl-*β*-D-galactopyranosyl)-(1→4)-2,6-di-*O*-benzoyl-*β*-D-glucopyranoside[13,15] (**4**) with 2.0 molar equivalents of **2*α*** in acetonitrile at −15° in the presence of DMTST and 3Å Molecular Sieves (MS) affords the desired *α*-glycoside **5** or **6** in about 50% yield, respectively, together with 2,3-dehydro derivative (**7**) of Neu5Ac (a major by-product) and unreacted acceptors. It is noteworthy that neither the *β*-glycoside nor a positional isomer is isolated in the reactions. This method has been successfully extended[16] to the large-scale preparation of *α*-sialyl-(2→3)- and *α*-sialyl-(2→6)-Gal derivatives by using **2*α*** or even the anomeric mixture (**2*α*,*β***) obtained almost quantitatively in one step from methyl 5-acetamido-2,4,7,8,9-penta-*O*-acetyl-3,5-dideoxy-D-*glycero*-D-*galacto*-2-nonulopyranosonate (**1**).

Iodonium ion-promoted glycosylation of **3** or **4** with **2*α*,*β*** was then examined[17] using NIS–TfOH as a promoter, and the desired *α*-glycoside **5** or **6** is obtained in 60–70% yield under milder reaction conditions than those of the DMTST-promoted glycosylation described before. Therefore, taking into account the ready availability, lower toxicity, and easy handling of glycosyl promoter, the NIS–TfOH method seems to be superior to the DMTST method. A proposed reaction mechanism[17] of the DMTST- or NIS–TfOH-promoted glycosylation of Neu5Ac using its 2-thioglycoside in acetonitrile is shown in Fig. 1.

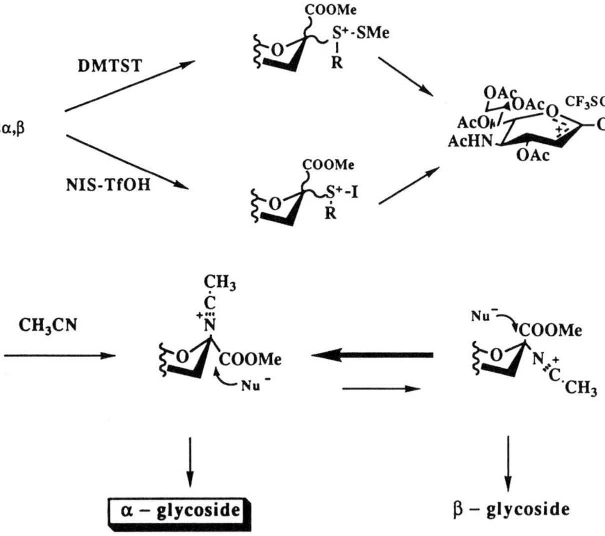

FIG. 1. Proposed reaction mechanism of the DMTST- or NIS–TfOH-promoted glycosylation of Neu5Ac using the 2-thioglycoside in acetonitrile.

Acetylation of **6** with acetic anhydride in pyridine (94%) and selective removal[24] of the SE group in **8** with $BF_3 \cdot OEt_2$ give the corresponding 1-hydroxy derivative **9** (~90%), which is then treated with trichloroacetonitrile and 1,8-diazabicyclo[5.4.0]undec-7-ene (DBU) to afford the trichloroacetimidate **10** (94%). Glycosylation of (2*S*,3*R*,4*E*)-2-azido-3-*O*-benzoyl-4-octadecene-1,3-diol[25–27] (**11**) by **10** yields the expected β-glycoside **12** in high yield (92%). Selective reduction of the azide group with H_2S in 5 : 1 pyridine–water gives the amine **13**, which is converted stepwise, by condensation with fatty acids in the presence of 1-ethyl-3-(3-dimethylaminopropyl)carbodiimide hydrochloride (DAC) in CH_2Cl_2 followed by O-deacylation and saponification of the methyl ester, to the desired ganglioside G_{M3}[13] (Scheme 2).

Detailed Procedures

*Preparation of 2-Thioglycosides of Neu5Ac (**2α,β**)*

Methyl 2-Thioglycoside of Neu5Ac. The methyl α- and methyl β-2-thioglycosides of Neu5Ac are each prepared[28] by methylation of the corresponding sodium salt of methyl 5-acetamido-4,7,8,9-tetra-*O*-acetyl-3,5-dideoxy-2-thio-D-*glycero*-α- or -β-D-*galacto*-2-nonulopyranosonate. A large amount of the anomeric mixture (37.7 g, α : β ~1 : 1) is obtained[16] in one step from the acetate[29] **1** (40 g) by treatment with (methylthio)trimethylsilane and trimethylsilyltrifluoromethanesulfonate in 1,2-dichloroethane.

Phenyl 2-Thioglycoside of Neu5Ac. The phenyl 2-thioglycoside of Neu5Ac is readily prepared[30] by treatment of **1** with thiophenol in the presence of boron trifluoride etherate in dichloromethane. Other 2-thioglycosides are also obtainable by this method.

*Preparation of "Lightly Protected" Lactose Acceptor (**4**)*

Partial benzoylation[13] of 2-(trimethylsilyl)ethyl *O*-(3-*O*-benzyl-β-D-galactopyranosyl)-(1→4)-β-D-glucopyranoside[13] (**14**) or 2-(trimethylsilyl) ethyl *O*-(3,4-*O*-isopropylidene-β-D-galactopyanosyl)-(1→4)-β-D-gluco-

[24] K. Jansson, S. Ahlfors, T. Frejd, J. Kihlberg, G. Magnusson, J. Dahmen, G. Noori, and K. Stenvall, *J. Org. Chem.* **53,** 5629 (1988).
[25] R. R. Schmidt and P. Zimmerman, *Angew. Chem., Int. Ed. Engl.* **25,** 725 (1986).
[26] M. Kiso, A. Nakamura, Y. Tomita, and A. Hasegawa, *Carbohydr. Res.* **158,** 101 (1986).
[27] Y. Itoh, M. Kiso, and A. Hasegawa, *J. Carbohydr. Chem.* **8,** 285 (1989).
[28] O. Kanie, M. Kiso, and A. Hasegawa, *J. Carbohydr. Chem.* **7,** 501 (1988).
[29] N. Baggett and B. J. Marsden, *Carbohydr. Res.* **110,** 11 (1982).
[30] A. Marra and P. Sinaÿ, *Carbohydr. Res.* **187,** 35 (1989).

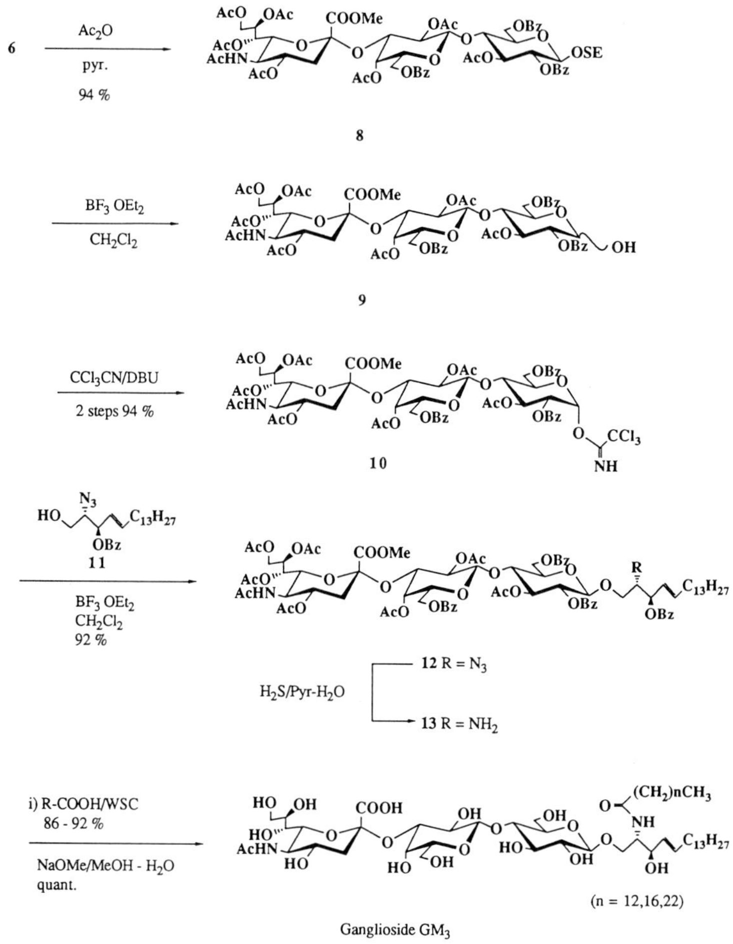

SCHEME 2. Synthesis of ganglioside G_{M3}.

pyranoside[31] (**15**) gives the corresponding 2,6,6′-tri-*O*-benzoyl derivative as a major product, which is then converted, either by hydrogenolytic removal of the benzyl group or by hydrolytic cleavage of the isopropylidene group, to the title compound **4** as needles, mp 106°–108°, $[\alpha]_D$ + 11° (*c* 0.6, chloroform).

[31] A. Hasegawa, K. Fushimi, H. Ishida, and M. Kiso, *J. Carbohydr. Chem.* **12,** 1203 (1993).

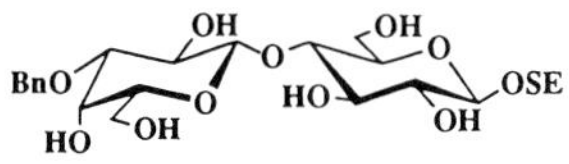

14

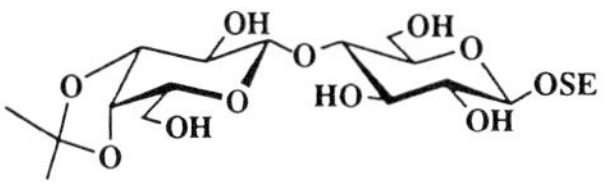

15

Glycosylation Promoters

The DMTST[32] ($CF_3SO_3^- \cdot Me_2S^+SMe$) reagent is prepared by treatment of methyl trifluoromethanesulfonate (CF_3SO_3Me) and a small excess of dimethyl disulfide (MeSSMe) in dichloromethane for 48 hr at 20°. Anhydrous ether is added, and the resulting crystals of DMTST are filtered with powdered MS-3Å. The DMTST content is adjusted to 60% by weight with MS-3Å, and the mixture is used for glycosylation. Both *N*-iodosuccinimide (NIS) and trifluoromethanesulfonic acid (TfOH) are commercially available.

Synthesis of α-Sialyl-(2→3)-Lactose Derivative **6**

DMTST Method.[15–17] The DMTST-promoted glycosylation of **4** (2.6 mmol) with the methyl 2-thioglycoside of Neu5Ac (**2α,β**, R = Me; 5.2 mmol) is achieved in acetonitrile (20 ml) at −15° in the presence of DMTST (15 mmol; see glycosylation promoters above) and additional MS-3Å (3 g) as described,[13,16,17] the reaction being monitored by thin-layer chromatography (TLC; silica gel 60 F_{254}, 15 : 1 to 25 : 1 CH_2Cl_2–methanol). The desired α-glycoside **6** (50% average yield) is isolated, on a column of silica gel with 4 : 1 (v/v) ethyl acetate–hexane, as an amorphous mass, $[\alpha]_D$ + 11° (*c* 1.74, chloroform).

NIS–TfOH Method. To a stirred mixture of **4** (1.9 mmol), the methyl or phenyl 2-thioglycoside of Neu5Ac (**2α,β**, R = Me or Ph; 3.2 mmol), and MS-3Å (3 g) in 10 : 1 acetonitrile–dichloromethane (15 ml) is added, at −40°, powdered NIS (6.4 mmol) and TfOH (0.64 mmol).[17,18] The mixture is stirred overnight at −40° and then neutralized by triethylamine. The precipitate is filtered off and washed with dichloromethane. The filtrate and washings are combined, successively washed with 5% $Na_2S_2O_3$, saturated $NaHCO_3$, and water, dried (Na_2SO_4), and concentrated. The residue is chromatographed on a column of silica gel to give **6** in 60–70% yield. In both methods, the 2,3-dehydro derivative of Neu5Ac (**7**) is formed as a major by-product, but neither the β-glycoside nor any positional isomers

[32] M. Ravenscroft, R. M. G. Roberts, and J. G. Tillett, *J. Chem. Soc. Perkin Trans. 2*, 1569 (1982).

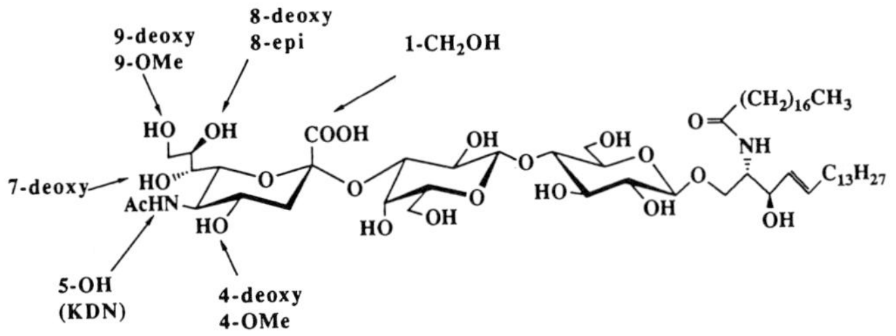

FIG. 2. Ganglioside G_{M3} analogs containing modified sialic acids.

are isolated. An alternative approach using phosphites and phosphates of Neu5Ac as glycosyl donors has also been reported.[33]

*Synthesis of Ganglioside G_{M3} from **6***

The conversion of **6** into ganglioside G_{M3} is carried out[13] by acetylation and removal of the SE group (**6** → **9**, 94%), imidate formation (94%) and coupling with azidosphingosine derivative **11** (92%), reduction of the azide group and N-acylation (86–92%), and final deprotection reactions (quantitative), yielding a series of natural ganglioside G_{M3} molecules containing three kinds of *N*-fatty acyl groups (n = 12, 16, and 22) (Scheme 2; for experimental details, see Ref. 13).

Synthesis of Ganglioside G_{M3} Analogs Containing Modified Sialic Acids

Various G_{M3} analogs containing modified sialic acids (Fig. 2) have been synthesized[34–37] by employing the same procedure just described for the natural G_{M3}. Chemical modifications of Neu5Ac are performed starting from methyl [2-(trimethylsilyl)ethyl 5-acetamido-3,5-dideoxy-D-*glycero*-α-D-*galacto*-2-nonulopyranosid]onate[38] (**16**) or methyl (methyl 5-acetamido-3,5-dideoxy-2-thio-D-*glycero*-α-D-*galacto*-2-nonulopyranosid) onate[39] (**17**), and the corresponding methyl 2-thioglycoside derivatives of

[33] T. J. Martin, R. Brescello, A. Toepfer, and R. R. Schmidt, *Glycoconjugate J.* **10,** 16 (1993).
[34] A. Hasegawa, T. Murase, K. Adachi, M. Morita, H. Ishida, and M. Kiso, *J. Carbohydr. Chem.* **9,** 181 (1990).
[35] A. Hasegawa, K. Adachi, M. Yoshida, and M. Kiso, *J. Carbohydr. Chem.* **11,** 95 (1992).
[36] A. Hasegawa, K. Adachi, M. Yoshida, and M. Kiso, *Carbohydr. Res.* **230,** 273 (1992).
[37] A. Hasegawa, K. Adachi, M. Yoshida, and M. Kiso, *Biosci. Biotech. Biochem.* **56,** 445 (1992).
[38] A. Hasegawa, Y. Ito, H. Ishida, and M. Kiso, *J. Carbohydr. Chem.* **8,** 125 (1989).
[39] A. Hasegawa, T. Murase, M. Ogawa, H. Ishida, and M. Kiso, *J. Carbohydr. Chem.* **9,** 415 (1990).

modified Neu5Ac (C_7- and C_8-Neu5Ac,[34] 4-*O*- or 9-*O*-methyl-Neu5Ac,[35] 8-epi-Neu5Ac,[35] and a series of deoxy-Neu5Ac[40] derivatives) are used as the glycosyl donors for **4**. The neutral G_{M3} analog containing 1-hydroxymethyl Neu5Ac is synthesized[37] by a mild reduction of the methoxycarbonyl group in **6** with sodium borohydride. De-*N*-acetyl-G_{M3} and some analogs are synthesized[41] by a similar approach using methyl [methyl 4,7,8,9-tetra-*O*-acetyl-5-(*tert*-butoxycarbonylamino)-2-nonulopyranosid]onate (**18**).

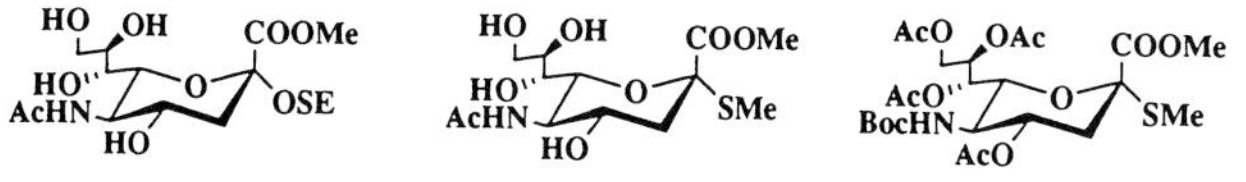

16 **17** **18**

The KDN-ganglioside G_{M3} derivative,[42] which contains 3-deoxy-D-*glycero*-α-D-*galacto*-2-nonulopyranosylonic acid (KDN) in place of Neu5Ac, was first isolated[43] from rainbow trout sperm. The KDN glycosyl donor **19** is prepared by treatment of methyl 3-deoxy-2,4,5,7,8,9-hexa-*O*-acetyl-D-*glycero*-D-*galacto*-2-nonulopyranosonate[44] with thiophenol in 92% yield as described for Neu5Ac. Coupling of **19** and **4** is achieved[42] in acetonitrile in the presence of NIS and a catalytic amount of trimethylsilyl trifluoromethanesulfonate (TMS·OTf), and the resulting trisaccharide **20** (46% isolated yield) is converted[42] to KDN-ganglioside G_{M3} as described for Neu5Ac ganglioside G_{M3}.

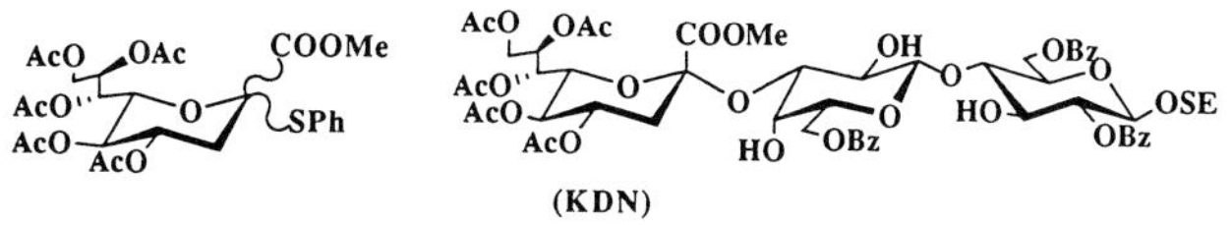

19 **20**

Synthesis of Ganglioside G_{M3} Analogs Containing Modified Ceramides

A series of G_{M3} analogs containing modified ceramides (Fig. 3) are synthesized[45] either by N-acylation of the spingosine derivative[13] **13** (analogs A) or by glycosylation of 3-benzyloxycarbonylamino-1-propanol,

[40] A. Hasegawa, K. Adachi, M. Yoshida, and M. Kiso, *Carbohydr. Res.* **230,** 257 (1992).
[41] S. Fujita, M. Numata, M. Sugimoto, K. Tomita, and T. Ogawa, *Carbohydr. Res.* **228,** 347 (1992).
[42] T. Terada, M. Kiso, and A. Hasegawa, *J. Carbohydr. Chem.* **12,** 425 (1993).
[43] S. Yu, K. Kitajima, S. Inoue, and Y. Inoue, *J. Biol. Chem.* **266,** 21929 (1991).
[44] R. Shirai and H. Ogura, *Tetrahedron Lett.* **30,** 2263 (1989).
[45] A. Hasegawa, T. Murase, M. Morita, H. Ishida, and M. Kiso, *J. Carbohydr. Chem.* **9,** 201 (1990).

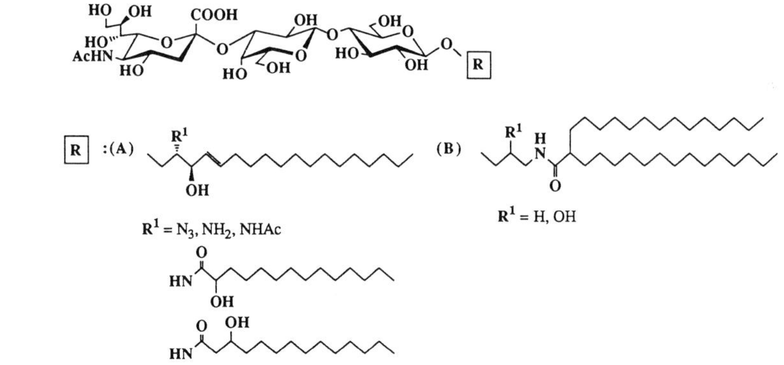

FIG. 3. Ganglioside G_{M3} analogs containing modified ceramides.

and (2*RS*)-3-benzyloxycarbonylamino-2-*O*-benzoyl-1,2-propanediol with trisaccharide imidate[13] **10** (analog B).

Synthesis of α-Neu5Ac-(2→6)-Isomer of Ganglioside G_{M3} (**22**)

The trisaccharide derivative[46] **21**, obtained by glycosylation of 2-(trimethylsilyl)ethyl *O*-(2-*O*-acetyl-3-*O*-benzyl-β-D-galactopyranosyl)-(1→4)-2,3,6-tri-*O*-acetyl-β-D-glucopyranoside[47] with the methyl α-2-thioglycoside of Neu5Ac (**2α**, R = Me), is converted[48] stepwise, by hydrogenolysis and O-acetylation (93%), removal of the SE group (96%), imidate formation and coupling with **11** (71%), reduction of the azide group and N-acylation (95%), and de-O-acylation and saponification of the methyl ester (quantitative), to the desired G_{M3} positional isomer **22**.

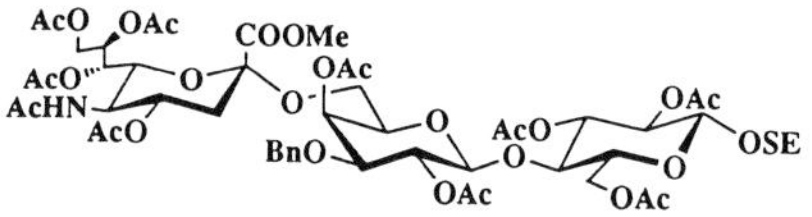

21

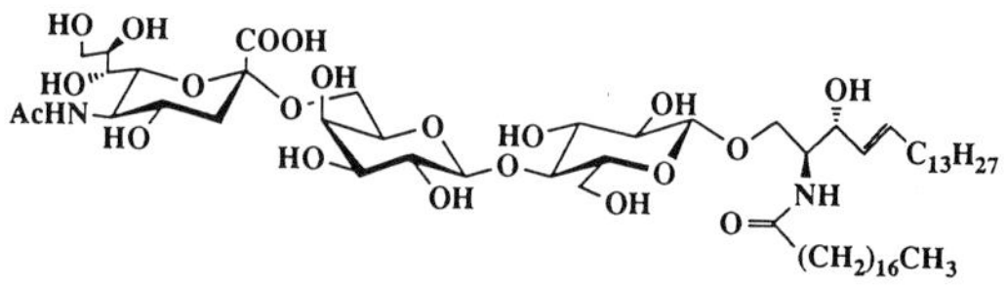

22

[46] A. Hasegawa, M. Ogawa, H. Ishida, and M. Kiso, *J. Carbohydr. Chem.* **9,** 393 (1990).
[47] K. P. R. Kartha, A. Kameyama, M. Kiso, and A. Hasegawa, *J. Carbohydr. Chem.* **8,** 145 (1989).
[48] A. Hasegawa, M. Ogawa, and M. Kiso, *Biosci. Biotech. Biochem.* **56,** 536 (1992).

Chemoenzymatic Synthesis of Ganglioside G_{M3}

The combined chemical enzymatic approach has been applied to the synthesis of natural ganglioside G_{M3}.[49,50] The key synthetic step is the regio- and α-stereoselective glycosylation of sialic acid to 3′-OH of the lactose moiety by use of CMP-Neu5Ac and α-2,3-sialyltransferase. However, this method is usually restricted to the synthesis of natural gangliosides, and in addition both the sugar nucleotide and glycosyltransferase are still not readily available.

[49] Y. Ito and J. C. Paulson, *J. Am. Chem. Soc.* **115,** 1603 (1993).
[50] K. K.-C. Liu and S. J. Danishefsky, *J. Am. Chem. Soc.* **115,** 4933 (1993).

[17] Synthesis of Ganglioside Analogs Containing Sulfur in Place of Oxygen at the Linkage Positions

By HIDEHARU ISHIDA, MAKOTO KISO, and AKIRA HASEGAWA

Introduction

Chemical synthesis of naturally occurring gangliosides as well as a variety of ganglioside analogs is becoming stimulating and rewarding as more and more biological functions[1–6] of sialoglycoconjugates are reported. Among these analogs, thioglycosides are of interest as inhibitors of the corresponding glycosides (mainly via competitive inhibition).[7,8] For instance, ganglioside analogs containing the α-thioglycoside of sialic acid are expected to affect the activity of sialidases, and analogs containing thioglycosidically linked ceramide could be an inhibitor of ceramide glycanase.[9] In this chapter we describe the established methods for the synthesis

[1] G. Walz, A. Aruffo, W. Kolanus, M. Bevilacqua, and B. Seed, *Science* **250,** 1132 (1990).
[2] M. L. Phillips, E. Nudelman, F. C. A. Graeta, M. Perez, A. K. Singhal, S. Hakomori, and J. C. Paulson, *Science* **250,** 1130 (1990).
[3] J. B. Lowe, L. M. Stoolman, R. P. Nair, R. D. Larsen, T. L. Berhend, and R. M. Marks, *Cell (Cambridge, Mass.)* **63,** 475 (1990).
[4] M. J. Polley, M. L. Phillips, E. Wayner, E. Nudelman, A. K. Singhal, S. Hakomori, and J. C. Paulson, *Proc. Natl. Acad. Sci. U.S.A.* **88,** 6224 (1991).
[5] H. Nojiri, M. Stroud, and S. Hakomori, *J. Biol. Chem.* **266,** 4531 (1991).
[6] I. Eggens, B. Fenderson, T. Toyokuni, B. Dean, M. Stroud, and S. Hakomori, *J. Biol. Chem.* **264,** 9476, (1989).
[7] D. Horton and J. D. Wander, *in* "The Carbohydrates, Chemistry/Biochemistry" (W. Pigman and D. Horton, eds.), 2nd Ed., Vol. 1B, p. 803. Academic Press, New York, 1980.
[8] P. J. Deschavanne, O. M. Viratelle, and J. M. Yon, *J. Biol. Chem.* **253,** 833 (1978).
[9] M. Ito and T. Yamagata, this series, Vol. 179, p. 488.

Copyright © 1994 by Academic Press, Inc.
All rights of reproduction in any form reserved.

of ganglioside analogs containing thioglycosidic linkages in the following positions: (1) between the sialic acid and hexopyranoside (compounds **A–D**); (2) between the oligosaccharide chain and ceramide (compounds **E–H**); (3) between both the sialic acid–hexopyranoside and oligosaccharide chain–ceramide (compounds **I–K**); and (4) between sialic acid and sialic acid (compound **L**).

Synthesis of *S*-(α-Sialyl)-(2→6)-β-D-hexopyranosyl and *S*-(α-Sialyl)-(2→6′)-β-D-lactosyl Ceramides (**A–D**)

In this section, the synthesis of *S*-(α-sialyl)-(2→6)-β-D-lactosyl ceramide (**D**), shown in Fig. 1, is described as an example of the synthesis of the title compounds. Other compounds (**A–C**) are synthesized essentially by the same method described for **D**.[10,11]

For the synthesis of **D**, we employ the sodium salt of methyl 5-acetamido-4,7,8,9-tetra-*O*-acetyl-3,5-dideoxy-2-thio-D-*glycero*-α-D-*galacto*-2-nonulopyranosonate (**3**)[12] as glycosyl donor and per-O-acetylated 2-(trimethylsilyl)ethyl 6′-bromo-6′-deoxy-β-D-lactoside (**9**)[10] as glycosyl acceptor, respectively. The intermediates are coupled and converted, by introduction of ceramide moiety, to the end product.

Treatment of methyl 5-acetamido-4,7,8,9-tetra-*O*-acetyl-2-chloro-2,3,5-trideoxy-D-*glycero*-β-D-*galacto*-2-nonulopyranosonate (**1**)[13] with potassium thioacetate in dry dichloromethane gives α-2-*S*-acetyl derivative **2** in 90% yield. Selective S-deacetylation of **2** with the amount of sodium methoxide equivalent to **2** in dry methanol at −40° gives the sodium salt **3**, which is used for the next reaction without further purification. Treatment of 2-(trimethylsilyl)ethyl β-D-lactoside **4**[14] with 2-methoxypropene in *N,N*-dimethylformamide (DMF) in the presence of *p*-toluenesulfonic acid monohydrate and subsequent acetylation give **6** in 93% yield. The de-O-isopropylidenation of **6** by heating with 80% aqueous acetic acid for 10 hr at 45° gives crystalline **7** in 91% yield. Selective C-6′ bromination of **7** with *N*-bromosuccinimide in the presence of triphenylphosphine and subsequent acetylation afford compound **9** as crystals in good yield. Condensation of **3** with **9** in DMF under a nitrogen atmosphere leads to the

[10] A. Hasegawa, M. Morita, Y. Ito, H. Ishida, and M. Kiso, *J. Carbohydr. Chem.* **9,** 369 (1990).
[11] A. Hasegawa, T. Terada, H. Ogawa, and M. Kiso, *J. Carbohydr. Chem.* **11,** 319 (1992).
[12] A. Hasegawa, J. Nakamura, and M. Kiso, *J. Carbohydr. Chem.* **5,** 11 (1986).
[13] R. Kuhn, P. Lutz, and D. L. MacDonald, *Chem. Ber.* **99,** 611 (1966).
[14] K. P. R. Kartha, A. Kameyama, M. Kiso, and A. Hasegawa, *J. Carbohydr. Chem.* **8,** 145 (1989).

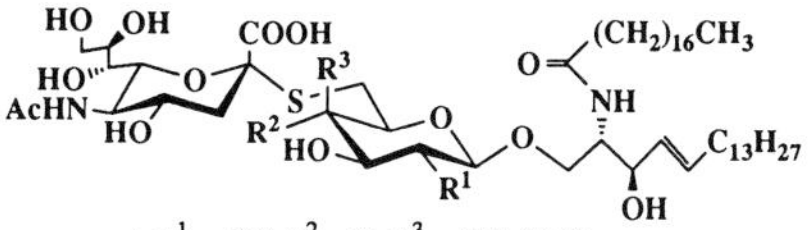

A $R^1 = OH, R^2 = H, R^3 = OH$ (Gal)
B $R^1 = OH, R^2 = OH, R^3 = H$ (Glc)
C $R^1 = NHAc, R^2 = OH, R^3 = H$ (GlcNAc)

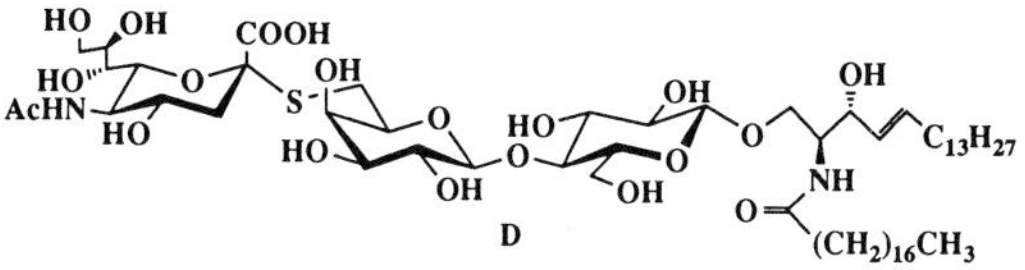

D

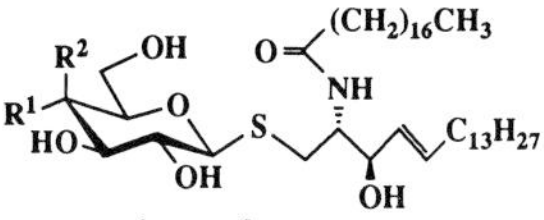

E $R^1 = H, R^2 = OH$ (Gal)
F $R^1 = OH, R^2 = H$ (Glc)

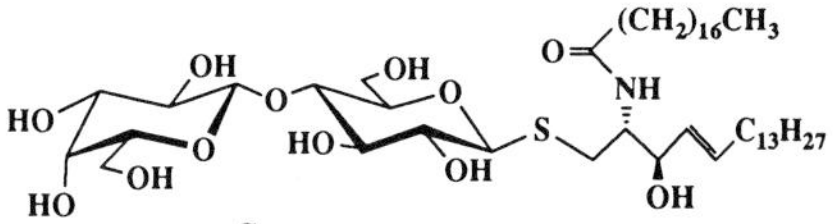

G

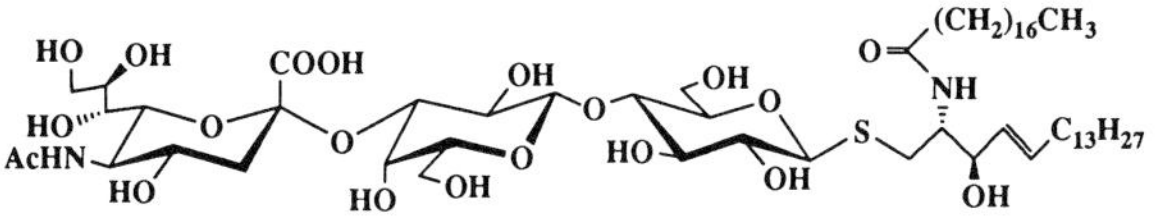

H

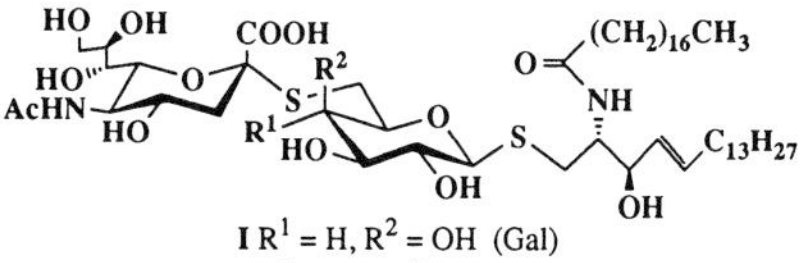

I $R^1 = H, R^2 = OH$ (Gal)
J $R^1 = OH, R^2 = H$ (Glc)

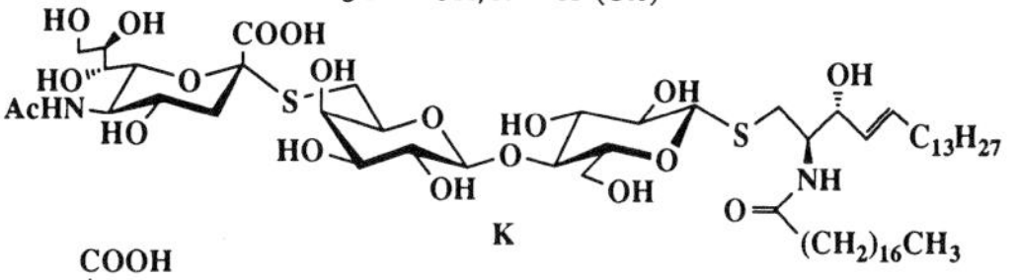

K

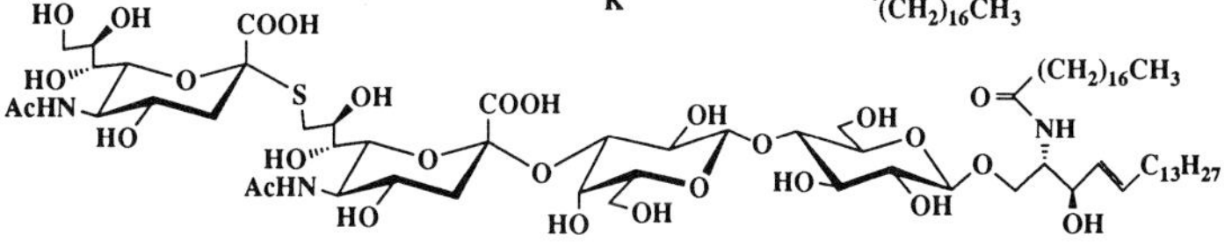

L

FIG. 1. Ganglioside analogs containing thioglycosides.

corresponding α-thioglycoside **10** in 87% yield. Selective removal[15,16] of the 2-(trimethylsilyl)ethyl group in **10** is performed by treatment of **10** with boron trifluoride etherate in dichloromethane for 2 hr at −20°, to give **11** in 84% yield. When treated with trichloroacetonitrile[15,17,18] in the presence of 1,8-diazabicyclo[5.4.0]undec-7-ene (DBU) for 2 hr at 0°, compound **11** gives the trichloroacetimidate **12** as the α-anomer in 92% yield, after column chromatography.

The glycosylation of (2*S*,3*R*,4*E*)-2-azido-3-*O*-benzoyl-4-octadecene-1,3-diol (**13**)[19,20] with **12** in the presence of boron trifluoride etherate for 6 hr at −20° yields only the expected β-glycoside **14** in 82% yield. Selective reduction[15,21] of the azide group in **14** with hydrogen sulfide in 5 : 1 (v/v) pyridine–water gives the amine **15**, which on condensation with octadecanoic acid using 1-(3-dimethylaminopropyl)-3-ethylcarbodiimide hydrochloride (DAC) in dichloromethane gives the *S*-(α-*N*-acetylneuraminyl)-(2→6′)-*O*-(6′-thio-β-D-lactosyl)-(1→1)-ceramide derivative **16** in 94% yield. Finally, de-O-acetylation of **16** with sodium methoxide in methanol, followed by saponification of the methyl ester group, yields almost quantitatively the end product **D**.

Compounds **A**, **B**, and **D** show potent inhibition against sialidases from several kinds of influenza virus, acting as competitive inhibitors. The order of decreasing inhibition is **D** > **A** > **B**.[22]

Detailed Procedures

Methyl 5-Acetamido-4,7,8,9-tetra-O-acetyl-2-S-acetyl-3,5-dideoxy-2-thio-D-glycero-α-D-galacto-2-nonulopyranosonate (2). To a solution of **1**[12] (3.3 g) in dry dichloromethane (30 ml) is added potassium thioacetate (3.5 g) (Scheme 1). The mixture is stirred overnight at room temperature and concentrated, the residue is extracted with chloroform, and the extract is washed with water, dried (Na_2SO_4), and evaporated to a syrup. The product is chromatographed on a column of silica gel (200 g) with 100 : 1 (v/v) chloroform–methanol, to give 3.2 g (90%) of compound **2** as an amorphous mass; mp 74°–76°, $[\alpha]_D$ −15.6° (*c* 0.75, chloroform).

*2-(Trimethylsilyl)ethyl O-(2,3-Di-O-acetyl-4,6-O-isopropylidene-β-D-galactopyranosyl)-(1→4)-2,3,6-tri-O-acetyl-β-D-glucopyranoside (**6**).* To a

[15] T. Murase, H. Ishida, M. Kiso, and A. Hasegawa, *Carbohydr. Res.* **188,** 71 (1989).
[16] K. Jansson, T. Frejd, J. Kihlberg, and G. Magnussun, *Tetrahedron Lett.* **27,** 753 (1986).
[17] M. Numata, M. Sugimoto, K. Koike, and T. Ogawa, *Carbohydr. Res.* **163,** 209 (1987).
[18] R. R. Schmidt and J. Michel, *Angew. Chem., Int. Ed. Engl.* **19,** 731 (1980).
[19] M. Kiso, A. Nakamura, Y. Tomita, and A. Hasegawa, *Carbohydr. Res.* **158,** 101 (1986).
[20] R. R. Schmidt and P. Zimmermann, *Angew. Chem., Int. Ed. Engl.* **25,** 726 (1986).
[21] T. Adachi, Y. Yamada, I. Inoue, and M. Saneyoshi, *Synthesis*, 45 (1977).
[22] Y. Suzuki, K. Sato, M. Kiso, and A. Hasegawa, *Glycoconjugate J.* **7,** 349 (1990).

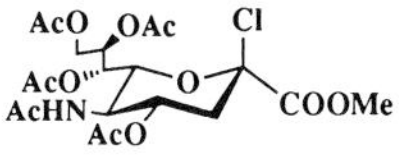

1

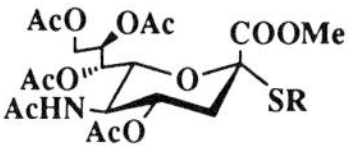

2 R = Ac
3 R = Na

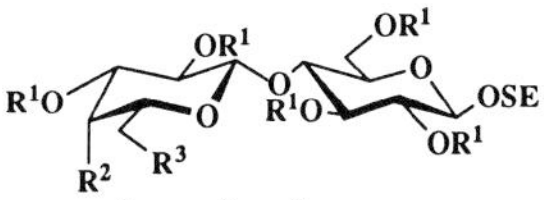

4 R^1 = H, R^2 = R^3 = OH
5 R^1 = H, R^2,R^3 = $OCMe_2O$
6 R^1 = Ac, R^2,R^3 = $OCMe_2O$
7 R^1 = Ac, R^2 = R^3 = OH
8 R^1 = Ac, R^2 = OH, R^3 = Br
9 R^1 = Ac, R^2 = OAc, R^3 = Br

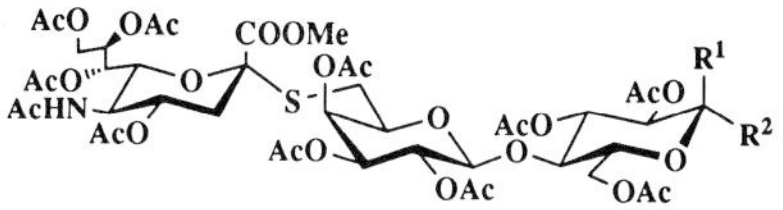

10 R^1 = OSE, R^2 = H
11 R^1, R^2 = H, OH
12 R^1 = H, R^2 = $OC(=NH)CCl_3$

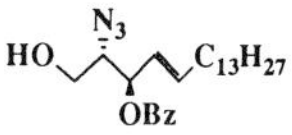

13

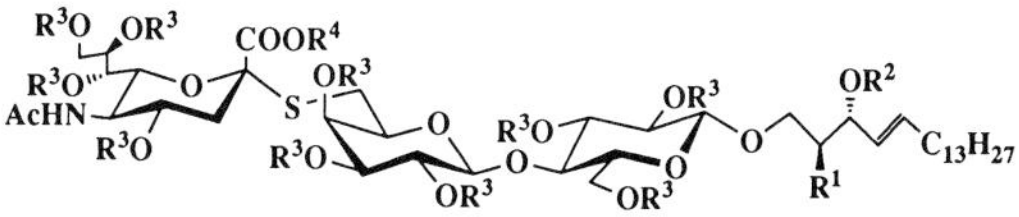

14 R^1 = N_3, R^2 = Bz, R^3 = Ac, R^4 = Me
15 R^1 = NH_2, R^2 = Bz, R^3 = Ac, R^4 = Me
16 R^1 = $NHCO(CH_2)_{16}Me$, R^2 = Bz, R^3 = Ac, R^4 = Me
D R^1 = $NHCO(CH_2)_{16}Me$, R^2 = R^3 = R^4 = H

SE = 2-(trimethylsilyl)ethyl, Bz = benzoyl

SCHEME 1. Synthesis of ganglioside analogs containing the α-thioglycoside of sialic acid.

solution of **4**[10] (1.5 g, 3.39 mmol) in DMF (15 ml), cooled to 0°, are added, with stirring, 2-methoxypropene (0.6 ml) and *p*-toluenesulfonic acid monohydrate (30 mg), and the mixture is stirred for 2 hr at 0°, the progress of the reaction being monitored by thin-layer chromatography (TLC). Acetic anhydride (7 ml) and pyridine (10 ml) are added to the mixture, and this is stirred for 5 hr at room temperature and concentrated to a syrup, which is chromatographed on a column of silica gel (150 g) with 1 : 4 (v/v) ethyl acetate–hexane to give **6** (2.14 g, 93%) as an amorphous mass, $[\alpha]_D$ +14.0° (*c* 1.0, chloroform).

*2-(Trimethylsilyl)ethyl O-(2,3-Di-O-acetyl-β-D-galactopyranosyl)-(1→4)-2,3,6-tri-O-acetyl-β-D-glucopyranoside (**7**).* A solution of **6** (2.34 g, 3.38 mmol) in 80% aqueous acetic acid (30 ml) is kept for 10 hr at 45°, then concentrated to a syrup, which is chromatographed on a column of silica gel (200 g) with 1 : 1 (v/v) ethyl acetate–hexane to give **7** (2.0 g, 91%) as crystals. Recrystallization from ether–hexane gives needles: mp 197°, $[\alpha]_D$ −8.1° (*c* 0.6, chloroform).

*2-(Trimethylsilyl)ethyl O-(2,3-Di-O-acetyl-6-bromo-6-deoxy-β-D-galactopyranosyl)-(1→4)-2,3,6-tri-O-acetyl-β-D-glucopyranoside (**8**).* To a solution of **7** (500 mg, 0.77 mmol) in DMF (10 ml), cooled to 0°, are added, with stirring, *N*-bromosuccinimide (267 mg, 1.5 mmol) and triphenylphosphine (262 mg, 1 mmol), and the mixture is stirred for 2 days at 0°. Methanol (1 ml) is added to the mixture, and this is stirred for 10 min, then concentrated. The residue is chromatographed on a column of silica gel (60 g) with 2 : 1 (v/v) ethyl acetate–hexane to give **8** (380 mg, 69%) as an amorphous mass, $[\alpha]_D$ −2.8° (*c* 0.86, chloroform).

*2-(Trimethylsilyl)ethyl O-(2,3,4-Tri-O-acetyl-6-bromo-6-deoxy-β-D-galactopyranosyl)-(1→4)-2,3,6-tri-O-acetyl-β-D-glucopyranoside (**9**).* Acetylation of **8** (329 mg, 0.46 mmol) with acetic anhydride (1 ml) in pyridine (2 ml) overnight at room temperature gives **9** (345 mg, quantitative) as crystals. Recrystallization from ether–hexane gives needles: mp 208°–209°, $[\alpha]_D$ −11.9° (*c* 1.0, chloroform).

*2-(Trimethylsilyl)ethyl S-(Methyl 5-Acetamido-4,7,8,9-tetra-O-acetyl-3,5-dideoxy-D-glycero-α-D-galacto-2-nonulopyranosylonate)-(2→6)-O-(2,3,4-tri-O-acetyl-6-thio-β-D-galactopyranosyl)-(1→4)-2,3,6-tri-O-acetyl-β-D-glucopyranoside (**10**).* A solution of **9** (280 mg, 0.37 mmol) and **3** (400 mg, 0.76 mmol) in dry DMF (3 ml) is stirred overnight at room temperature under a nitrogen atmosphere, the course of the reaction being monitored by TLC. Acetic anhydride (2 ml) and pyridine (4 ml) are added to the solution, and the mixture is stirred overnight at room temperature. Dichloromethane (200 ml) is added, and the solution is successively washed with 1 *M* sodium carbonate, 2 *M* hydrochloric acid, and water, dried (Na_2SO_4), and concentrated. The residue is chromatographed on a column of silica gel (100 g) with 1 : 1 (v/v) ethyl acetate–hexane to give **10** (380 mg, 86.5.%) as an amorphous mass, $[\alpha]_D$ +1.5° (*c* 0.67, chloroform).

*S-(Methyl 5-Acetamido-4,7,8,9-tetra-O-acetyl-3,5-dideoxy-D-glycero-α-D-galacto-2-nonulopyranosylonate)-(2→6)-O-(2,3,4-tri-O-acetyl-6-thio-β-D-galactopyranosyl)-(1→4)-2,3,6-tri-O-acetyl-D-glucopyranose (**11**).* To a solution of **10** (904 mg, 0.76 mmol) in dry dichloromethane (10 ml), cooled to −20°, is added boron trifluoride etherate (1.2 ml), and the mixture is stirred for 2 hr at −20°, the progress of the reaction is monitored by TLC. Dichloromethane (100 ml) is added to the mixture, and the solution is successively washed with 1 *M* sodium carbonate and water, dried (Na_2SO_4), and concentrated to a syrup, which is chromatographed on a column of silica gel (250 g) with 100 : 1 (v/v) dichloromethane–methanol to give **11** (695 mg, 84%) as an amorphous mass, $[\alpha]_D$ +2.1° (*c* 1.9, chloroform).

*S-(Methyl 5-Acetamido-4,7,8,9-tetra-O-acetyl-3,5-dideoxy-D-glycero-α-D-galacto-2-nonulopyranosylonate)-(2→6)-O-(2,3,4-tri-O-acetyl-6-thio-β-D-galactopyranosyl)-(1→4)-2,3,6-tri-O-acetyl-D-glucopyranosyl Trichloroacetimidate (**12**).* To a stirred solution of **11** (403 mg, 0.37 mmol) in dry dichloromethane (5 ml), cooled to 0°, are added trichloroacetonitrile (0.075 ml) and 1,8-diazabicyclo[5.4.0]undec-7-ene (DBU, 0.03 ml). The mixture is stirred for 2 hr at 0° and then concentrated. The residue is chromatographed on a column of silica gel (20 g) with 120 : 1 (v/v) dichloromethane–methanol, to give **12** (420 mg, 92%) as an amorphous mass, $[\alpha]_D$ +52.4° (*c* 0.9, chloroform).

*S-(Methyl 5-Acetamido-4,7,8,9-tetra-O-acetyl-3,5-dideoxy-D-glycero-α-D-galacto-2-nonulopyranosylonate)-(2→6)-O-(2,3,4-tri-O-acetyl-6-thio-β-D-galactopyranosyl)-(1→4)-O-(2,3,6-tri-O-acetyl-β-D-glucopyranosyl)-(1→1)-(2S,3R,4E)-2-azido-3-O-benzoyl-4-octadecene-1,3-diol (**14**).* To a solution of **12** (453 mg, 0.37 mmol) and **13** (320 mg, 0.74 mmol) in dry dichloromethane (15 ml) is added Molecular Sieves 4Å powder (MS-4Å; 600 mg), and the mixture is stirred for 1 hr at room temperature, then cooled to −20°. Boron trifluoride etherate (0.06 ml) is added to the cooled mixture, and this is stirred for 6 hr at −20°, the progress of the reaction being monitored by TLC. The precipitate is filtered off and washed with chloroform. The filtrate and washings are combined, and the solution is successively washed with 1 *M* sodium carbonate and water, dried (Na_2SO_4), and concentrated. The residue is chromatographed on a column of silica gel (60 g) with 120 : 1 (v/v) dichloromethane–methanol to give **14** (452 mg, 82%) as an amorphous mass, $[\alpha]_D$ +0.9° (*c* 0.56, chloroform).

*S-(Methyl 5-Acetamido-4,7,8,9-tetra-O-acetyl-3,5-dideoxy-D-glycero-α-D-galacto-2-nonulopyranosylonate)-(2→6)-O-(2,3,4-tri-O-acetyl-6-thio-β-D-galactopyranosyl)-(1→4)-O-(2,3,6-tri-O-acetyl-β-D-glucopyranosyl)-(1→1)-(2S,3R,4E)-3-O-benzoyl-2-octadecanamido-4-octadecene-1,3-diol (**16**).* Hydrogen sulfide is bubbled through a solution of **14** (100 mg, 0.069 mmol) in pyridine (2.5 ml) and water (0.5 ml) for 2 days while the solution is stirred at room temperature. The mixture is concentrated to give the syrupy amine **15**, which is used for the next reaction without further purification. To a solution of **15** in dry dichloromethane (2.5 ml) are added octadecanoic acid (40 mg, 0.14 mmol) and DAC (35 mg, 0.19 mmol), and the mixture is stirred overnight at room temperature. After completion of the reaction, dichloromethane (15 ml) is added to the mixture, and the solution is washed with water, dried (Na_2SO_4), and concentrated to a syrup that is chromatographed on a column of silica gel (5 g) with 90 : 1 (v/v) dichloromethane–methanol, to give **16** (106 mg, 94%) as an amorphous mass, $[\alpha]_D$ +8.0° (*c* 0.4, chloroform).

*S-(5-Acetamido-3,5-dideoxy-D-glycero-α-D-galacto-2-nonulopyranosylonic acid)-(2→6)-O-(6-thio-β-D-galactopyranosyl)-(1→4)-O-(β-D-glucopyranosyl)-(1→1)-(2S,3R,4E)-2-octadecanamido-4-octadecene-1,3-diol (**D**).* To a solution of **16** (121 mg, 0.072 mmol) in dry methanol (5 ml) is added sodium methoxide (25 mg), and the mixture is stirred overnight at room temperature. After completion of the reaction, water (0.2 ml) is added to the mixture at 0°, and this is stirred for 1 hr, then treated with Amberlite IR-120 (H^+) resin to remove the base. The solution is concentrated, and the residue is chromatographed on a column of Sephadex LH-20 (100 g) with 1 : 1 (v/v) chloroform–methanol to give compound **D** (79.6 mg, 97%) as an amorphous mass, $[\alpha]_D$ +16.5° (*c* 1.5, chloroform–methanol).

Synthesis of Cerebroside, Lactosyl Ceramide, and Ganglioside G_{M3} Analogs Containing β-Thioglycosidically Linked Ceramide

In this section, we describe the synthesis of a ganglioside G_{M3} analog carrying β-thioglycosidically linked ceramide (**H**). Other thioglycolipids (**E–G**) were synthesized by the same procedure.[23] Compound **H** is prepared by the coupling of the sodium salt of *O*-(α-sialyl)-(2→3′)-β-D-1-thiolactose (**19**) and the tosylate derivative of azidosphingosine (**20**), and the coupled product is readily converted to the target compound **H**.

Treatment of per-O-acetylated *O*-(α-sialyl)-(2→3′)-α-D-lactosyl trichloroacetimidate (**17**)[15] with thioacetic acid in the presence of boron trifluoride etherate gives the required β-thioacetate derivative **18** in 91% yield. (2*S*,3*R*,4*E*)-2-Azido-3-*O*-benzoyl-1-*O*-(p-toylysulfonyl)-4-octadecene-1,3-diol (**20**) is derived by *p*-tolylsulfonylation of (2*S*, 3*R*, 4*E*)-2-azido-3-*O*-benzoyl-4-octadecene-1,3-diol (**13**),[19,20] which is used as the glycosyl acceptor. Treatment of sodium salt **19**, freshly prepared from **18** by selective S-deacetylation with sodium methoxide, with **20** in DMF under nitrogen affords the desired β-thioglycoside (**21**) in 41% yield. The ^1H nuclear magnetic resonance (NMR) data for **21** demonstrate it to be a fully blocked glycoside. The intermediate **21** is converted, by selective reduction of the

[23] A. Hasegawa, M. Morita, Y. Kojima, H. Ishida, and M. Kiso, *Carbohydr. Res.* **214,** 43 (1991).

azide group, condensation with octadecanoic acid, de-O-acylation, and saponification of the methyl ester group, to the end product **H** as described for the conversion of **14** to **D**.

Bär and Schmidt[24] synthesized β-lactosyl 1-thioceramide (**G**) in an alternative manner, namely, by the coupling of lactosyl bromide with the sodium salt of the 1-thio derivative of azidosphingosine.

Detailed Procedure

*O-(Methyl 5-Acetamido-4,7,8,9-tetra-O-acetyl-3,5-dideoxy-*D*-glycero-α-*D*-galacto-2-nonulopyranosylonate)-(2→3)-O-(2,4-di-O-acetyl-6-O-benzoyl-β-*D*-galactopyranosyl)-(1→4)-3-O-acetyl-1-S-acetyl-2,6-di-O-benzoyl-1-thio-β-*D*-glucopyranose (**18**).* To a solution of **17** (100 mg, 71.5 μmol) in dry dichloromethane (5 ml) are added thioacetic acid (10 μl) and boron trifluoride etherate (10 μl), and the mixture is stirred overnight at room temperature (Scheme 2). The mixture is successively washed with 1 *M* sodium carbonate and water, dried (Na_2SO_4), and concentrated to a syrup that is chromatographed on a column of silica gel (50 g) with 90 : 1 (v/v) dichloromethane–methanol to give **18** (85 mg, 91%) as an amorphous mass, $[\alpha]_D$ +0.5° (*c* 0.2, chloroform).

*O-(Methyl 5-Acetamido-4,7,8,9-tetra-O-acetyl-3,5-dideoxy-*D*-glycero-α-*D*-galacto-2-nonulopyranosylonate)-(2→3)-O-(2,4-di-O-acetyl-6-O-benzoyl-β-*D*-galactopyranosyl)-(1→4)-S-(3-O-acetyl-2,6-di-O-benzoyl-β-*D*-glucopyranosyl)-(1→1)-(2R,3R,4E)-2-azido-3-benzoyloxy-4-octadecene-1-thiol (**21**).* To a stirred solution of **18** (30 mg, 0.023 mmol) in dry methanol (0.5 ml) and chloroform (0.1 ml), cooled to −20°, is added a solution of sodium metal (5.0 mg) in dry methanol (0.2 ml). Stirring is continued for 5 min at −20°, and the mixture is concentrated to give **19** as an amorphous mass, which is used for the next reaction without purification. A solution of **19** and **20** (26 mg, 0.045 mmol) in dry DMF (2 ml) is stirred for 24 hr at 40° under a nitrogen atmosphere, the course of the reaction being monitored by TLC. Acetic anhydride (0.5 ml) and pyridine (1 ml) are added to the mixture, which is stirred overnight at room temperature, then concentrated. The residue is dissolved in dichloromethane (20 ml), and the solution is successively washed with 2 *M* hydrochloric acid, 1 *M* sodium carbonate, and water, and then dried (Na_2SO_4) and evaporated to a syrup that is chromatographed on a column of silica gel (10 g) with 6 : 1 (v/v) hexane–ethyl acetate to afford **21** (15.5 mg, 41%) as an amorphous mass, $[\alpha]_D$ −64.5° (*c* 0.73, chloroform).

[24] T. Bär and R. R. Schmidt, *Liebigs Ann. Chem.*, 185 (1991).

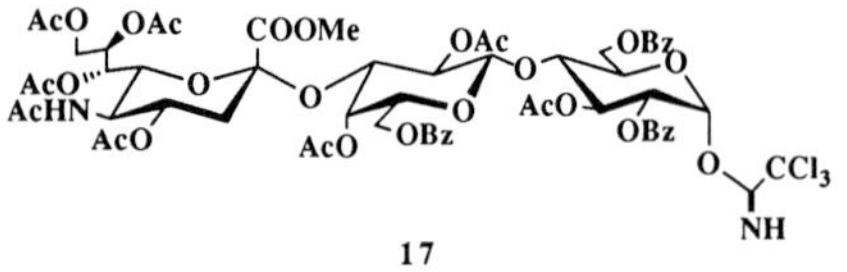

17

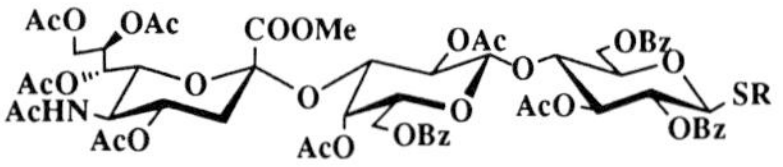

18 R = Ac
19 R = Na

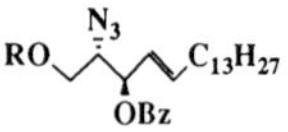

13 R = H
20 R = Ts

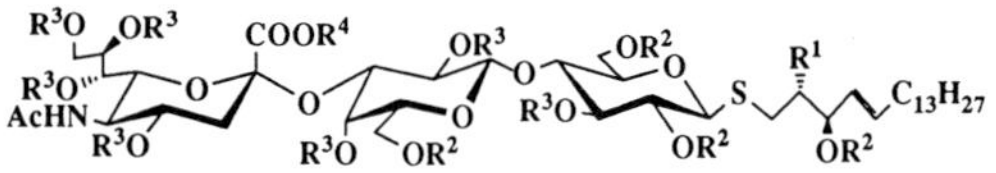

21 $R^1 = N_3$, $R^2 = Bz$, $R^3 = Ac$, $R^4 = Me$
22 $R^1 = NH_2$, $R^2 = Bz$, $R^3 = Ac$, $R^4 = Me$
23 $R^1 = NHCO(CH_2)_{16}Me$, $R^2 = Bz$, $R^3 = Ac$, $R^4 = Me$
H $R^1 = NHCO(CH_2)_{16}Me$, $R^2 = R^3 = R^4 = H$

Ts = *p*-tolylsulfonyl

SCHEME 2. Synthesis of ganglioside analogs containing thioglycosically linked ceramide.

Synthesis of *S*-(α-Sialyl)-(2→6)-β-hexopyranosyl and *S*-(α-Sialyl)-(2→6′)-β-lactosyl Ceramides Containing β-Thioglycosidically Linked Ceramides (**I–K**)

In this section, the synthesis of *S*-(α-sialyl)-(2→6′)-β-lactosyl 1-thioceramide (**K**) is described as an example of the synthesis of the title compounds (**I–K**), all of which are synthesized in the same manner.[25] For the synthesis of the target thio analog of ganglioside, we employ the sodium salt of the per-O-acetylated *S*-α-sialyl-(2→6′)-1,6′-dithio-β-D-lactose (**26**) as the glycosyl donor, for coupling with (2*S*,3*R*,4*E*)-2-azido-3-*O*-benzoyl-

[25] A. Hasegaw, H. Ogawa, and M. Kiso, *J. Carbohydr. Chem.* **10,** 1009 (1991).

1-*O*-(*p*-tolylsulfonyl)-4-*O*-octadecene-1,3-diol (**20**). Treatment of *S*-(methyl 5-acetamido-4,7,8,9-tetra-*O*-acetyl-3,5-dideoxy-D-*glycero*-α-D-*galacto*-2-nonulopyranosylonate)-(2→6)-*O*-(2,3,4-tri-*O*-acetyl-6-thio-β-D-galactopyranosyl)-(1→4)-2,3,6-tri-*O*-acetyl-D-glucopyranose (**11**) with methane sulfonyl chloride in dichloromethane in the presence of 2,4,6-collidine for 1 hr at −15° gives the 1-chloro derivative **24**, which is converted to the desired β-thioacetate **25** in 53% yield. Coupling of the sodium salt **26**, freshly derived from **25** by selective S-deacetylation with sodium methoxide at −30°, with the tosylate of azidosphingosine **20** in DMF under a nitrogen atmosphere overnight at 45° yields the expected β-glycoside **27** (38%) after column chromatography. Compound **27** is converted to the final compound **K** by the set reaction sequence, that is, selective reduction of the azide group, condensation with octadecanoic acid, and final deprotection, as described for the preparation of **D**.

Detailed Procedures

*S-(Methyl 5-Acetamido-4,7,8,9-tetra-O-acetyl-3,5-dideoxy-D-glycero-α-D-galacto-2-nonulopyranosylonate)-(2→6)-O-(2,3,4-tri-O-acetyl-6-thio-β-D-galactopyranosyl)-(1→4)-2,3,6-tri-O-acetyl-1-S-acetyl-1-thio-β-D-glucopyranose (**25**).* To a solution of **11** (500 mg, 0.46 mmol) in dichloromethane (6 ml) is added 2,4,6-collidine (0.6 ml), and the mixture is cooled to −15° (Scheme 3). Methane sulfonyl chloride (0.3 ml, 3.88 mmol) is added to the mixture, and the mixture is stirred for 20 min at −15°, then for 1 hr at room temperature; progress of the reaction is monitored by TLC. Dichloromethane (60 ml) is added, and the solution is successively washed with 2 *M* hydrochloric acid and water, dried (Na_2SO_4), and concentrated to the crude **24**, which is used for the next reaction without further purification. To a solution of **24** in acetone (10 ml) is added Drierite (2 g), and the mixture is stirred for 2 hr at room temperature, after which potassium thioacetate (315 mg, 2.76 mmol) is added. The mixture is stirred overnight at 45°, then filtered, and the precipitate is washed with dichloromethane. The filtrate and washings are combined and concentrated to a syrup. The residue is chromatographed on a column of silica gel with 120 : 1 (v/v) dichloromethane–methanol, to give **25** (279 mg, 53%) as an amorphous mass, $[\alpha]_D$ +0.26° (*c* 0.75, chloroform).

*S-(Methyl 5-Acetamido-4,7,8,9-tetra-O-acetyl-3,5-dideoxy-D-glycero-α-D-galacto-2-nonulopyranosylonate)-(2→6)-O-(2,3,4-tri-O-acetyl-6-thio-β-D-galactopyranosyl)-(1→4)-S-(2,3,6-tri-O-acetyl-β-D-glucopyranosyl)-(1→1)-(2R,3R,4E)-2-azido-3-benzoyloxy-4-octadecene-1-thiol (**27**).* To a stirred solution of **25** (250 mg, 0.22 mmol) in dry methanol (2 ml), cooled to −30°, is added a solution of sodium metal (15 mg) in dry methanol (0.2

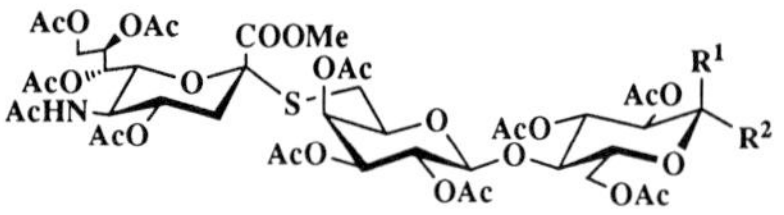

11 $R^1, R^2 = H, OH$
24 $R^1 = Cl, R^2 = H$
25 $R^1 = H, R^2 = SAc$
26 $R^1 = H, R^2 = SNa$

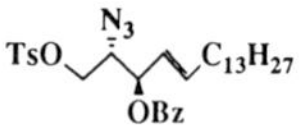

20

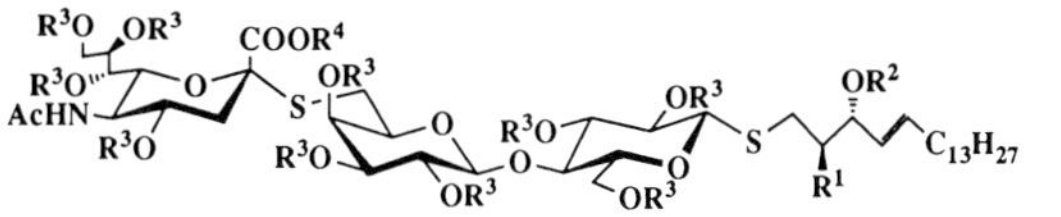

27 $R^1 = N_3, R^2 = Bz, R^3 = Ac, R^4 = Me$
28 $R^1 = NH_2, R^2 = Bz, R^3 = Ac, R^4 = Me$
29 $R^1 = NHCO(CH_2)_{16}Me, R^2 = Bz, R^3 = Ac, R^4 = Me$
K $R^1 = NHCO(CH_2)_{16}Me, R^2 = R^3 = R^4 = H$

SCHEME 3. Synthesis of ganglioside analogs containing both the α-thioglycoside of sialic acid and thioglycosidically linked ceramide.

ml). Stirring is continued for 5 min at −30°, and the mixture is concentrated to give **26** as an amorphous mass, which is used for the next step without purification. A solution of **26** and **20** (410 mg, 0.7 mmol) in dry DMF (4.5 ml) is stirred overnight at 45° under nitrogen. Acetic anhydride (2 ml) and pyridine (4 ml) are added to the mixture, which is stirred overnight at room temperature, then concentrated. The residue is chromatographed on a column of silica gel (50 g) with 100 : 1 (v/v) dichloromethane–methanol, to give **27** (124 mg, 38%) as an amorphous mass, $[\alpha]_D$ −6.4° (*c* 0.95, chloroform).

Synthesis of S-(α-Sialosyl)-(2→9)-O-(α-sialosyl)-(2→3′)-β-lactosyl Ceramide (*L*)

For the synthesis of the target ganglioside analog,[26] we set out to synthesize the per-O-acylated 2-(trimethylsilyl)ethyl *S*-(methyl α-*N*-ace-

[26] A. Hasegawa, H. Ogawa, H. Ishida, and M. Kiso, *J. Carbohydr. Chem.* **10,** 1009 (1991).

tylneuraminyl)-(2→9)-*O*-(methyl α-*N*-acetyl-9-thioneuraminyl)-(2→3′)-β-lactoside (**39**) as the intermediate. Compound **39** could then, by introduction of the ceramide moiety, be transformed to the end product.

Treatment of methyl (methyl 5-acetamido-3,5-dideoxy-2-thio-D-*glycero*-α-D-*galacto*-2-nonulopyranosid)onate (**30**) with 2,2-dimethoxypropane in the presence of *p*-toluenesulfonic acid monohydrate gives the 8,9-*O*-isopropylidene derivative **31** in 70% yield, which is acetylated with acetic anhydride in pyridine to give **32** (Scheme 4). Removal of the isopropylidene group from **32** with 80% aqueous acetic acid for 3.5 hr at 45° gives compound **33** in 89% yield, without acetyl migration. In the NMR spectrum, H-7 appears as a doublet of doublets at δ 5.08 ($J_{6,7}$ = 2.0, $J_{7,8}$ = 9.3 Hz). Selective bromination[27] of **33** with carbon tetrabromide–triphenylphosphine in pyridine gives methyl (methyl 5-acetamido-4,7-di-*O*-acetyl-9-bromo-3,5,9-trideoxy-2-thio-D-*glycero*-α-D-*galacto*-2-nonulopyranosid)onate (**34**) in 72% yield, which is converted to the 4,7,8-tri-*O*-acetyl derivative **35** quantitatively.

The glycosylation[15,28] of 2-(trimethylsilyl)ethyl *O*-(6-*O*-benzoyl-β-D-galactopyranosyl)-(1→4)-2,6-di-*O*-benzoyl-β-D-glucopyranoside[15] (**36**) with **35** in acetonitrile for 20 hr at −15° in the presence of dimethyl(methylthio)-sulfonium triflate (DMTST) as the glycosyl promoter gives the expected α-glycoside **37** of the sialic acid at O-3′ in compound **36**, in 37% yield. Acetylation of **37** with acetic anhydride in pyridine gives the acetate **38** in 93% yield. Condensation of the sodium salt **3** with compound **38** in DMF under nitrogen affords the desired tetrasaccharide **39** in 80% yield. The intermediate **39** is transformed into the final product **I**, by the procedures described for the conversion of **10** to **D**.

Detailed Procedures

*Methyl (Methyl 5-Acetamido-3,5-dideoxy-8,9-O-isopropylidene-2-thio-D-glycero-α-D-galacto-2-nonulopyranosid)onate (**31**).* To a solution of **30** (500 mg, 1.4 mmol) in dry DMF (5 ml) are added 2,2-dimethoxypropane (0.87 ml), *p*-toluenesulfonic acid monohydrate (15 mg), and Drierite (1.0 g), and the mixture is stirred for 1 hr at 0°, then neutralized with solid sodium bicarbonate (Scheme 4). The precipitate is filtered off and washed with methanol. The filtrate and washings are combined, then concentrated. The residue is chromatographed on a column of silica gel (50 g) with 40 : 1 (v/v) dichloromethane–methanol, giving compound **31** (390 mg, 70%) as an amorphous mass; $[\alpha]_D$ +12.6° (*c* 0.58, chloroform).

[27] A. Kashem, M. Anisuzzaman, and R. L. Whistler, *Carbohydr. Res.* **61,** 511 (1978).

[28] T. Murase, H. Ishida, M. Kiso, and A. Hasegawa, *Carbohydr. Res.* **188,** 71 (1989).

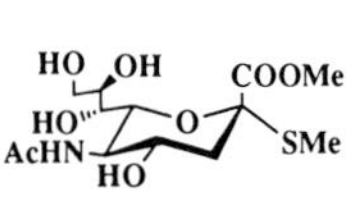

30

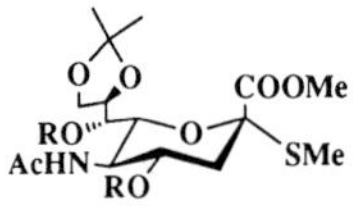

31 R = H
32 R = Ac

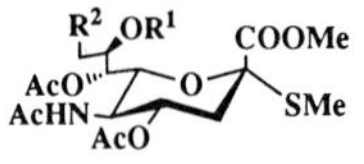

33 R^1 = H, R^2 = OH
34 R^1 = H, R^2 = Br
35 R^1 = Ac, R^2 = Br

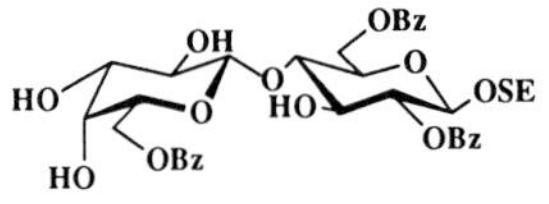

36

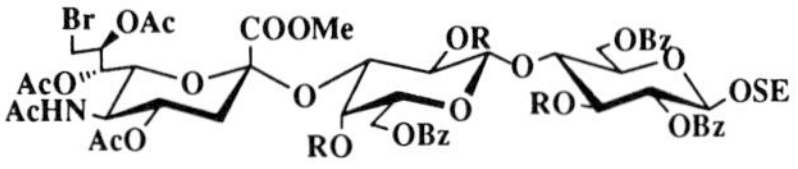

37 R = H
38 R = Ac

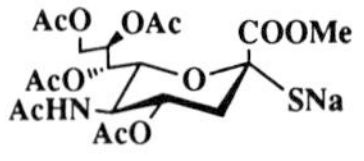

3

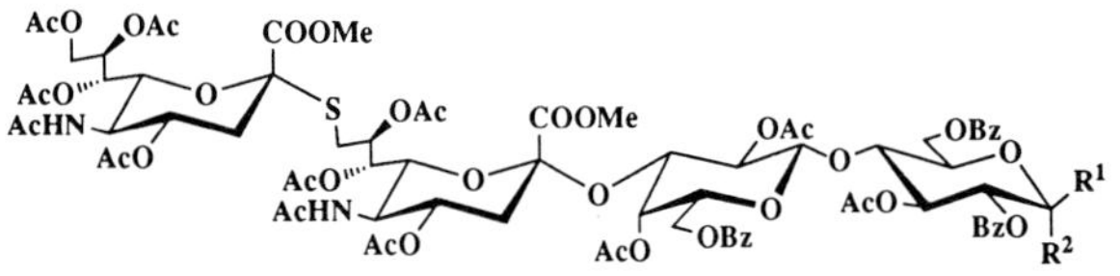

39 R^1 = OSE, R^2 = H
40 R^1, R^2 = H, OH
41 R^1 = H, R^2 = OC(=NH)CCl_3

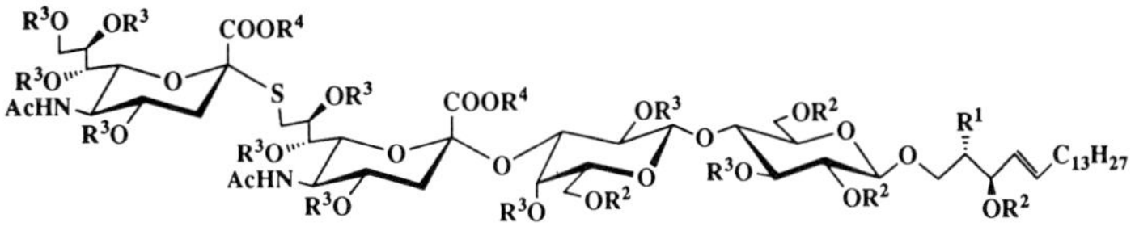

42 R^1 = N_3, R^2 = Bz, R^3 = Ac, R^4 = Me
43 R^1 = NH_2, R^2 = Bz, R^3 = Ac, R^4 = Me
44 R^1 = NHCO$(CH_2)_{16}$Me, R^2 = Bz, R^3 = Ac, R^4 = Me
L R^1 = NHCO$(CH_2)_{16}$Me, R^2 = R^3 = R^4 = H

SCHEME 4. Synthesis of ganglioside G_{D3} analogs containing the α-thioglycoside of sialic acid.

Methyl (Methyl 5-Acetamido-4,7-di-O-acetyl-3,5-dideoxy-8,9-O-isopropylidene-2-thio-D-glycero-α-D-galacto-2-nonulopyranosid)onate (**32**). Compound **31** (390 mg, 0.99 mmol) is acetylated with acetic anhydride (2 ml) in pyridine (4 ml) overnight at room temperature. The product is purified by chromatography on silica gel (60 g) with 60 : 1 (v/v) dichloromethane–methanol, to give **32** (470 mg, quantitative) as an amorphous mass, $[\alpha]_D$ +14.6° (*c* 0.8, chloroform).

*Methyl (Methyl 5-Acetamido-4,7-di-O-acetyl-3,5-dideoxy-2-thio-D-glycero-α-D-galacto-2-nonulopyranosid)onate (**33**).* A solution of **32** (2.12 g, 4.44 mmol) in 80% aqueous acetic acid (20 ml) is heated for 3.5 hr at 45°, then concentrated. The residue is chromatographed on a column of silica gel (100 g) with 80 : 1 (v/v) dichloromethane–methanol to give **33** (1.73 g, 89%) as an amorphous mass, $[\alpha]_D$ +33.8° (*c* 0.7, chloroform).

*Methyl (Methyl 5-Acetamido-4,7-di-O-acetyl-9-bromo-3,5,9-trideoxy-2-thio-D-glycero-α-D-galacto-2-nonulopyranosid)onate (**34**).* To a solution of **33** (2.0 g, 4.57 mmol) in pyridine (40 ml) is added carbon tetrabromide (3.03 g, 9.14 mmol), and the solution is cooled to 0°. Triphenylphosphine (2.4 g, 9.15 mmol) is added, and the mixture is stirred for 5 hr at room temperature. Methanol (1 ml) is added to the mixture, which is concentrated to a syrup and then chromatographed on a column of silica gel (100 g) with 2 : 1 (v/v) ethyl acetate–hexane, giving **34** (1.65, 72%) as an amorphous mass, $[\alpha]_D$ +38.5° (*c* 0.94, chloroform).

*Methyl (Methyl 5-Acetamido-4,7,8-tri-O-acetyl-9-bromo-3,5,9-trideoxy-D-glycero-α-D-galacto-2-nonulopyranosid)onate (**35**).* Compound **34** (1.35 g, 2.7 mmol) is acetylated with acetic anhydride (5 ml) in pyridine (10 ml) overnight at room temperature. The product is purified by column chromatography on silica gel (60 g) with 2 : 1 (v/v) ethyl acetate–hexane, to give **35** (1.45 g, quantitative) as an amorphous mass, $[\alpha]_D$ +28.0° (*c* 0.88, chloroform).

*2-(Trimethylsilyl)ethyl O-(Methyl 5-Acetamido-4,7,8-tri-O-acetyl-9-bromo-3,5,9-trideoxy-D-glycero-α-D-galacto-2-nonulopyranosylonate)-(2→3)-O-(6-O-benzoyl-β-D-galactopyranosyl)-(1→4)-2,6-di-O-benzoyl-β-D-glucopyranoside (**37**).* To a solution of **36**[15,29] (800 mg, 1.06 mmol) and **35** (1.15 g, 2.12 mmol) in dry acetonitrile (10 ml) is added MS-3Å (2.5 g). The mixture is stirred for 10 hr at room temperature and then cooled to −25°. To the cooled mixture is added, with stirring, a mixture (2.94 g; 74.6% DMTST by weight) of DMTST[7,19] and MS-3Å, and the stirring is continued for 20 hr at −15°. The precipitate is filtered off, then washed thoroughly with dichloromethane. The filtrate and washings are combined, and the solution is successively washed with 1 *M* sodium carbonate and water, dried (Na_2SO_4), and evaporated to a syrup that is chromatographed on a column of silica gel (80 g) with 3 : 1 (v/v) ethyl acetate–hexane, to give **37** (490 mg, 37%) as an amorphous mass, $[\alpha]_D$ +9.1°, (*c* 0.78, chloroform).

2-(Trimethylsilyl)ethyl O-(Methyl 5-Acetamido-4,7,8-tri-O-acetyl-9-

[29] P. Fügedi and P. J. Garegg, *Carbohydr. Res.* **149,** c9 (1986).

*bromo-3,5,9-trideoxy-*D*-glycero-α-*D*-galacto-2-nonulopyranosylonate)-(2→3)-O-(2,4-di-O-acetyl-6-O-benzoyl)-β-*D*-galactopyranosyl)-(1→4)-3-O-acetyl-2,6-di-O-benzoyl-β-*D*-glucopyranoside (**38**).* Compound **37** (540 mg, 0.43 mmol) is acetylated with acetic anhydride (3 ml) in pyridine (6 ml) overnight at room temperature. The product is purified by chromatography on silica gel (50 g) with 2 : 1 (v/v) ethyl acetate–hexane, giving **38** (550 mg, 93%) as an amorphous mass, $[\alpha]_D$ +5.0° (*c* 0.6, chloroform).

*2-(Trimethylsilyl)ethyl S-(Methyl 5-Acetamido-4,7,8,9-tetra-O-acetyl-3,5-dideoxy-*D*-glycero-α-*D*-galacto-2-nonulopyranosylonate)-(2→9)-O-(methyl 5-acetamido-4,7,8-tri-O-acetyl-3,5-dideoxy-9-thio-*D*-glycero-α-*D*-galacto-2-nonulopyranosylonate)-(2→3)-O-(2,4-di-O-acetyl-6-O-benzoyl-β-*D*-galactopyranosyl)-(1→4)-3-O-acetyl-2,4-di-O-benzoyl-β-*D*-glucopyranoside (**39**).* A solution of **38(** (400 mg, 0.29 mmol) and the sodium salt **3** (308 mg, 0.58 mmol) in DMF (5 ml) is stirred overnight at room temperature under a nitrogen atmosphere, then concentrated. The residue is chromatographed on a column of silica gel (60 g) with ethyl acetate to give **39** (421 mg, 80%) as an amorphous mass, $[\alpha]_D$ +9.0° (*c* 0.58, chloroform).

[18] Replacement of Glycosphingolipid Ceramide Residues by Glycerolipid for Microtiter Plate Assays

By RENÉ ROY, ANNA ROMANOWSKA, and FREDRIK O. ANDERSSON

Introduction

Glycosphingolipids and gangliosides in particular constitute well-defined antigenic carbohydrate structures which have been implicated in a wide range of biological and immunological activities. When monoclonal antibodies are used for determining the fine chemical specificities involved in protein–carbohydrate interactions, solid-phase enzyme-linked immunosorbent assays (ELISA) are frequently used as a standard technique.[1,2] The prime event in the ELISA is a nonspecific lipophilic adsorption of coating antigens on the surface on the plastic wells, which are generally made of either polystyrene or polyvinyl chloride (PVC). Although the exact structures of the ceramide portions have been suggested as being involved in monoclonal antibody–ganglioside (G_{M3}, **2** in Scheme 1) interactions,[3] the structure of the lipid component is not critically important for

[1] P. Fredman, S. Jeansson, E. Lycke, and L. Svennerholm, *FEBS Lett.* **189,** 23 (1985).
[2] J. Miyoshi, Y. Fujii, and M. Naiki, *J. Biochem.* (*Tokyo*) **92,** 89 (1982).
[3] S. Itonori, K. Hidari, Y. Sanai, M. Tanigushi, and Y. Nagai, *Glycoconjugate J.* **6,** 551 (1989).

Copyright © 1994 by Academic Press, Inc.
All rights of reproduction in any form reserved.

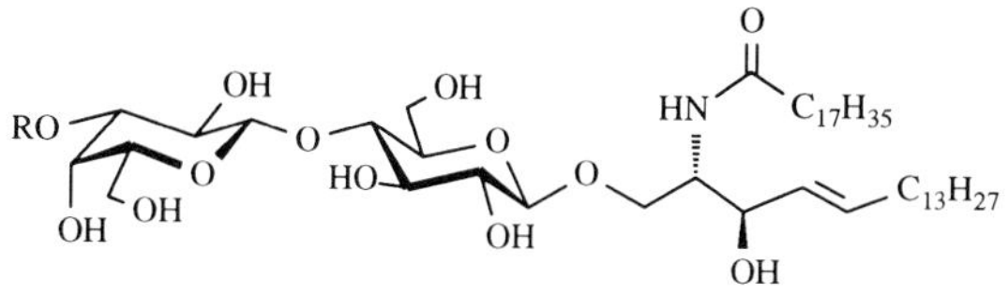

1 R = H
2 R = Sialic acid

SCHEME 1

the purpose of coating microtiter plates to be used, for example, in the determination of the carbohydrate-binding specificities of lectins.[4]

Because the large-scale synthesis of glycosphingolipids containing intact ceramide residues is nontrivial[5] and requires ample amounts of the latter, the replacement of sphingolipids by glycerolipids for ELISA (or similar methods) represents a useful technical advantage. This chapter describes chemical syntheses of glycosylglycerolipids **6** and **11** as illustrated in Scheme 2 and their use in model ELISA for the detection of plant lectins. Of particular interest in the synthesis is the utilization of the stable, crystalline sialosyl donor **8** obtained under phase-transfer catalysis (PTC) conditions from the unstable chloride **7**.

Materials

Sialic acid is isolated from edible bird's nest as previously described.[6,7] 1,2-Di-*O*-tetradecyl-*sn*-glycerol (**4**) is prepared from D-mannitol as previously described.[8] Silver triflate and all other chemicals described here are reagent grade and used without purification (Aldrich Chemical Co., Milwaukee, WI). Dimethyl(methylthio)sulfonium triflate (DMTST) is prepared according to Ravenscroft *et al.*[9] and kept dry under nitrogen in the cold. Linbro enzyme immunoassay (EIA) microtiter plates are from Flow

[4] I. J. Goldstein and R. D. Poretz *in* "The Lectins: Properties, Functions and Application in Biology and Medicine" (I. E. Liener, N. Sharon, and I. J. Goldstein, eds.). Academic Press, Orlando, Florida, 1986.

[5] See, for example, S. Fujita, M. Sugimoto, K. Tomita, Y. Nakahara, and T. Ogawa, *Agric. Biol. Chem.* **55,** 2561 (1991), and references cited therein.

[6] M. F. Czarniecki and E. R. Thornton, *J. Am. Chem. Soc.* **99,** 8273 (1977).

[7] R. Roy and C. A. Laferrière, *Can. J. Chem.* **68,** 2045 (1990).

[8] R. Roy, M. Letellier, D. Fenske, and H. C. Jarrell, *J. Chem. Soc., Chem. Commun.*, 378 (1990).

[9] M. Ravenscroft, R. M. G. Roberts, and J. G. Tillett, *J. Chem. Soc., Perkin Trans. 2,* 1569 (1982).

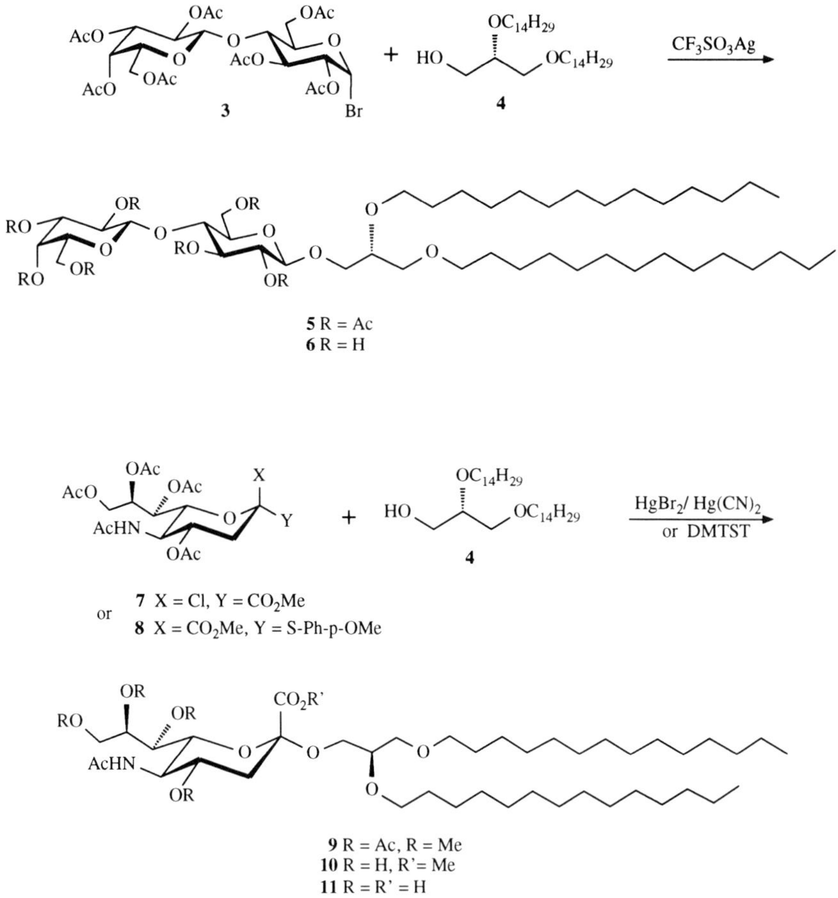

SCHEME 2

Laboratories (Mississauga, ON, Canada). Horseradish peroxidase-labeled peanut lectin from *Arachis hypogaea* (L-7759, 20 units/mg protein) and labeled wheat germ agglutinin from *Triticum vulgaris* (L-3892, 60 units purpurogallin/mg protein) are from Sigma Chemical Co. (St. Louis, MO). The peroxidase substrate, 2,2′-azinobis(3-ethylbenzothiazoline-6-sulfonic acid) (ABTS), is obtained from Aldrich Chemical Co.

General Methods

Melting points are determined on a Gallenkamp apparatus and are uncorrected. The ^{1}H nuclear magnetic resonance (NMR) and ^{13}C NMR spectra are recorded on a Varian (Palo Alto, CA) XL-300 instrument at 300 and 75.4 MHz, respectively. The proton chemical shifts (δ) are given relative to internal chloroform (7.24 ppm) for $CDCl_3$ solutions and to the HOD signal (4.75 ppm) for D_2O solutions. The analyses are done as a first-order approximation, and assignments are based on correlated spectroscopy (COSY) and heteronuclear correlated spectroscopy (HETCOR) experiments. Optical rotations are measured on a Perkin-Elmer (Norwalk, CT) 241 polarimeter and are run in chloroform at 23° for 1% solutions unless stated otherwise. Thin-layer chromatography (TLC) is performed using silica gel 60F-254 and column chromatography on silica gel 60 (230–400 mesh, E. Merck, Darmstadt, Germany, No. 9385). Optical densities in enzyme-linked lectin assays are measured at 410 nm on a Dynatech (Alexandria, VA) MR-600 spectrophotometer.

Experimental Procedures

*Synthesis of Lactosylglycerolipid **6***

Per-O-acetylated lactosyl bromide (**3**) (acetobromolactose) (300 mg, 0.43 mmol)[10] and 1,2-di-*O*-tetradecyl-*sn*-glycerol (**4**) (249 mg, 0.52 mmol)[8,11] are dissolved in a mixture of dichloromethane (2.2 ml) and toluene (1.4 ml). Dried 4 Å molecular sieves (500 mg) and silver triflate (135 mg, 0.6 mmol) are then added. The reaction mixture is stirred at room temperature for 45 min. After filtration over a Celite pad, the organic phase and washings (CH_2Cl_2) are successively washed with sodium sulfite, 1 *M* HCl, saturated $NaHCO_3$, and brine. The organic phase is dried (Na_2SO_4), and the crude residue obtained after evaporation of the solvent is purified by silica gel column chromatography using toluene–ethyl acetate (3 : 1, v/v) as eluant. Pure **5** is obtained in 65% yield (302 mg). The amorphous solid obtained on cooling has mp 45°–50°. The ^1H NMR and ^{13}C NMR data agree with literature values.[11]

The peracetylated lactosylglycerolipid (**5**) (73.4 mg) is de-O-acetylated in methanol (2 ml) containing a catalytic amount of 0.2 *M* sodium methoxide (25 μl) to provide a quantitative yield (53.5 mg) of the unprotected glycoside **6**, mp 215°–220° (methanol–H_2O).[11] The solid is used as such for the microtiter plate assays.

*Synthesis of Sialylglycerolipid **11***

Glycosylation of the glycerol derivative **4** with appropriate sialyl donors can be accomplished using a classic method [$HgBr_2/Hg(CN)_2$] from

[10] C. P. Stowell and Y. C. Lee, this series, Vol. 83, p. 282.

[11] T. Ogawa and K. Beppu, *Agric. Biol. Chem.* **46,** 255 (1982).

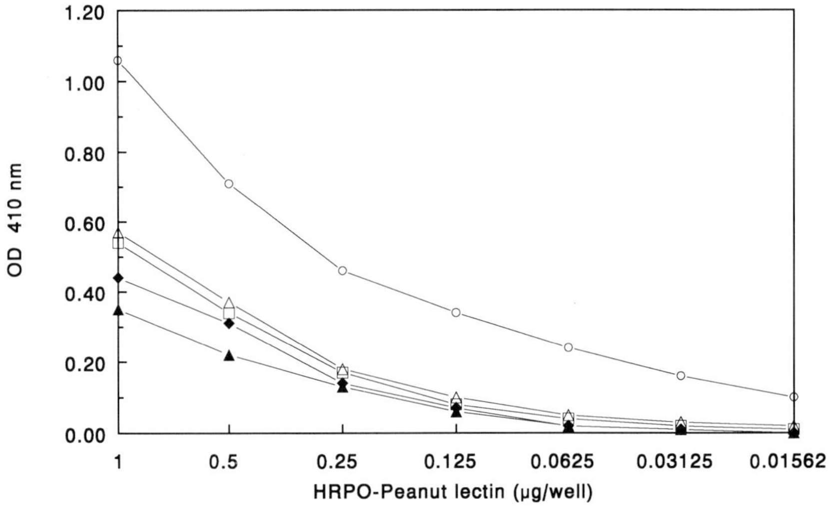

FIG. 1. Microtiter plate enzyme-linked lectin assays (ELLA) of lactosylglycerolipid **6** as coating antigen with horseradish peroxidase (HRPO)-labeled peanut lectin in serial 2-fold dilutions using ABTS–H_2O_2 as enzyme substrates. The coating of **6** was 0.01 (▲), 0.1 (◆), 1 (□), 10 (△), and 100 (○) μg/well.

the chloride **7**[7,12] or, more effectively, with the "active"[13] thiosialosyl donor **8** using DMTST as catalyst. The latter reaction is shorter (30 min instead of 48 hr), gives a slightly higher yield, and provides better α-selectivity.

Method A from **7**. To a solution of acetochloroneuraminic acid **7** (1 mmol)[7,12] and **4** (1.5 mmol) dissolved in 1,2-dichloroethane (5 ml) is added 4 Å molecular sieves (300 mg), mercuric bromide (1.2 mmol), and mercuric cyanide (1.2 mmol). The reaction mixture is stirred for 48 hr at room temperature and then filtered through a Celite pad which is thoroughly washed with dichloromethane. The organic phase is washed with sodium sulfite, saturated $NaHCO_3$, and water. The dried organic phase is evaporated to dryness, and the residue is separated by silica gel column chromatography using hexane–ethyl acetate (1 : 1, v/v) as eluant. The β anomer (R_f 0.66, $CHCl_3$–methanol, 15 : 4, v/v) is obtained in 25% yield, whereas the desired α anomer **9** (R_f 0.53) is obtained in 38% yield, $[\alpha]_D$ −8.6° (*c* 0.92, $CHCl_3$). The ^1H NMR and ^{13}C NMR data agree with those published.[8,14]

[12] R. Kuhn, P. Lutz, and D. L. McDonald, *Chem. Ber.* **99,** 611 (1966).
[13] R. Roy, F. O. Andersson, and M. Letellier, *Tetrahedron Lett.* **33,** 6053 (1992).
[14] T. Ogawa and M. Sugimoto, *Carbohydr. Res.* **128,** C1 (1984).

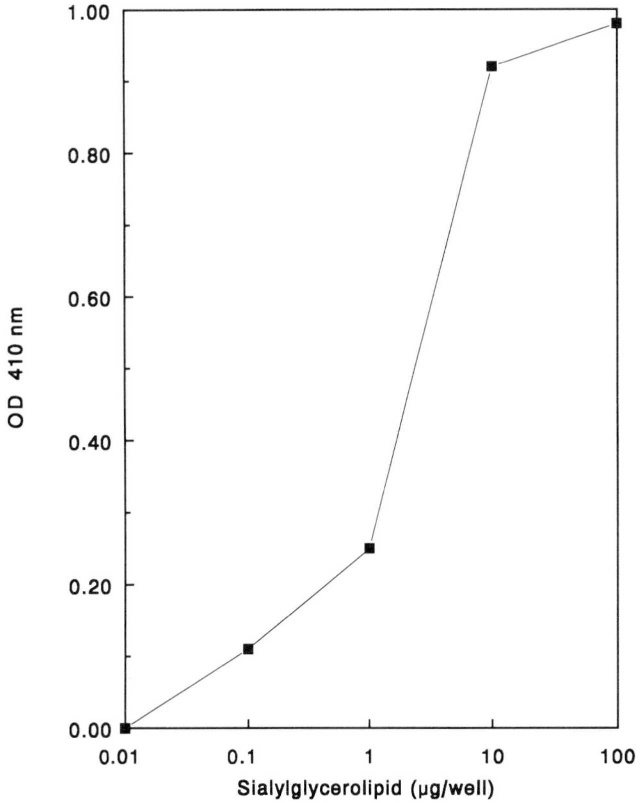

FIG. 2. Microtiter plate ELLA of sialylglycerolipid **11** as coating antigen with HRPO–WGA at 1 μg/well.

The α anomer **9** is de-O-acetylated as above (sodium acetate, methanol, 25°, 4.5 hr) to provide **10** in 95% yield, mp 139°–140°, $[\alpha]_D$ +0.4° (*c* 1.0, methanol). The methyl ester function in **10** is hydrolyzed in a mixture of tetrahydrofuran (THF)–0.1 *M* NaOH (1 : 4, v/v) at room temperature for 4.5 hr, followed by neutralization with cation-exchange resin (H^+), back titration to pH 7.0 with 0.1 *M* NaOH, and evaporation of the solvent to give **11** in 90% yield, mp 138°–138.5°, $[\alpha]_D$ +10.4°.[8,14] The product **11** is used as the sodium salt in the microtiter plate assays.

Method B from ***8***. To a solution of freshly prepared β-acetochloroneuraminic acid **7** (510 mg, 1 mmol), tetrabutylammonium hydrogen sulfate (340 mg, 1 mmol), and 4-methoxybenzenethiol (185μl, 1.5 mmol) in ethyl acetate (5 ml) is added 1 *M* sodium carbonate (5 ml). The reaction mixture

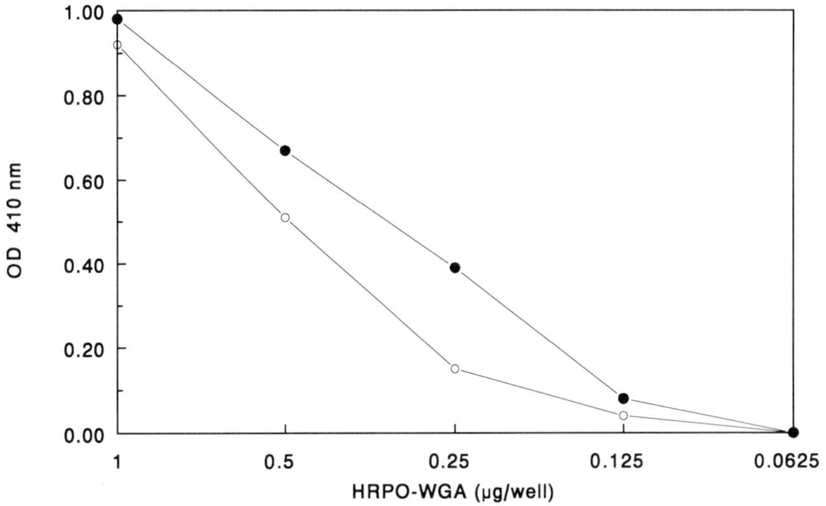

FIG. 3. Effect of HRPO–WGA concentration on the detection of sialylglycerolipid **11** used as coating antigen at 100 (●) and 10 (○) μg/well in a microtiter plate ELLA.

is vigorously stirred at room temperature for 15 min (TLC monitoring in ethyl acetate). The reaction mixture is then diluted with ethyl acetate (10 ml), and the organic phase is washed successively with 1 *M* sodium carbonate (3 times), water (2 times), and saturated brine. The dried organic phase (Na_2SO_4) is evaporated under reduced pressure. The resulting residue is purified by silica gel column chromatography using ethyl acetate : toluene (3 : 1, v/v) as eluant. Compound **8** crystallizes from benzene–hexane in 81% yield, mp 136°–137°, $[\alpha]_D$ +20.7°.

The *p*-methoxyphenylthio-α-sialoside **8** (32 mg, 52 μmol), which is a stable crystalline sialosyl donor, and glycerolipid **4** (37.8 mg, 78 μmol) dissolved in a mixture of CH_3CN–CH_2Cl_2 (1 : 1, v/v; 400 μl) are stirred at room temperature for 3 hr in the presence of 4 Å molecular sieves to ensure complete dryness. Freshly prepared DMTST (47 mg, 209 μmol) is added, and the reaction mixture is stirred for 30 min at room temperature. The reaction mixture is diluted with CH_2Cl_2 and filtered through a Celite pad. The solution is evaporated to dryness, and the residue is purified on silica gel column as above. The desired α anomer **9** is obtained in 52% yield, whereas the β anomer is obtained in 22% yield.

Enzyme-Linked Lectin Assays

The lactosyl- or sialylglycerolipids (**6** or **11**) are prepared as stock solutions in ethanol (1 mg/ml). In the case of **6**, gentle warming of the ethanol solution is necessary to ensure complete dissolution. To coat the wells of the microtiter plates, 100 μl of the stock glycosylglycerolipid solutions is added to each well. Coating with lower amounts of the glycosylglycerolipids is done by adding 100 μl of 10-fold dilutions of the stock solutions down to a final concentration of 0.01 μg/well. The plates are sealed with Parafilm and are left overnight at room temperature. Ethanol then evaporates at room temperature. The plates are blocked with 300 μl of 0.2% bovine serum albumin (BSA) in phosphate-buffered saline (PBS) for 1 hr at room temperature. After washing the plates with PBS (4 times), the wells are filled with 100 μl of horseradish peroxidase (HRPO)-labeled peanut lectin (for **6**) or wheat germ agglutinin (WGA) (for **11**) [100-fold diluted stock solution (1 mg/ml) of lectins]. The plates are incubated at room temperature for 3 hr and then washed with PBS. The enzyme substrate (ABTS, 1 mg/2 ml) (50 μl/well) in citrate–phosphate buffer (0.2 *M*, pH 4.0) containing 0.01% H_2O_2 is added. After 15 min, the absorbance is measured at 410 nm.

In the first set of experiments, different amounts of lactosylglycerolipid **6** are used in coating of the wells to determine the detection limit by enzyme-linked lectin assays (ELLA) using HRPO–peanut lectin. Figure 1 shows that as little as 0.1 μg/well of lipid **6** is sufficient for detection with 1 μg/well of HRPO–lectin.

In the case of the sialylglycerolipid **11**, the effect of coating the microtiter plates with various concentrations of **11** is illustrated in Fig. 2. The optimum coating occurs at 10–100 μg/well. The concentration effect of the HRPO–WGA on binding to plates coated with either concentration of **11** (10 and 100 μg/well) is illustrated on Fig. 3. As also seen from Fig. 2, there is only a slight difference in the absorbance obtained from 10 or 100 μg/well of **11**. In both cases, the coating sialylglycerolipid could be reliably detected with 0.5 μg/well of HRPO–WGA.

[19] Neoglycolipids: Probes in Structure/Function Assignments to Oligosaccharides

By T. Feizi and R. A. Childs

Introduction

The rationale for developing "*the neoglycolipid technology*" as a micromethod to determine antigenicities and ligand functions of oligosac-

Copyright © 1994 by Academic Press, Inc.
All rights of reproduction in any form reserved.

charides was elaborated in 1985.[1] There was an increasing awareness that carbohydrate chains of glycoproteins and proteoglycans of animal cells behave as oncodevelopmental antigens, as attachment sites for infective agents, and as ligands for endogenous proteins. However, there was a lack of microscale binding methods, analogous to the "blotting" methods for nucleic acids and proteins, applicable to oligosaccharides after their release from carrier proteins. Glycoproteins typically contain multiple oligosaccharide chains with differing structures, and only small amounts of the individual oligosaccharides can be isolated in the pure state. Moreover, the loss of cooperative effects of multivalence following the release of clustered oligosaccharides from glycoproteins such as mucins means that the amounts of individual components required for conventional immuno- or bioassays such as inhibition assays are usually prohibitive. It was reasoned that, ideally, the released oligosaccharides should be analyzed in a clustered, multivalent state, and as single species (or as far as possible separated species) rather than mixtures. Knowing that sensitive binding experiments can be performed with glycolipids resolved on chromatograms,[2] it was decided to develop a microtechnology whereby oligosaccharides released from glycoproteins are conjugated by reductive amination to an aminophospholipid.[1] Initial efforts were focused on rendering the technology applicable both to reducing oligosaccharides and to reduced oligosaccharides such as those released by alkaline borohydride degradation from O-glycosylated proteins.[1,3–7] More recently oligosaccharides released under mild nonreductive conditions from O-glycosylated proteins have been successfully investigated as neoglycolipids.[8]

There was precedent for preparing lipid-linked oligosaccharides by reductive amination. In two relatively large-scale experiments, involving sialyllactose[9] and dextran oligomers,[10] and in another experiment (scale

[1] P. W. Tang, H. C. Gooi, M. Hardy, Y. C. Lee, and T. Feizi, *Biochem. Biophys. Res. Commun.* **132,** 474 (1985).

[2] J. L. Magnani, D. F. Smith, and V. Ginsburg, *Anal. Biochem.* **109,** 399 (1980).

[3] M. S. Stoll, T. Mizuochi, R. A. Childs, and T. Feizi, *Biochem. J.* **256,** 661 (1988).

[4] T. Mizuochi, R. W. Loveless, A. M. Lawson, W. Chai, P. J. Lachmann, R. A. Childs, S. Thiel, and T. Feizi, *J. Biol. Chem.* **264,** 13834 (1989).

[5] M. S. Stoll, E. F. Hounsell, A. M. Lawson, W. Chai, and T. Feizi, *Eur. J. Biochem.* **189,** 499 (1990).

[6] W. Chai, G. C. Cashmore, R. A. Carruthers, M. S. Stoll, and A. M. Lawson, *Biol. Mass Spectrom.* **20,** 169 (1991).

[7] W. Chai, M. S. Stoll, G. C. Cashmore, and A. M. Lawson, *Carbohydr. Res.* **239,** 107 (1993).

[8] C.-T. Yuen, A. M. Lawson, W. Chai, M. Larkin, M. S. Stoll, A. C. Stuart, F. X. Sullivan, T. J. Ahern, and T. Feizi, *Biochemistry* **31,** 9126 (1992).

[9] H. Wiegandt and W. Ziegler, *Hoppe-Seyler's Physiol. Chem.* **355,** 11 (1974).

[10] C. Wood and E. A. Kabat, *J. Exp. Med.* **154,** 432 (1981).

not specified) involving bacterial polysaccharides,[11] the oligosaccarides were made into neoglycolipids, though not referred to as such. In the latter two experiments, the resulting neoglycolipids were successfully used as immunogens for the raising of monoclonal antibodies[10] and for evaluation as potential antibacterial vaccines.[11] Special features of the technology as developed in our laboratory are that it is microscale and is applicable to (1) as little as a few micrograms of purified oligosaccharide material, (2) unfractionated or partially fractioned oligosaccharides released from glycoproteins, and (3) desired structurally defined or chemically synthesized oligosaccharides. The technology has proved to be a powerful means of singling out the ligands for carbohydrate-binding proteins and evaluating the influence of carrier proteins or lipids in presentation and recognition of oligosaccharide ligands. It also constitutes a microsequencing strategy for oligosaccharides, as the neoglycolipids have unique ionization properties in mass spectrometry.[12] Details of the conjugation procedures for reducing and reduced oligosaccharides, analyses of the resulting neoglycolipids by liquid secondary ion mass spectrometry (LSIMS), and, where necessary, purification of neoglycolipids from reaction mixtures, for ligand-binding studies, have been described in another volume of the series.[13] In this chapter we summarize the new structure/function assignments that have been made using the neoglycolipid technology.

Neoglycolipids in Microsequencing of Glycoprotein Oligosaccharides

Analysis by LSIMS of neoglycolipids derived from O-glycosidic oligosaccharides has proved to be a powerful sequencing approach.[12–14] The discovery, by this means, of novel sulfated oligosaccharide ligands for selectins is highlighted in Table I. Here we consider results of the application of the neoglycolipid technology to analyses of the core and backbone domains of O-glycosidic oligosaccharides released by alkaline borohydride degradation from mucin-type glycoproteins. As the reduced oligosaccharides contain no reactive aldehydes, mild periodate oxidation conditions are used such that overoxidation does not occur and a reactive aldehyde is generated at the reduced end.[1,5] The lipids used most extensively in our

[11] H. Snippe, J. E. G. van Dam, A. J. van House, J. M. N. Willers, J. P. Kamerling, and J. F. G. Vliegenthart, *Infect. Immun.* **42,** 842 (1983).

[12] A. M. Lawson, W. Chai, G. C. Cashmore, M. S. Stoll, E. F. Hounsell, and T. Feizi, *Carbohydr. Res.* **200,** 47 (1990).

[13] T. Feizi, M. S. Stoll, C.-T. Yuen, W. Chai, and A. M. Lawson, this series, Vol. 230, p. 484.

[14] A. M. Lawson, E. F. Hounsell, M. S. Stoll, J. Feeney, W. Chai, J. R. Rosankiewicz, and T. Feizi, *Carbohydr. Res.* **221,** 191 (1991).

laboratories for generating neoglycolipids have been L-1,2-dipalmitoyl-*sn*-glycero-3-phosphoethanolamine (DPPE):

$$\begin{array}{l} CH_2OCO(CH_2)_{14}CH_3 \\ | \\ CHOCO(CH_2)_{14}CH_3 \\ | \\ CH_2OP(O)(OH)OCH_2CH_2NH_2 \end{array}$$

and L-1,2-dihexadecyl-*sn*-glycero-3-phosphoethanolamine (DHPE):

$$\begin{array}{l} CH_2O(CH_2)_{15}CH_3 \\ | \\ CHO(CH_2)_{15}CH_3 \\ | \\ CH_2OP(O)(OH)OCH_2CH_2NH_2 \end{array}$$

Chromatographic and mass spectrometric studies of the products of oxidation/conjugation of a large number of oligosaccharide alditols[5,7] have shown that no oxidative changes occur along the chain except on sialic acid, which nevertheless is preserved to the extent of 60%[7] allowing isolation of neoglycolipids containing intact sialic acid.

The mechanism of oxidative cleavage under the mild periodate conditions used has been analyzed by LSIMS, and it has been shown[5] that cleavage occurs specifically at the *threo*-diol C-4–C-5 bond of the core *N*-acetylgalactosaminitol, giving rise to two cleavage products with differing molecular masses (Scheme 1). Although this cleavage would be a drawback in carbohydrate-recognition systems that require the intact core monosaccharide, it does have the advantage of allowing unambiguous assignment by LSIMS of positions of linkage (at C-3 or C-6) of oligosaccharide sequences to the core *N*-acetylgalactosamine. For example, from an oligosaccharide disubstituted at the core, two lipid-linked oligosaccharides are obtained corresponding to the branches C-3- and C-6-linked to the *N*-acetylgalactosaminitol as depicted below:

```
Galβ1-4GlcNAcβ1                    Galβ1-4GlcNAcβ1
               \                                  \
                6                                  6—OY
                 GalNAcol
                3                                  3—OX
               /                                  /
          Galβ1                              Galβ1
```

where OX = −OCH[CH(NHAc)CH_2OH]CH_2—DPPE and OY = −OCH_2-CH_2—DPPE.

Another advantage is that oligosaccharide alditol mixtures that are difficult to separate in the free state by existing chromatographic proce-

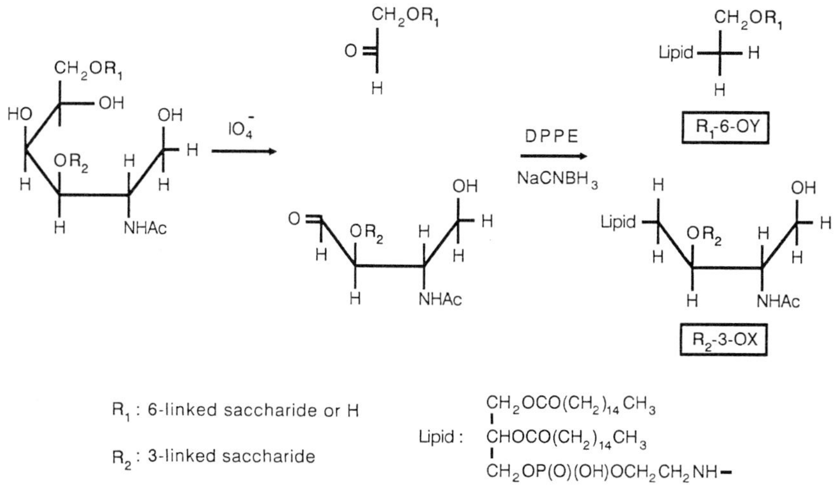

SCHEME 1. Proposed reaction scheme for periodate cleavage and conjugation of the resulting fragments to the lipid DPPE by reductive amination of C-3 and/or C-6-linked *N*-acetylgalactosaminitol. (Adapted from Stoll *et al.*[5])

dures can often be resolved into clearly separated bands after the controlled oxidation and conjugation to lipid, and they can be identified and sequenced by LSIMS using submicrogram amounts of carbohydrate material.[5] This approach has been crucial for deriving sequence information on the core and backbone sequences of oligosaccharide mixtures that were available in only limited amounts from human fetal gastrointestinal glycoproteins (meconium) as well as those that are too complex for characterization by conventional mass spectrometry and nuclear magnetic resonance (NMR) alone.[5,12,14] The method has been applied with success to profiling the core sequences among partially fractionated oligosaccharide alditols derived from human amniotic fluid[15] and from human gastric mucins.[16]

Neoglycolipids in Assignments of Carbohydrate-Binding Specificities

New assignments of combining specificities of antibodies, animal lectins, microbial adhesins, glycosyltransferases, and glycosidases made by the neoglycolipid technology are summarized in Table I. Of special note

[15] F.-G. Hanisch and J. Peter-Katalinic, *Eur. J. Biochem.* **205,** 527 (1992).

[16] F.-G. Hanisch, W. Chai, J. R. Rosankiewicz, A. M. Lawson, M. S. Stoll, and T. Feizi, *Eur. J. Biochem.* **217,** 645 (1993).

TABLE I
ASSIGNMENTS OF SACCHARIDE RECOGNITION MADE BY THE NEOGLYCOLIPID TECHNOLOGY[a]

Recognition proteins	Oligosaccharide sources/types	Lipids[b]	Salient observations
Antibodies			
Anti-I (Ma)	O-Glycosidic from sheep gastric mucins; chemically synthesized	DPPE	The C-5–C-6 part of oxidized and cleaved *N*-acetylgalactosaminitol giving rise to 6-OY[c] found to contain essential recognition motif in epitope Galβ1-4GlcNAcβ1-6GalNAc/Gal recognized by antibody (*1, 2*)
Anti-Le[a] and anti-SSEA-1	O- and N-glycosidic from human milk galactosyltransferase	DPPE	Le[a] and Le[x] antigens demonstrated on galactosyltransferase (*3*)
Anti-blood group A (TH-1)	Human fetal gastrointestinal tract	DPPE	Minor oligosaccharide detected containing repetitive blood group A (*4*)
Anti-i and anti-keratan sulfate	Bovine corneal keratan sulfate released using endo-β-galactosidase	DPPE	Reciprocal expression of keratan sulfate antigen and i antigen demonstrated (*5*)
Anti-human CD24 (VIB-E3)	O-Glycosidic from bovine submaxillary mucin; human milk; chemically synthesized	DPPE	Epitope characterized as NeuAcα2-6GalNAc/Gal (*6*)
Anti-*Candida albicans*	Mannooligosaccharides obtained by partial acid hydrolysis, β-elimination, and acetolysis of cell wall polysaccharides	4HDA	Immunogenicity of mannooligosaccharides demonstrated in mice; monoclonal antibodies with different binding patterns produced (*7–9*)
Sponge (*Microciona prolifera*) aggregation-blocking antibody (Block 1)	Oligosaccharides obtained by mild acid hydrolysis of adhesive proteoglycan	DPPE	Epitope characterized as 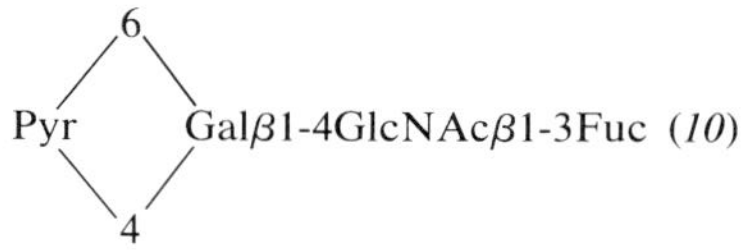
Anti-human colon cancer (BW494)	Human milk; *O*-glycosidic from human amniotic fluid	DPPE	Binding detected to oligosaccharides terminating in Galβ1-3/4HexNAc (*11*)
Anti-mouse L1 cell adhesion molecule (L3 and L4)	N-Glycosidic from neural glycoprotein AMOG and from RNase B; chemically synthesized	DPPE	Binding detected to high-mannose-type oligosaccharides (*12*)

Anti-human amniotic fluid mucins (FW6)	O-Glycosidic from human amniotic mucins	DPPE	Epitope proposed to contain galactose and fucose but distinct from Le^x and Le^y (*13*)
Anti-human gastric mucosa (2B5)	Human milk; O-glycosidic from human gastric mucins	DPPE	Epitope identified as GlcNAcβ1-3Galβ1-4GlcNAcβ1-6GalNAcol (*14*)
Antibodies to selectin ligands	Human milk; chemically synthesized	DPPE	Five patterns of binding activities detected toward Le^a- and Le^x-related sequences (*15*)
C-type lectins			
Conglutinin, bovine	Human milk; chitin oligosaccharides obtained by partial acid hydrolysis; mannobiose	DPPE	Multiple binding specificities demonstrated both for conglutinin and mannan-binding proteins toward N-linked oligosaccharides terminating in mannose or *N*-acetylglucosamine and toward fucooligosaccharides of Le^a type (*16–19*); for human mannan-binding protein, binding to Le^x-type sequence also (*18*); for both human mannan-binding protein and conglutinin binding to high-mannose-type oligosaccharides derived from complement glycoprotein iC3b; binding for conglutinin only to Man_8/Man_9 oligosaccharides as presented on iC3b protein (*20*)
	N-Glycosidic high-mannose, hybrid, and complex types from various glycoproteins	DPPE	
	High-mannose-type oligosaccharides from C-3 glycoprotein and from α and β chains of glycoprotein	DPPE	
Mannan-binding proteins of rat and human	Human milk and N-glycosidic (see conglutinin); di- and trisaccharides from acid-hydrolyzed dextran	DPPE	
	N-Glycosidic oligosaccharides from envelope glycoprotein gp120 of human immunodeficiency virus	DPPE	For human mannan-binding protein, interactions shown with Man_8 and Man_9 oligosaccharides of gp120, as well as with intact glycoprotein (*21*)
Pulmonary surfactant protein A, human and canine	Human milk; N-glycosidic (see conglutinin); monosaccharides	DPPE	Specificity demonstrated for galactose or glucose or lactose linked to specific ceramides and not as neoglycolipids (*22*)
E-selectin, human	Human milk; human fetal gastrointestinal tract	DPPE	Dependence of E-selectin binding on density of surface expression; requirement of threshold density for binding to sialyl-Le^a/Le^x type sequences; requirement of high surface density for binding to asialo Le^a/Le^x sequences (*23*)

(*continued*)

TABLE I (*continued*)

Recognition proteins	Oligosaccharide sources/types	Lipids[b]	Salient observations
	O-Glycosidic oligosaccharides from ovarian cystadenoma glycoprotein	DPPE	Sulfated Le^a/Le^x type sequences identified as strong supporters of E-selectin binding (*24*)
	Chemically synthesized	DHPE	Sulfated Le^a penta- and tetrasaccharides shown to be most potent E-selectin ligands so far (*25*)
L-selectin of rat	Human milk oligosaccharides; O-glycosidic from ovarian cystadenoma glycoprotein	DPPE	Sulfated Le^a/Le^x type sequences shown to be strongly bound by L-selectin; no absolute requirement for fucose in contrast to E-selectin (*26*)
Other carbohydrate-binding proteins			
Natural killer cell protein NKR-P1 of rat	Milk oligosaccharide lacto-*N*-neotetraose (from human milk), β-galactosidase-treated	DPPE	NKR-P1 binding shown to sequence GlcNAcβ1-3Galβ1-4Glc (*27*)
Serum amyloid P protein, human	N-Glycosidic (as with conglutinin); pentamannose phosphate and dephosphorylated form from *Hansenula holsteii*	DPPE	Binding specificity revealed toward mannose-6-phosphorylated oligosaccharides, as on lysosomal hydrolases (*28*)
14-kDa β-galactoside-binding protein (galectin 1)[d]	Complex type bi-, tri-, tetraantennary; milk and human fetal gastrointestinal tract; chemically synthesized	DPPE	Minimum lipid-linked sequence bound shown to be tetrasaccharide with terminal Galβ1-3/4GlcNAc sequence (*29*)
30-kDa β-galactoside-binding protein (galectin 2)	As with galectin 1	DPPE	Minimum lipid-linked sequence bound shown to be pentasaccharide with terminal or subterminal Galβ1-3/4GlcNAc sequence; preference for blood group A and blood group B termini (*30*)
Neural adhesion protein NCAM	Human milk; N-glycosidic oligosaccharides from neural adhesion protein AMOG, ovalbumin, RNase B, and fetuin	DPPE	Binding observed (attributed to domain 4 of NCAM) to high-mannose-type oligosaccharides (*31*)
Signaling systems			
Rat glioma cells (recognition proteins unknown)	Oligosaccharide moiety of G_{M1} ganglioside	DPPE	G_{M1} neoglycolipids shown not only to be taken up and metabolized by cells in manner similar to natural glycolipid, but also to serve as functional receptors that effectively mediate rise in cAMP levels in cells exposed to cholera toxin (*32, 33*)

Liposome clearance systems			
Murine (recognition proteins unknown)	Oligosaccharide moiety of G_{M1}	DOPE	Incorporation of G_{M1} sequence into liposomes shown to prolong liposome circulation time *in vivo;* evidence provided for oligosaccharide sequence-specific effect (*34*)
Sialyltransferases			
Rat sialyltransferase	Milk oligosaccharides	DHPE, DPPE	Lactose and *N*-acetyllactosamine neoglycolipids shown to be effective substrates for sialyltransferase activities in liver Golgi vesicles (*35*[a])
Glycosidases			
Human lysosomal *N*-acetylgalactosaminidase	Urinary oligosaccharide	DHPE	Blood group A neoglycolipid shown to be effective substrate for lysosomal β-*N*-acetylgalactosaminidase activity (*36*)
β-Galactosidase (jack bean)	N-Glycosidic complex type; human milk	DPPE	β-Galactosyl-terminating oligosaccharides shown to be effective substrates (*17, 37*)
β-*N*-Acetylhexosaminidase (jack bean)	N-Glycosidic complex type	DPPE	β-*N*-Acetylglucosaminyl-terminating neoglycolipids shown to be effective substrates (*17*)
Microbial adhesins[e]			
Escherichia coli type 1 fimbriated	N-Glycosidic (as with conglutinin); human milk	DPPE	Direct demonstration of preferential binding to Man_5 oligosaccharides; novel binding specificity to lipid-linked lactose (*38*)
Escherichia coli S fimbriated	Human milk	DPPE	Extension of earlier observations of others on preferential binding to α2-3 linked sialic acid (*39*)
Pseudomonas aeruginosa, cystic fibrosis patients	Human milk and urine	4HDA	Binding specificity toward lacto/neolacto type sequences demonstrated (*37, 40*)
Fusobacterium nucleatum, human gingiva	N-Glycosidic oligosaccharides from salivary glycoprotein	DPPE	Binding specificity demonstrated toward β-galactosyl-terminating oligosaccharides (*41*)
Rotavirus, murine	Chemically synthesized	DPPE	Binding specificity demonstrated toward the HexNAcβ1-3/4 sequence (*42*)

(*continued*)

TABLE I (*continued*)

Recognition proteins	Oligosaccharide sources/types	Lipids[b]	Salient observations
Rotavirus, simian	Synthetic oligosaccharide; O-glycosidic from bovine submaxillary mucins	DPPE	Binding specificity demonstrated toward NeuAcα2-3Galβ1-4GlcNAc (*43*)

[a] *Key to references:* (*1*) P. W. Tang, H. C. Gooi, M. Hardy, Y. C. Lee, and T. Feizi, *Biochem. Biophys. Res. Commun.* **132,** 474 (1985); (*2*) M. S. Stoll, E. F. Hounsell, A. M. Lawson, W. Chai, and T. Feizi, *Eur. J. Biochem.* **189,** 499 (1990); (*3*) P. W. Tang and T. Feizi, *Carbohydr. Res.* **161,** 133 (1987); (*4*) M. S. Stoll, T. Mizuochi, R. A. Childs, and T. Feizi, *Biochem. J.* **256,** 661 (1988); (*5*) P. W. Tang, P. Scudder, H. Mehmet, E. F. Hounsell, and T. Feizi, *Eur. J. Biochem.* **160,** 537 (1986); (*6*) M. Larkin, W. Knapp, M. S. Stoll, H. Mehmet, and T. Feizi, *Clin. Exp. Immunol.* **85,** 536 (1991); (*7*) C. Faille, J. C. Michalski, G. Strecker, D. W. R. MacKenzie, D. Camus, and D. Poulain, *Infect. Immun.* **58,** 3537 (1990); (*8*) M. P. Hayette, G. Strecker, C. Faille, D. Dive, D. Camus, D. W. R. MacKenzie, and D. Poulain, *J. Clin. Microbiol.* **30,** 411 (1992); (*9*) P.-A. Trinel, C. Faille, P. M. Jacquinot, J.-C. Cailliez, and D. Poulain, *Infect. Immun.* **60,** 3845 (1992); (*10*) D. Spillmann, K. Hård, J. Thomas-Oates, J. F. G. Vliegenthard, G. Misevic, M. M. Burger, and J. Finne, *J. Biol. Chem.* **268,** 13378 (1993); (*11*) F. G. Hanisch, B. Auerbach, K. Bosslet, K. Kolbe, U. Karsten, Y. Nakahara, T. Ogawa, and G. Uhlenbruck, *Biol. Chem. Hoppe-Seyler* **374,** 1083 (1993); (*12*) B. Schmitz, J. Peter-Katalinic, H. Egge, and M. Schachner, *Glycobiology* **3,** 609 (1993); (*13*) F. G. Hanisch, G. Heimbüchel, S. E. Baldus, G. Uhlenbruck, R. Schmits, M. Pfreundschuh, M. Schwonzen, M. Vierbuchen, J. Bara, and J. Peter-Katalinic, *Cancer Res.* **53,** 4367 (1993); (*14*) F. G. Hanisch, U. Koldovsky, and F. Borchard, *Cancer Res.* **53,** 4791 (1993); (*15*) P. J. Green, C.-T. Yuen, and T. Feizi, *in* "Leucocyte Typing V" (S. Schlossman, ed.), in press. Oxford Univ. Press, Oxford, 1994; (*16*) R. W. Loveless, T. Feizi, R. A. Childs, T. Mizuochi, M. Stoll, R. G. Oldroyd, and P. J. Lachmann, *Biochem. J.* **258,** 109 (1989); (*17*) T. Mizuochi, R. W. Loveless, A. M. Lawson, W. Chai, P. J. Lachmann, R. A. Childs, S. Thiel, and T. Feizi, *J. Biol. Chem.* **264,** 13834 (1989); (*18*) R. A. Childs, K. Drickamer, T. Kawasaki, S. Thiel, T. Mizuochi, and T. Feizi, *Biochem. J.* **262,** 131 (1989); (*19*) R. A. Childs, T. Feizi, C.-T. Yuen, K. Drickamer, and M. S. Quesenberry, *J. Biol. Chem.* **265,** 20770 (1990); (*20*) D. Solís, T. Feizi, C.-T. Yuen, A. M. Lawson, R. A. Harrison, and R. W. Loveless, *J. Biol. Chem.* **269,** 11555 (1994); (*21*) M. Larkin, R. A. Childs, T. J. Matthews, S. Thiel, T. Mizuochi, A. M. Lawson, J. S. Savill, C. Haslett, R. Diaz, and T. Feizi, *AIDS* (*London*) **3,** 793 (1989); (*22*) R. A. Childs, J. R. Wright, G. F. Ross, C.-T. Yuen, A. M. Lawson, W. Chai, K. Drickamer, and T. Feizi, *J. Biol. Chem.* **267,** 9972 (1992); (*23*) M. Larkin, T. J. Ahern, M. S. Stoll, M. Shaffer, D. Sako, J. O'Brien, C.-T. Yuen, A. M. Lawson, R. A. Childs, K. M. Barone, P. R. Langer-Safer, A. Hasegawa, M. Kiso, G. R. Larsen, and T. Feizi, *J. Biol. Chem.* **267,** 13661 (1992); (*24*) C.-T. Yuen, A. M. Lawson, W. Chai, M. Larkin, M. S. Stoll, A. C. Stuart, F. X. Sullivan, T. J. Ahern, and T. Feizi, *Biochemistry* **31,** 9126 (1992); (*25*) C.-T. Yuen, K. Bezouska, J. O'Brien, M. Stoll, R. Lemoine, A. Lubineau, M. Kiso, A. Hasegawa, N. J. Bockovich, K. C. Nicolaou, and T. Feizi, *J. Biol. Chem.* **269,** 1595 (1994); (*26*) P. J. Green, T. Tamatani, T. Watanabe, M. Miyasaka, A. Hasegawa, M. Kiso, C.-T. Yuen, M. S. Stoll, and T. Feizi, *Biochem. Biophys. Res. Commun.* **188,** 244 (1992); (*27*) K. Bezouska, G. Vlahas, O. Horváth, G. Jinochová, A. Fiserová, R. Giorda, W. H. Chambers, T. Feizi, and M. Pospísil, *J. Biol. Chem.* **269,** 16945–16952 (1994); (*28*) R. W. Loveless, G. Floyd-O'-Sullivan, J. G. Raynes, C.-T. Yuen, and T. Feizi, *EMBO J.* **11,** 813 (1992); (*29*) J. C. Solomon, M. S. Stoll, P. Penfold, W. M. Abbott, R. A. Childs, P. Hanfland, and T. Feizi, *Carbohydr. Res.* **213,** 293 (1991); (*30*) T. Feizi, J. C. Solomon,

C.-T. Yuen, K. C. G. Jeng, L. G. Frigeri, D. K. Hsu, and F.-T. Liu, *Biochemistry* **33,** 6342–6349 (1994); (*31*) R. Horstkorte, M. Schachner, J. P. Magyar, T. Vorherr, and B. Schmitz, *J. Cell Biol.* **121,** 1409 (1993); (*32*) T. Pacuszka and P. H. Fishman, *Biochim. Biophys. Acta* **1083,** 153 (1991); (*33*) T. Pacuszka, R. M. Bradley, and P. H. Fishman, *Biochemistry* **30,** 2563 (1991); (*34*) Y. S. Park and L. Huang, *Biochim. Biophys. Acta* **1166,** 105 (1993); (*35*) G. Pohlentz, S. Schlemm, and H. Egge, *Eur. J. Biochem.* **203,** 387 (1992); (*35*[a]) G. Pohlentz and H. Egge, this volume [13]; (*36*) B. Klima, G. Pohlentz, D. Schindler, and H. Egge, *Biol. Chem. Hoppe-Seyler* **373,** 989 (1992); (*37*) I. J. Rosenstein, C.-T. Yuen, M. S. Stoll, and T. Feizi, *Infect. Immun.* **60,** 5078 (1992); (*38*) I. J. Rosenstein, M. S. Stoll, T. Mizuochi, R. A. Childs, E. F. Hounsell, and T. Feizi, *Lancet* **2,** 1327 (1988); (*39*) F. G. Hanisch, J. Hacker, and H. Schroten, *Infect. Immun.* **61,** 2108 (1993); (*40*) R. Ramphal, C. Carnoy, S. Fievre, J.-C. Michalski, N. Houdret, G. Lamblin, G. Strecker, and P. Roussel, *Infect. Immun.* **59,** 700 (1991); (*41*) B. L. Gillece-Castro, A. Prakobphol, A. L. Burlingame, H. Leffler, and S. J. Fisher, *J. Biol. Chem.* **266,** 17358 (1991); (*42*) C. A. Srnka, M. Tiemeyer, J. H. Gilbert, M. Moreland, H. Schweingruber, B. W. de Lappe, P. G. James, T. Gant, R. E. Willoughby, R. H. Yolken, M. A. Nashed, S. A. Abbas, and R. Laine, *Virology* **190,** 794 (1992); (*43*) R. E. Willoughby, *Glycobiology* **3,** 437 (1993); (*44*) S. H. Barondes, V. Castronovo, D. N. W. Cooper, *et al., Cell* (*Cambridge, Mass.*) **76,** 597 (1994).

[b] DPPE, L-1,2-Dipalmitoyl-*sn*-glycero-3-phosphoethanolamine; DHPE, L-1,2-dihexadecyl-*sn*-glycero-3-phosphoethanolamine; 4HDA, 4-hexadecyl aniline; DOPE, dioleoyl-*sn*-glycero-3-phosphoethanolamine.

[c] For definition of 6-OY, see section on neoglycolipids in microsequencing of glycoprotein oligosaccharides.

[d] Galectin is the new nomenclature for soluble, S-type, β-galactoside-binding animal lectins (*44*).

[e] Microbial adhesins were assayed as whole microorganisms.

is the information accumulating on the animal lectins. A picture is emerging (particularly among proteins involved in inflammation and host defense) of ways in which biological specificities may arise through different modes of carbohydrate recognition, often involving oligosaccharides which are not unique (see also Refs. 17–19): (1) recognition of different motifs in the backbones and core regions of N-glycosidic oligosaccharides by two mannan-binding proteins, (2) recognition of core regions of N-linked oligosaccharides in the context of carrier proteins by serum conglutinin, (3) recognition of saccharides in the context of specific lipids by pulmonary surfactant protein A, (4) recognition of oligosaccharide sequences common to the backbone and peripheral regions of clustered O- and N-glycosidic oligosaccharides as well as glycolipids, among them the major blood groups and related antigens, variously recognized by the selectins and by the galectins, and (5) recognition of 6-phosphorylated mannose by amyloid P protein.

Comments

The initial hopes that the neoglycolipid technology would prove of value in immunochemical analyses, particularly in establishing whether antibodies are directed at carbohydrates, identifying new epitopes, and detecting minor antigenic oligosaccharides, have been realized. Still awaiting wide exploitation is the use of neoglycolipids as immunogens. The combined use of sequence-specific monoclonal antibodies and glycosidases to achieve microimmunosequencing, in conjunction with LSIMS, is another area for development.

At the inception of the neoglycolipid technology, great potential was envisaged[1,20–22] in the broader aspects of biological chemistry, namely, for identifying new oligosaccharide recognition systems, involving new endogenous lectins and adhesins of microbial agents. The emergence of several biologically important and interesting carbohydrate-binding proteins through molecular biological approaches, particularly the C-type lectins, has led our laboratory to focus efforts on oligosaccharide ligand identification. Thus, the potential of the technology to isolate hitherto unknown carbohydrate-binding proteins is largely untapped. Areas for

[17] T. Feizi, *Trends Biochem. Sci.* **16,** 84 (1991).

[18] T. Feizi, *Curr. Opin. Struct. Biol.* **1,** 766 (1991).

[19] T. Feizi, *Curr. Opin. Struct. Biol.* **3,** 701 (1993).

[20] T. Feizi and R. A. Childs, *Biochem. J.* **245,** 1 (1987).

[21] T. Feizi, *Biochem. Soc. Trans.* **16,** 930 (1988).

[22] T. Feizi, *in* "Carbohydrate Recognition in Cellular Function" (G. Bock and S. Harnett, eds.), p. 62. Wiley, Chichester, 1989.

future technical development include the design of high-yield conjugation conditions for glycosaminoglycan chains and micromethods whereby the reducing-end (core) monosaccharide ring structures are preserved. The required chemical principle has long been established[23] and applied,[24–27] and it will need development to become a simple "one test-tube experiment" comparable to the present neoglycolipid procedure.

[23] L. M. Likhosherstov, O. S. Novikova, V. A. Derevitskaja, and N. K. Kochetkov, *Carbohydr. Res.* **146,** C1 (1986).

[24] E. Kallin, H. Lonn, and T. Norberg, *Glycoconjugate J.* **3,** 311 (1986).

[25] I. D. Manger, S. Y. C. Wong, T. W. Rademacher, and R. A. Dwek, *Biochemistry* **31,** 10733 (1992).

[26] G. Arsequell, R. A. Dwek, and S. Y. C. Wong, *Anal. Biochem.* **216,** 165 (1994).

[27] E. Kallin, this volume [20].

Section III

Synthetic Polymers

[20] Use of Glycosylamines in Preparation of Oligosaccharide Polyacrylamide Copolymers

By ELISABET KALLIN

Introduction

Reducing oligosaccharides can be converted to neoglycoconjugates through their glycosylamines. The oligosaccharides are transformed to N-acylated glycosylamine derivatives to obtain derivatives suitable for further conjugaton. In this chapter the N-acryloylation of glycosylamines and subsequent polymerization of the glycosylamides into polyacrylamide copolymers are described.

Polyacrylamide copolymers are nontoxic, high molecular weight, water-soluble conjugates where the degree of incorporation of oligosaccharide can be chosen simply by altering the proportions of the reactants. These polyvalent structures have been successfully used as coating antigens in immunoassays. An advantage of these types of conjugates, over protein conjugates, is that they have a simple and well-characterized structure that reduces the risk for undesired immunological reactions. By using a combination of neoglycoproteins and oligosaccharide polyacrylamide copolymers in immunization and screening, antibodies directed to parts of an antigen other than the sugar epitope can be excluded.

Glycosylamines can be obtained through treatment of a reducing saccharide with saturated ammonia in methanol.[1] This procedure, however, is generally not applicable to complex oligosaccharides for several reasons. First, the reaction is low-yielding and time-consuming. Second, the reaction with methanolic ammonia produces both the α and β forms of the glycosylamines, causing separation problems. Third, and most importantly, larger oligosaccharides are nearly insoluble in methanol, and the addition of water causes the reaction to proceed even slower.

A better way to obtain pure β-glycosylamines has been described for *N*-acetylglucosamine and *N,N'*-diacetylchitobiose.[2] The method consists of treatment of the free saccharide with saturated aqueous ammonium bicarbonate for a period of 6 days at 30°. Pure β-glycosylamines are

[1] H. Paulsen and K. W. Pflughaupt, *in* "The Carbohydrates" (W. Pigman and D. Horton, eds.), Vol. 1B, p. 881. Academic Press, New York, 1980; H. S. Isbell and H. L. Frush, *Methods Carbohydr. Chem.* **8,** 255 (1980).

[2] L. M. Likhosherstov, O. S. Novikova, V. A. Derevitskaja, and N. K. Kochetkov, *Carbohydr. Res.* **146,** c1 (1986).

METHODS IN ENZYMOLOGY, VOL. 242
Copyright © 1994 by Academic Press, Inc.
All rights of reproduction in any form reserved.

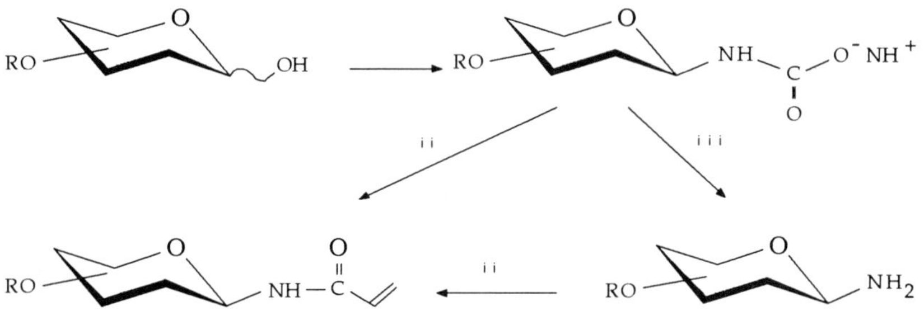

SCHEME 1. Synthesis of glycosylamines and *N*-acryloylglycosylamines. Conditions: i, NH_4HCO_3; ii, acryloyl chloride/Na_2CO_3; iii, cation exchange.

obtained after repeated evaporations and ion-exchange chromatography. This method has been shown to be extendable to sugars other than those terminating with *N*-acetylglucosamine. Several mono-, di-, and oligosaccharides, including fucosylated and sialylated structures, gave β-glycosides in good yields.[3,4]

The rate of the reactions is highly dependent on temperature, and on repeated additions of solid ammonium bicarbonate to compensate for evaporated ammonia and carbon dioxide. The possibility for liberated gases to pass out freely is important for a successful reaction.[2] Later results have shown, however, that the pH is the important factor. By adjusting the pH to 9 through the addition of a few drops of concentrated ammonia, the vessel could be closed and the reaction time lowered.

Nuclear magnetic resonance (NMR) experiments have shown that the formation of glycosylamine occurs via the ammonium carbamate form of the amine (Scheme 1).[3] Such carbamates are known to be stable in alkaline solution or as solid salts, but they decarboxylate rapidly on acidification, and this was also shown to be the case during the cation-exchange workup procedure. This procedure, performed both to obtain the free primary glycosylamine and to purify it from salt and reducing oligosaccharide, implies subjecting the glycosylamine to acidic conditions. It has to be performed with great care and might explain the lower yields previously reported.[2]

Omitting the cation-exchange step in the workup procedure gave the oligosaccharide β-carbamate derivatives in good yields. The amount of reducing sugar in the products was less than 10%, according to NMR. Contamination by free ammonium bicarbonate was minimized by three

[3] E. Kallin, H. Lönn, T. Norberg, and M. Elofsson, *J. Carbohydr. Chem.* **8,** 597 (1989).
[4] R. Roy and C. Laferrier, *J. Chem. Soc., Chem. Commun.* 1709 (1990).

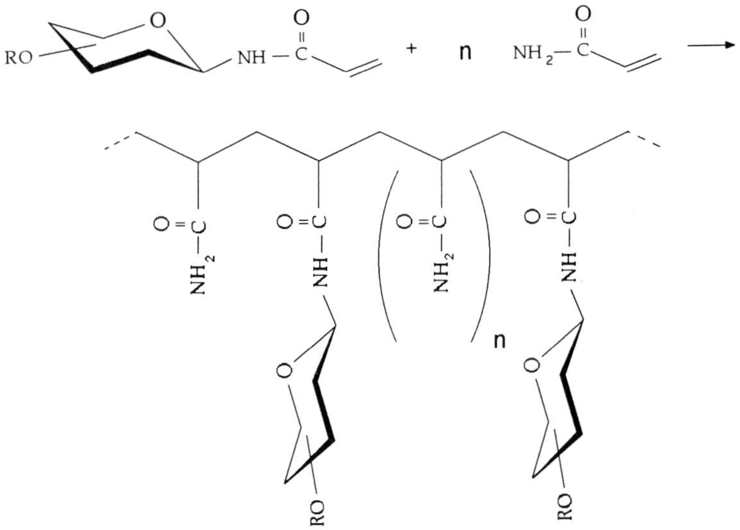

SCHEME 2. Copolymerization of *N*-acryloylglycosylamines with acrylamide.

successive evaporations of highly diluted aqueous solutions, followed by lyophilization. Whether free or in the form of carbamates, the glycosylamines are stable in aqueous solutions above pH 8 for several days or weeks, and as lyophilized powders for long periods. They hydrolyze rapidly around pH 5 but are stable again at low pH (aqueous HCl) owing to the stabilizing effect caused by protonation of the amino group.

Acylation of glycosylamines with acryloyl chloride proceeds smoothly in water-containing media provided a sufficient amount of base is added to prevent acidification of the reaction mixture. It is convenient to use a water-soluble salt as base.[5] Besides the advantage of giving a buffering effect to the system, the water-soluble salt is easily removed by a solid-phase extraction procedure. Thus, treatment of β-glycosylamines with acryloyl chloride in aqueous solution in the presence of sodium carbonate gives N-acryloylated β-glycosylamines (Scheme 1).[3,4] Acylation of glycosyl-*N*-carbamates is done as easily as when using the free glycosylamines.[3]

The N-acryloylated glycosylamines are copolymerized with acrylamide to form high molecular weight, linear copolymers (Scheme 2).[6] By adding a cross-linking reagent to the reaction mixture insoluble gels can

[5] P. H. Weigel, R. L. Schnaar, S. Roseman, and Y. C. Lee, this series, Vol. 83, p. 294.
[6] V. Hořejší, P. Smolek, and J. Kocourek, *Biochim. Biophys. Acta* **538,** 293 (1978).

be prepared.[5,7] The degree of oligosaccharide incorporation in the copolymers is about 70% of the theoretical value. This is higher than what is the case when, for example, allyl glycosides are copolymerized with acrylamide. Then the lower reactivity of the allyl group compared to the acryloyl function necessitates the use of a larger amount of glycoside.[8]

The molecular weight distribution of the copolymers is usually in the range of 100,000–500,000, centered around 300,000 although molecular weights of over 1,000,000 can be obtained. The molecular weight increases with decreasing concentrations of initiator and increasing concentrations of monomers. However, too high a concentration of monomers may result in an insoluble product. The dominant factors influencing the molecular weight distribution are the purity and concentration of the monomers and maintenance of an oxygen-free reaction mixture.

General Methods

All reactions except the preparation of glycosylamines are performed under nitrogen. Concentrations are performed at a bath temperature below 30°. Optical rotations are recorded at 21° with a Perkin-Elmer (Norwalk, CT) 241 polarimeter. The NMR spectra are recorded in D_2O with a Bruker AM 500 instrument. Acrylamide (enzyme grade, Eastman Kodak Co., Rochester, NY) is used without further purification. Gel-permeation chromatography is performed on Fractogel TSK HW-55(F) (Merck, Darmstadt, Germany) with water as eluent. Dextran standards are from Pharmacosmos (Viby, Denmark). To prevent self-polymerization 2,6-di-*tert*-butyl-4-methylphenol (0.5%, w/v) in tetrahydrofuran is used as inhibitor solution. Organic solvents are of analytical grade. Other methods are the same as described before.[9]

Preparation of Glycosylamines

Solid ammonium bicarbonate is added until saturation to a solution of oligosaccharide (50 mg) in water (2.5 ml). Concentrated aqueous ammonia is added to bring the pH to 9. The mixture is stirred at room temperature for 2–5 days. Ammonium bicarbonate is added at intervals, assuring saturation by keeping a portion of solid salt constantly present in the mixture. When thin-layer chromatography (TLC) indicates no further conversion,

[7] V. Hořejší and J. Kocourek, *Biochim. Biophys. Acta* **297,** 346 (1973); V. Hořejší and J. Kocourek, this series, Vol. 34, p. 361.

[8] A. Y. Chernyak, K. V. Antonov, N. D. Kochetkov, L. N. Padyukov, and N. V. Tsvetkova, *Carbohydr. Res.* **141,** 199 (1985).

[9] E. Kallin, this volume [12].

the mixture is diluted with water (100 ml) and concentrated to half the volume. This procedure is repeated twice, and the residue is lyophilized. The obtained crude glycosylamine, present in carbamate form, is used without further purification. It should be noted that TLC analysis often gives a false picture of how the reactions proceed, owing to the lability of the product under the chromatographic conditions.

Preparation of N-Acryloylated Glycosylamines

Sodium carbonate (100 mg) and methanol (1.0 ml) are added to a solution of the crude glycosylamine (0.14 mmol) in water (1.0 ml). The mixture is stirred at 0° while acryloyl chloride (60 ml, 0.74 mmol) in tetrahydrofuran (0.5 ml) is added dropwise. After 10 min, the solution is diluted with water (3 ml) and concentrated to 2 ml. The solution is again diluted with water (2 ml), inhibitor solution (0.2 ml) is added, and the solution is concentrated to 1–2 ml and applied to a Bond Elut C_{18} cartridge (5 g gel), equilibrated in water. Elution with water gives salts, unreacted glycosylamine, and reducing sugar in the first fractions, and the desired product in the later fractions. In some cases, elution of the product is preferably speeded up by adding methanol to the eluant. Product-containing fractions are combined, mixed with a few drops of inhibitor solution, and concentrated to 2 ml. This solution is purified by gel filtration on a BioGel P-2 (Bio-Rad, Richmond, CA) column. Appropriate fractions are combined and lyophilized.

The *N*-acryloylated glycosylamines are, as predicted, much more stable toward hydrolysis than the glycosylamines. However, the presence of the acryloyl group introduces a tendency to self-polymerization; therefore, addition of small amounts of inhibitor solution is necessary during some operations.

Preparation of Polyacrylamide Copolymers

A solution of the *N*-acryloylglycosylamine (52 mmol) and acrylamide (210 mmol, 15 mg) in distilled water (0.4 ml) is deaerated by flushing with nitrogen for 20 min. The solution is then stirred at 0°, and *N,N,N′,N′*-tetramethylethylenediamine (TEMED, 0.002 ml) and ammonium persulfate (1.0 mg) are added. The mixture is slowly stirred at 0° for 2 hr, then at room temperature overnight. The viscous solution is diluted with water (1 ml) and fractionated by gel filtration on Fractogel HW 55(F). Fractions containing polymer are combined and lyophilized.

The polyacrylamide copolymers are characterized by optical rotation and NMR. ^{1}H NMR spectroscopy at 50° allows a good estimation of the degree of substitution of the polymer, through integration of the sugar anomeric signals and the CH and CH_2 groups in the polymer backbone. The molecular weight distribution of the copolymer is determined by gel filtration on Fractogel HW 55(F), using dextran standards for calibration.

[21] Synthesis of Poly(*N*-acetyl-β-lactosaminide-carrying Acrylamide): Chemical–Enzymatic Hybrid Process

By KAZUKIYO KOBAYASHI, TOSHIHIRO AKAIKE, and TAICHI USUI

Introduction

Increasing attention is being paid to glycopolymers[1] which are synthetic polymers substituted with pendant carbohydrate moieties. Several different types of glycopolymers have been used as biomedical materials such as cell-specific culture substrata,[2–6] artificial antigens,[7,8] and targeted drug delivery agents.[9] They are also useful as tools for investigating biological recognition phenomena[10–17] using lectins and anticarbohydrate monoclonal

[1] R. Roy, F. D. Tropper, and A. Romanowska, *Bioconjugate Chem.* **3,** 256 (1992).
[2] P. H. Weigel, R. L. Schnaar, M. S. Kuhlenschmidt, E. Schmell, R. T. Lee, Y. C. Lee, and S. Roseman, *J. Biol. Chem.* **254,** 10830 (1979).
[3] A. Kobayashi, T. Akaike, K. Kobayashi, and H. Sumitomo, *Makromol. Chem., Rapid Commun.* **7,** 645 (1986).
[4] S. Tobe, Y. Takei, K. Kobayashi, and T. Akaike, *Biochem. Biophys. Res. Commun.* **184,** 225 (1992).
[5] A. Kobayashi, K. Kobayashi, S. Tobe, and T. Akaike, *J. Biomater. Sci. Polym. Ed.* **3,** 499 (1992).
[6] K. Kobayashi, A. Kobayashi, S. Tobe, and T. Akaike, *in* "Neoglycoconjugates" (Y. C. Lee and R. T. Lee, eds.) p. 261. Academic Press, San Diego, California, 1994.
[7] N. K. Kochetkov, *Pure Appl. Chem.* **56,** 923 (1984).
[8] A. Rozalski, L. Brade, H.-M. Kuhn, J. Brade, P. Kosma, B. J. Appelmek, S. Kusumoto, and H. Paulsen, *Carbohydr. Res.* **193,** 257 (1989).
[9] R. Duncan, P. Kopeckova-Rojmanova, J. Strohalm, I. Hume, H. C. Cable, J. Pohl, J. B. Lloyd, and J. Kopecek, *Br. J. Cancer* **55,** 165 (1987).
[10] L. A. Carpino, H. Ringsdorf, and H. Ritter, *Makromol. Chem.* **177,** 1631 (1976).
[11] R. Roy and F. D. Tropper, *J. Chem. Soc., Chem. Commun.*, 1058 (1988).
[12] R. Roy, F. D. Tropper, and A. Romanowska, *J. Chem. Soc., Chem. Commun.*, 1611 (1992).
[13] R. Roy, F. D. Tropper, T. Morrison, and J. Boratynski, *J. Chem. Soc., Chem. Commun.*, 536 (1991).
[14] S. Nishimura, K. Matsuoka, and K. Kurita, *Macromolecules* **23,** 4182 (1990).
[15] S. Nishimura, K. Matsuoka, T. Furuike, S. Ishii, K. Kurita, and K. Nishimura, *Macromolecules* **24,** 4236 (1991).

Copyright © 1994 by Academic Press, Inc.
All rights of reproduction in any form reserved.

antibodies. Attempts have been made to introduce more biologically important, complex oligosaccharides as well as to design more simple synthetic routes.

This chapter concerns mainly *N*-acetyllactosamine residues, which are major components of core oligosaccharides of glycoproteins and glycolipids and can serve as differentiation antigens, tumor-associated antigens, and components of receptor systems.[18,19] Glycopolymers carrying pendant *N*-acetyllactosamine residues are expected to exhibit interesting functions in various biological recognition events. For this reason, a simple and large-scale preparation of *N*-acetyllactosamine must be established.

N-Acetyllactosamine was found in bovine colostrum as a free component[20] and has been isolated from partial hydrolyzates of porcine gastric mucin[21] and human milk.[22] Because the yield of the disaccharide from natural products is insufficient, several organic syntheses of *N*-acetyllactosamine have been developed by way of glycosylations between galactose and *N*-acetylglucosamine derivatives and chemical modifications of disaccharides.[23,24] However, these tedious multistep procedures involving protections and deprotections are not well suited for large-scale productions. To overcome these difficulties, synthetic strategies using enzymes as catalysts have been introduced.[25–29] Approaches using glycosyltransferases[30] are promising, but, at present, the transferases and their cofactors as well as UDP-galactose are not readily available in large quantities.

More practical approaches take advantage of the transglycosylation activity of glycosidases.[25–29] Usui *et al.*[29] succeeded in regioselective syn-

[16] K. Hatanaka, Y. Ito, A. Maruyama, Y. Watanabe, and T. Akaike, *Macromolecules* **26,** 1483 (1993).

[17] J. Klein and A. H. Bergli, *Makromol. Chem.* **190,** 2527 (1989).

[18] J. C. Paulson, *Trends Biochem. Sci.* **14,** 272 (1983).

[19] T. Feizi, *Trends Biochem. Sci.* **16,** 84 (1991).

[20] T. Saito, T. Ito, and S. Adachi, *Biochim. Biophys. Acta* **801,** 147 (1984).

[21] R. J. Tomarelli, J. B. Hassinen, E. R. Eckhardt, R. H. Clark, and F. W. Bernhart, *Arch. Biochem. Biophys.* **48,** 225 (1954).

[22] R. Kuhn, H. H. Baer, and A. Gauhe, *Chem. Ber.* **87,** 1553 (1954).

[23] R. T. Lee and Y. C. Lee, *Carbohydr. Res.* **77,** 270 (1979).

[24] J. Alais and A. Veyrieres, *Carbohydr. Res.* **93,** 164 (1981).

[25] K. G. I. Nilsson, *Trends Biotech.* **6,** 256 (1988).

[26] S. David, C. Auge, and C. Gautheron, *Adv. Carbohydr. Chem. Biochem.* **49,** 175 (1991).

[27] E. J. Toone, E. S. Simon, M. D. Bednarski, and G. M. Whitesides, *Tetrahedron* **45,** 5365 (1989).

[28] D. G. Drueckhammer, W. J. Hennen, R. L. Pederson, C. F. Barbas III, C. M. Gautheron, T. Krach, and C. H. Wong, *Synthesis* 499 (1991).

[29] T. Usui, *Trends Glycosci. Glycotech.* **4** (No. 15), 116 (1992).

[30] C. H. Wong, S. L. Haynie, and G. M. Whitesides, *J. Org. Chem.* **47,** 5418 (1982).

thesis of several oligosaccharide derivatives on a preparative scale using chitinase,[31,32] lysozyme,[33] amylase,[34] and galactosidase[35,36] all of which are readily available and inexpensive. This chapter describes an example of such synthetic procedures. *p*-Nitrophenyl *N*-acetyl-β-lactosaminide (**1**),[36] an attractive starting material for glycopolymers, is synthesized from *p*-nitrophenyl *N*-acetyl-β-D-glucosaminide as glycosyl acceptor and lactose as glycosyl donor by one-step procedure using a β-D-galactosidase from *Bacillus circulans* (Scheme 1). Poly(*N*-acetyl-β-lactosaminide-carrying acrylamide) (**4**)[37] can be obtained by the following three steps: reduction of the nitro function to yield the *p*-aminophenyl glycoside (**2**), introduction of an acryloyl function to give the *p*-acryloylaminophenyl glycoside monomer (**3**), and finally radical homopolymerization of **3** to yield **4**.

Methods

The synthesis procedure follows previously published methods.[36,37]

Materials

β-D-Galactosidase from culture filtrates of *Bacillus circulans* is commercially available as Biolacta from Daiwa Kasei Co., Ltd. (Osaka, Japan). *p*-Nitrophenyl *N*-acetyl-β-D-glucosaminide (*p*NP = GlcNAc) is available from Yaizu Suisankagaku Co., Ltd. (Yaizu, Japan). GlcNAc-Sepharose CL-4B is prepared by coupling *p*-aminophenyl *N*-acetyl-β-D-glucosaminide to formyl-Sepharose CL-4B according to the method reported.[32] Palladium-on-carbon is obtained from Kishida Chemical Co. Ltd. (Osaka, Japan). 1-Ethyl-3-(3-dimethylaminopropyl)carbodiimide hydrochloride is obtained from Tokyo Chemical Industry Co. Ltd. (Tokyo, Japan). Azobisisobutyronitrile (AIBN; Tokyo Chemical Industry Co. Ltd.) is purified by recrystallization from ethanol. Dimethyl sulfoxide (DMSO) is distilled under reduced pressure.

[31] T. Usui, Y. Hayashi, F. Nanjo, K. Sakai, and Y. Ishido, *Biochim. Biophys. Acta* **923,** 302 (1987).

[32] T. Usui, H. Matsui, and K. Isobe, *Carbohydr. Res.* **203,** 65 (1990).

[33] T. Usui and T. Murata, *J. Biochem.* (*Tokyo*) **103,** 969 (1988).

[34] T. Usui, Y. Hayashi, F. Nanjo, and Y. Ishido, *Biochim. Biophys. Acta* **953,** 179 (1988).

[35] K. Sakai, R. Katsumi, H. Ohi, T. Usui, and Y. Ishido, *J. Carbohydr. Chem.* **11,** 553 (1992).

[36] T. Usui, S. Kubota, and H. Ohi, *Carbohydr. Res.* **244,** 315 (1993).

[37] K. Kobayashi, N. Kakishita, M. Okada, T. Akaike, and T. Usui, *J. Carbohydr. Chem.* **13,** 753 (1994).

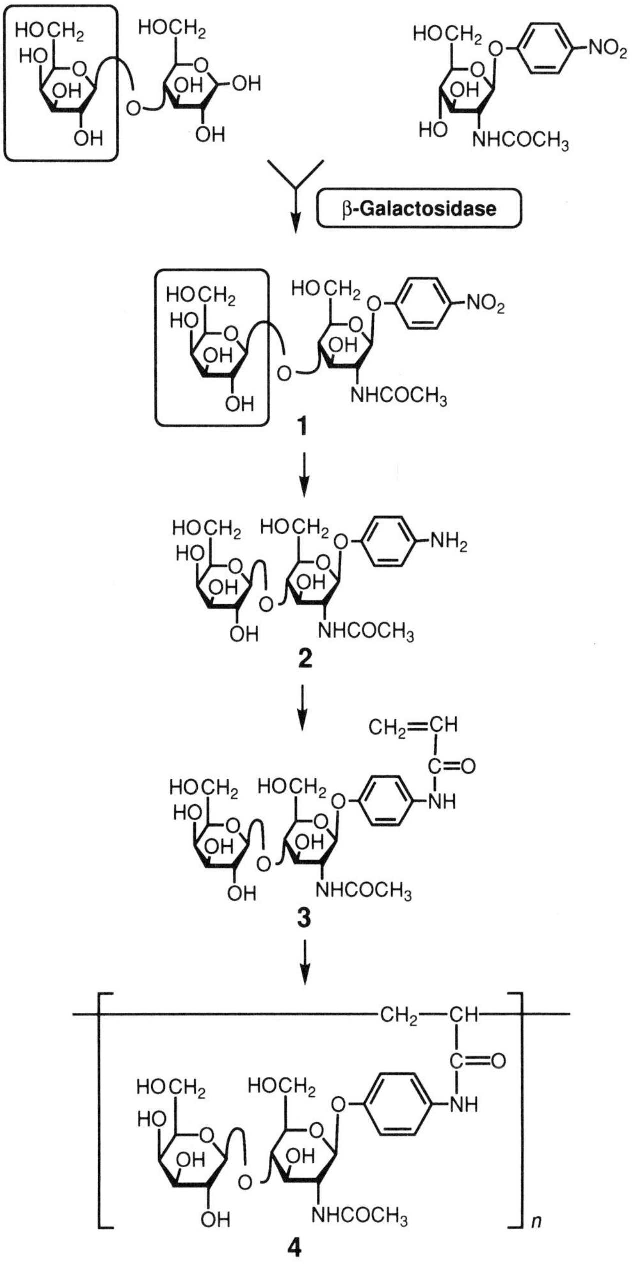

SCHEME 1. Synthesis of poly(*N*-acetyl-*β*-lactosaminide-carrying acrylamide) via a chemical–enzymatic hybrid process.

Enzyme Assay

An appropriate amount of the enzyme in a total volume of 0.1 ml is incubated with 2 m*M* *o*-nitrophenyl β-D-galactopyranoside in 0.9 ml of 50 m*M* sodium phosphate buffer (pH 7.0) at 30° for 10 min. The reaction is stopped with 0.1 *M* sodium carbonate (2 ml), and the *o*-nitrophenol liberated is determined colorimetrically at 420 nm. The amount of the enzyme releasing 1 μmol of *o*-nitrophenol per minute is defined as one unit of activity.

Partial Purification of Enzyme

Crude β-D-galactosidase (20 U) is dissolved in 4 ml of 60 m*M* phosphate buffer (pH 6.0), loaded onto a GlcNAc-Sepharose CL-4B column (1.3 × 4 cm), and eluted with the same buffer (84 ml). The eluate is concentrated to 2 ml with an Amicon (Danvers, MA) Diaflo ultrafilter equipped with a PM10 membrane operating at a pressure of 50 psi. The solution is lyophilized, and then the product is stored over $CaSO_4$ at 4°. This fraction exhibits most of the β-D-galactosidase activity but no β-*N*-acetylhexosaminidase (NAHase) activity. When the elution is continued after changing the eluent to 0.1 *M* acetic acid (36 ml) containing 1 *M* NaCl and 1% (w/v) GlcNAc, the resulting eluate exhibits most of the NAHase activity.

Synthesis Procedure

p-Nitrophenyl 2-Acetamide-2-deoxy-4-O-β-D-galactopyranosyl-β-D-glucopyranoside (pNP = β-LacNAc, ***1****).* Lactose (7.7 g) and *p*NP = β-GlcNAc (2.9 g) in a molar ratio of 1 : 0.4 are dissolved in a mixture of 20 m*M* phosphate buffer (pH 7.0) (50 ml) and acetonitrile (50 ml). The partially purified β-D-galactosidase (40 U) is added, and the solution is incubated at 30° for 8 hr. The solution is concentrated to a syrup in a rotary evaporator (0.1 torr) at 40°, the syrup is dissolved in 3 : 1 (v/v) water–methanol (150 ml), and the solution is centrifuged to remove insoluble substances. The solution in portions of 40–50 ml is chromatographed through a Toyopearl HW-40S (Tosoh Co., Tokyo, Japan) column (4.5 × 95 cm) with 3 : 1 (v/v) water–methanol as eluent. The elution is monitored by the absorbance at 300 nm (*p*-nitrophenyl group) and by the phenol–sulfuric acid method at 490 nm (carbohydrate content).[38]

A representative chromatogram is depicted in Fig. 1, which shows one main peak (fraction 3) and two minor peaks (fractions 1 and 2) as transglycosylation products. Fraction 3 is concentrated to dryness, and

[38] J. E. Hodge and B. T. Hofreiter, *in* "Methods in Carbohydrate Chemistry" (R. L. Whisler and M. L. Wolfrom, eds.), Vol. 1, p. 380. Academic Press, New York, 1962.

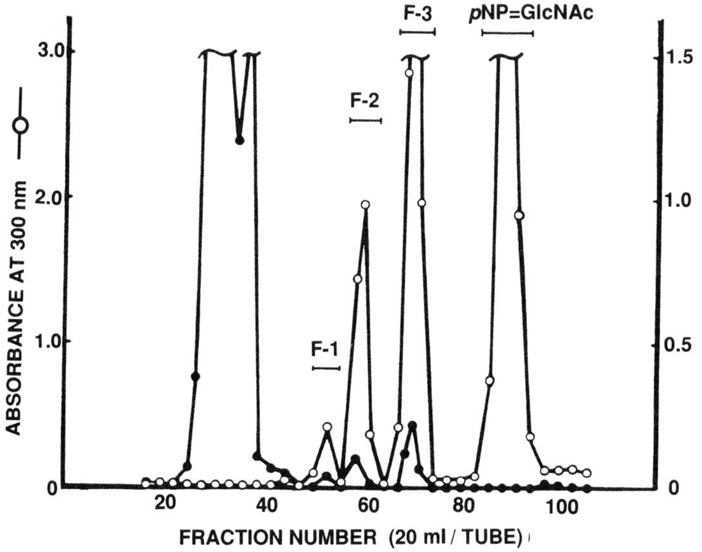

FIG. 1. Chromatograms of transglycosylation products obtained from lactose and pNP = GlcNAc by the action of *B. circulans* β-D-galactosidase. Elution through a Toyopearl HW-40S column (4.5 × 95 cm) was accomplished with 3 : 1 (v/v) water–methanol as eluent at 150 ml/hr.

the product is crystallized from methanol to afford the disaccharide **1** in a yield of 0.57 g, mp 213°; $[\alpha]_D^{25}$ −18.4° (*c* 1, water). Fractions 1 and 2 are separately concentrated and lyophilized to afford β(1→6)-linked disaccharide (*p*-nitrophenyl 2-acetamide-2-deoxy-6-*O*-β-D-galactopyranosyl-β-D-glucopyranoside) and β(1→4)-linked trisaccharide [*p*-nitrophenyl 2-acetamide-2-deoxy-4-*O*-(4-*O*-β-D-galactopyranosyl-β-D-galactopyranosyl)-β-D-glucopyranoside], respectively.

*p-Acryloylaminophenyl 2-Acetamido-2-deoxy-4-O-β-D-galactopyranosyl-β-D-glucopyranoside (p-acryloylaminophenyl = β-LacNAc, **3**).* The *p*NP = β-LacNAc (**1**, 1.50 g, 3 mmol) is dissolved in a mixture of methanol (75 ml) and water (75 ml). Palladium-on-carbon (270 mg) is added, and hydrogen gas is introduced from a bomb for 8.5 hr. Thin-layer chromatography (TLC) shows that the spot (ethanol) of R_f 0.51 (**1**) is converted to R_f 0.40 (**2**). The palladium catalyst is filtered off, and the filtrate is condensed under reduced pressure. The residue (**2**) is dissolved in 50 ml of water, and a trace of 2,6-di-*tert*-butyl-4-methylphenol is added as radical inhibitor. The flask is cooled at 5°, 1-ethyl-3-(3-dimethylaminopropyl)carbodiimide hydrochloride (0.58 g, 3.0 mmol) is added, and then acrylic acid (0.22 ml, 3.2 mmol) is added through a dropping funnel. The solution is stirred in a cold room at 5° for 24 hr, after which TLC (ethanol) shows **3** at R_f 0.48. Water is removed by freeze-drying, and the product is purified by silica gel chromatography (4 × 50 cm; eluent, ethanol). The yield is 1.1 g (69%); mp 227°–228°; $[\alpha]_D^{25}$ +7.0° (*c* 0.10, water–ethanol 1 : 1 by volume).

*Poly[p-acryloylaminophenyl 2-Acetamido-2-deoxy-4-O-β-D-galactopyranosyl-β-D-glucopyranoside] (**4**).* Monomer **3** (1.10 g) and AIBN (3.4 mg) are dissolved in DMSO (2.5 ml) in a polymerization ampoule. The solution is frozen and degassed three times under reduced pressure by placing the ampoule in a solid carbon dioxide–methanol bath. The ampoule is sealed under reduced pressure and kept in a thermostat at 60° for 12 hr. Polymerization is terminated by pouring the solution into an excess amount of cold methanol. After centrifugation, the polymeric product is redissolved in DMSO, precipitated by pouring it into methanol (repeated three times), and dialyzed in a cellulose tube (Nakarai Chemicals, Ltd., Kyoto, Japan; molecular weight cutoff, 8000) against water. A white powdery polymer is isolated by freeze-drying from an aqueous solution. The yield of **4** is 1.00 g (91%).

Summary of Synthesis of Glycopolymers

These types of glycopolymers can be synthesized from other *p*-nitrophenyl glycosides, some of which are commercially available and others can be prepared readily.[37] Three other glycopolymers, carrying α-lactose (**5**), *N*-acetyl-β-D-glucosamine (**6**), and α-D-glucopyranose (**7**) moieties (Fig. 2), have been prepared as described for **4**. Table I summarizes the representative data of the polymerization using 1 to 3 g of *p*-acryloylaminophenyl glycosides as monomers, together with some physical properties of both monomers and polymers. Because the solubility in DMSO is higher than that in water, the polymerization is carried out at rather high monomer concentrations in DMSO with azobisisobutyronitrile (1 mol%) as initiator at 60°. The polymerizations proceed homogeneously to give viscous solutions. All of the polymers are obtained as white powders and are soluble in DMSO and water. Weight-average molecular weights ($\overline{M}_w$) determined by light scattering range from 1.4 × 10^5 to 4.4 × 10^5.

Properties of Glycopolymers

Glycopolymers **4–7** showed some unique solution properties in water in terms of solubility, solution viscosities, nuclear magnetic resonance (NMR) spectra, gel-permeation chromatograms, and light scattering measurements as follows. (1) The homopolymer **6** was soluble in water, in

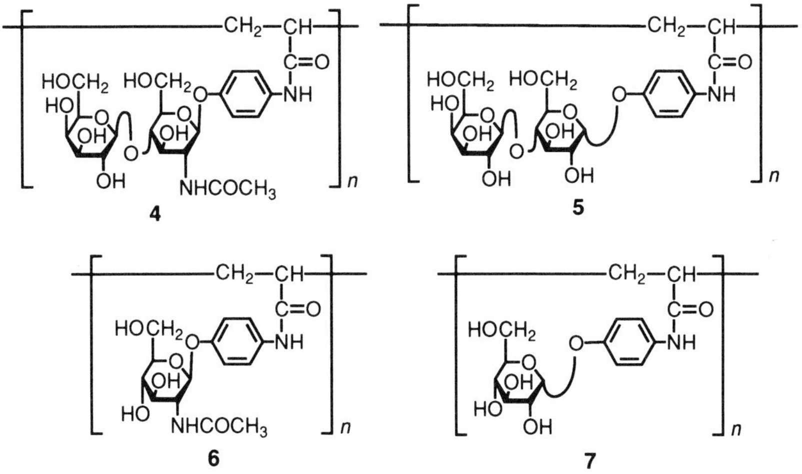

FIG. 2. Glycopolymers prepared by homopolymerization.

spite of the insolubility of the corresponding monomeric compound in water. (2) The limiting viscosity numbers of the polymers determined in water were smaller than the corresponding values determined in DMSO (Table I). (3) In the ^{13}C NMR spectra of the polymers in D_2O, some signals of the phenyl moiety became broad, and the main-chain methylene, methine, and carbonyl signals were too broad to be detected. (4) Average molecular weights estimated by gel-permeation chromatography using two column/eluent combinations were quite different from one another. For example, the polymer **4** had $\overline{M}_n = 1.9 \times 10^6$ ($\overline{M}_w/\overline{M}_n = 3.0$) (water eluent) and $\overline{M}_n = 2.4 \times 10^4$ ($\overline{M}_w/\overline{M}_n = 24$) (DMSO eluent). The column/eluent combinations used were Shodex OHpak KB-802 and KB-803 columns with water eluent and Shodex KF-803 and KF-804 columns with DMSO eluent. Several standard pullulan samples (M_n or $M_w = 5.3 \times 10^3$ to 2.36×10^6) were used as references for both combinations. (5) Zimm plots in light scattering measurements in water gave distorted rectilinear grids.

The solution properties (1)–(3) in water can be explained if one assumes that these polymers tend to take the following conformations owing to their amphiphilic structures. The hydrophobic phenyl moieties along the polymer chain are occluded in the inside of the molecule and the hydrophilic carbohydrate moieties protrude outside, which induces the polymer to form a tightly coiled conformation in water. The amphiphilic polymers also may have tendencies to be adsorbed to solid surfaces and to form molecular aggregates, which might be the reason for phenomena (4) and (5).

TABLE I

SYNTHESIS OF POLY[*p*-ACRYLOYLAMINOPHENYL GLYCOSIDE] POLYMERS

Glycoside	Monomer		Polymerization[a]				Polymer				
	Mp (°C)	$[\alpha]_D^{25}$	Feed (g)	DMSO (ml)	Time (hr)	Yield (%)	Structure	$\overline{M_w}$[b] ($\times 10^5$)	[η], dl/g[c] DMSO	[η], dl/g[c] Water	$[\alpha]_D^{25}$ [d]
β-LacNAc	227–228	+7.0° [e]	1.1	2.5	12	91	**4**	3.2	—	0.19	−5.2°
α-Lac	170–172	+40.6° [f]	1.5	3.0	9	50	**5**	3.0	0.38	0.20	+42.3°
α-Lac	—	—	3.0	8.0	12	45	**5**	1.4	0.20	0.10	+41.3°
β-GlcNAc	243–245	+9.0° [f]	1.8	5.0	8	85	**6**	4.4	0.66	0.21	−11.2°
α-Glc	170–172	+149.9° [d]	2.3	5.0	5.5	80	**7**	1.4	0.92	0.62	+136.6°

[a] Azobisisobutyronitrile, 1 mol%; 60°.
[b] Determined by light scattering method (in water at 25°).
[c] At 25°.
[d] *c* 1.0 in water.
[e] *c* 0.1 in water/ethanol (1/1).
[f] *c* 1.0 in DMSO.

N-Acetyl-β-lactosaminide and the other carbohydrate moieties of the polymers are not only attached to every repeating unit along the polymer chain, but are also crowded together outside the tightly coiled hydrophobic cores. Clustering of glycosignals is one of the most important factors in many biological recognition events.[39] It is likely, therefore, that these polymers have superior properties in biorecognition. These characteristics of the homopolymers are distinct from those of the copolymers with acrylamide and other vinyl-containing compounds, which is another interesting strategy in the design of biomaterials.

The amphiphilic properties of glycopolymers **4–7** are comparable to those of the corresponding glycopolymers of styrene derivatives previously reported,[40,41] although the hydrophobicity of the *p*-acryloylaminophenyl portion is not as strong as that of the *p*-vinylbenzyl portion of polystyrene derivatives because of the hydrophilic nature of the amide group. Among the polystyrene-type glycopolymers, poly[*N*-*p*-vinylbenzyl-*O*-β-D-galatopyranosyl-(1 → 4)-D-gluconamide] (lactose-carrying polystyrene, abbreviated as PVLA) has been reported[3–6,42] to be a useful substratum for culture of hepatocytes. This unique culture system is based on the carbohydrate-specific dynamic interactions between clustered galactose along the polymer chains and asialoglycoprotein receptors on the surface of the cells. We expect that glycopolymers **4–7** described in this chapter will exhibit effective biological recognitions owing to the high density of clustered glycosignals and will be applied for cell-specific culture substrata as well as other biomedical materials.

[39] Y. C. Lee, *Ciba Found. Symp.* **145,** 80 (1989).
[40] K. Kobayashi and T. Akaike, *Trends Glycosci. Glycotech.* **2** (No. 2), 26 (1990).
[41] K. Kobayashi, H. Sumitomo, and Y. Ina, *Polym. J.* **17,** 567 (1985).
[42] K. Kobayashi, A. Kobayashi, and T. Akaike, this series, Vol. 247 [30].

[22] Preparation of Glycoprotein Models: Pendant-Type Oligosaccharide Polymers

By SHIN-ICHIRO NISHIMURA, TETSUYA FURUIKE, and KOJI MATSUOKA

Introduction

This chapter covers advances in polymerizable glycosides which are useful for the preparation of macromolecules having pendant-type carbohydrates. Synthetic glycoconjugates have gained interest both as conve-

Copyright © 1994 by Academic Press, Inc.
All rights of reproduction in any form reserved.

nient tools for investigating the biological roles of natural glycoconjugates and as a new class of biomaterials in medical science.

The conversion of sugars to polymers with high molecular weight via a copolymerization reaction was originally based on the method first reported by Horejsi and Kocourek[1] and involved allyl glycosides of monosaccharides with acrylamide. Using this method, Kochetkov and co-workers[2] obtained a variety of synthetic antigens which showed the group specificity of the lipopolysaccharide epitopes of *Salmonella*. Furthermore, Lee and co-workers[3] demonstrated that polyacrylamide gels containing sugar residues exhibited a specific capacity to adhere hepatocytes. Some of the polymers containing *N*-acetylneuramic acid have been found to show potent inhibitor effects on hemagglutination by influenza virus.[4]

Progress in the chemical syntheses of complex oligosaccharides of glycoconjugates[5] has considerably enhanced our ability to design and synthesize not only naturally occurring bioactive oligosaccharides but simplified biomimetic molecules for investigating the nature of cell surface receptors that specifically interact with sugar molecules. Although a number of polymerizable glycosides are available as monomers for the preparation of carbohydrate-carrying macromolecules, judicious consideration must be given for establishment of sophisticated methods for syntheses of the functional glycosides that have, in addition to the polymerizable functions, appropriate spacer-arm moieties to provide the flexibility of carbohydrate branches.

n-Pentenyl Group as a Simple and Efficient Polymerizable Aglycon

Model Monomers from N-Acetyl-D-Glucosamine

Prior to the design of synthetic glycoconjugates containing complex oligosaccharide chains, careful evaluation of several aglycons having a C=C bond at the terminal end has been carried out with simple *N*-acetyl-D-glucosamine derivatives to clarify the effect of the chain length of the spacer arm on polymerization behavior. These model monomers (Scheme 1, **6–9**) were derived from the oxazoline derivative **1**, readily prepared by the trimethylsilyltrifluoromethane sulfonate (TMSOTf) method[6] from

[1] V. Horejsi and J. Kocourek, *Biochim. Biophys. Acta* **297,** 346 (1973).

[2] A. Y. Chernyak, A. B. Levinsky, B. A. Dmitriev, and N. K. Kochetkov, *Carbohydr. Res.* **128,** 269 (1984).

[3] P. H. Weigel, E. Schmell, Y. C. Lee, and S. Roseman, *J. Biol. Chem.* **253,** 330 (1978).

[4] A. Spaltenstein and G. M. Whiteside, *J. Am. Chem. Soc.* **113,** 686 (1991).

[5] O. Kanie and O. Hindsgaul, *Curr. Opin. Struct. Biol.* **2,** 674 (1992).

[6] S. Nakabayashi, C. D. Warren, and R. W. Jeanloz, *Carbohydr. Res.* **150,** c7 (1986).

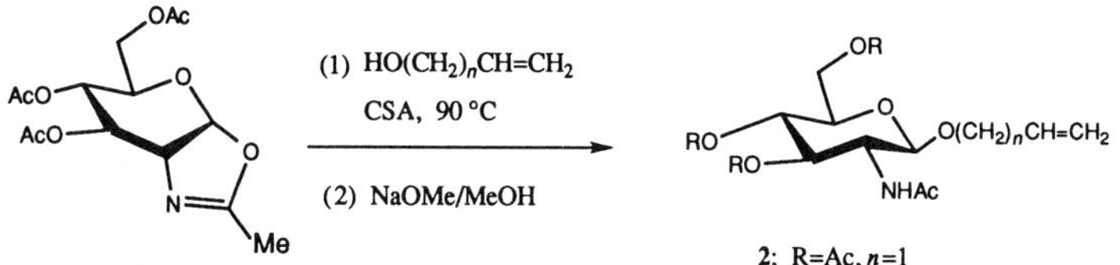

1

2; R=Ac, *n*=1
3; R=Ac, *n*=2
4; R=Ac, *n*=3
5; R=Ac, *n*=9
6; R=H, *n*=1
7; R=H, *n*=2
8; R=H, *n*=3
9; R=H, *n*=9

SCHEME 1. Monomers from *N*-acetyl-D-glucosamine. NaOMe, Sodium methoxide; MeOH, methanol.

2-acetamido-1,3,4,6-tetra-*O*-acetyl-2-deoxy-α-D-glucopyranose, by glycosidation reactions with commercially available alcohols such as 2-propen-1-ol, 3-buten-1-ol, 4-penten-1-ol, and 10-undecen-1-ol in the presence of 10-camphorsulfonic acid (CSA) as the promoter and subsequent de-O-acetylation by the Zemplen method (Scheme 1). As discussed in the following, the TMSOTf method seems to be one of the most efficient and versatile procedures for chemical conversion of oligosaccharides containing a Glc*p*NAc residue at the reducing end to glycosides through the oxazoline derivatives.

*Preparation of n-Pentenyl-2-Acetamido-2-deoxy-β-D-glucopyranoside (**8**).* To a solution of 2-acetamido-1,3,4,6-tetra-*O*-acetyl-2-deoxy-D-glucopyranose (α,β mixture of peracetate) (5.0 g, 12.8 mmol) in 1,2-dichloroethane (10 ml) is added TMSOTf (2.6 ml, 14.1 mmol), and the mixture is stirred at 50° for 5 hr.[7,8] After adjusting the solution to an alkaline pH with triethylamine (9.0 ml), the mixture is directly applied to a column (2.0 × 30 cm) of silica gel, then eluted with 100 : 200 : 1 (v/v/v) toluene–ethyl acetate–triethylamine to give oxazoline **1** in a quantitative yield, as examined by thin-layer chromatography (TLC) [R_f 0.58, 5 : 4 : 1 (v/v) chloroform–ethyl acetate–methanol].

A solution of oxazoline derivative **1** (4.2 g) in dry 4-penten-1-ol (6.6 ml, 64.2 mmol) containing CSA (79.1 mg, 0.34 mmol) is stirred under a nitrogen atmosphere for 5 hr at 90°. The solution is cooled to room temperature, diluted with chloroform, and poured into ice–water. The extract with chloroform is washed successively with aqueous sodium hydrogen carbonate and water, dried over magnesium sulfate, filtered, and evaporated. The residual syrup is subjected to chromatography on

[7] P. Stangier, W. Treder, and J. Tieme, *Glycoconjugate J.* **10,** 26 (1993).
[8] S.-I. Nishimura, K. Matsuoka, and K. Kurita, *Macromolecules* **23,** 4182 (1990).

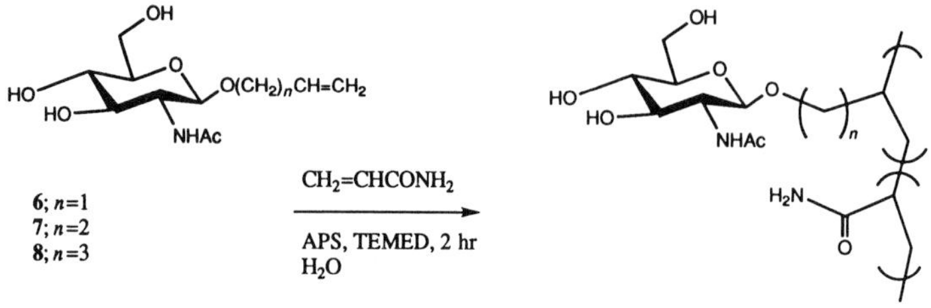

SCHEME 2. Copolymerization of monomers. APS, Ammonium peroxodisulfate; TEMED, *N,N,N′,N′*-tetramethylethylenediamine.

silica gel with 30 : 1 (v/v) chloroform–methanol to give pure **4** (3.8 g, 72%): mp 124°–125° (from ethyl acetate–hexane as a recrystallization system); $[\alpha]_D$ −14.1° (*c* 0.35, chloroform).

Compound **4** (3.0 g, 7.22 mmol) is treated with sodium methoxide (120 mg, 2.24 mmol) in dry methanol (30 mL) for 3 hr at room temperature. It is made neutral with Dowex 50W-X8 (H^+) resin, and the suspension is filtered. The filtrate is evaporated to give a quantitative yield of **8** (2.1 g): mp 184° (from ethanol); $[\alpha]_D$ −27.3° (*c* 0.13, water); ^1H nuclear magnetic resonance (NMR): δ 1.61 (m, 2H, OCH_2CH_2), 2.00 (s, 3 H, $NHCOCH_3$), 2.05 (m, 2 H, $CH_2CH{=}CH_2$), 3.35–3.73 (m, 5 H, H-2, H-3, H-4, H-5, H-6a, and H-6b), 3.86 (m, 2 H, OCH_2), 4.46 (d, J = 8.3 Hz, H-1), 5.01 (m, 2 H, $CH{=}CH_2$), and 5.85 (m, 1 H, $CH{=}CH_2$).

Analysis: Calculated for $C_{13}H_{23}O_6N \cdot 1.1H_2O$; C, 50.51; H, 8.22; N, 4.53. Found: C, 50.48; H, 8.04; N, 4.40.

Copolymerization of N-Acetyl-D-glucosamine Monomers

Copolymerization of the simple monomers[8,9] with acrylamide in water has been examined by a method based on that of Kochetkov *et al.*[10] (Scheme 2). The results of copolymerization of the monomers are shown in Table I with some physical data. As can be seen from the data, the effect of the spacer-arm length on polymerization behavior is evident, and the degrees of Glc*p*NAc incorporation clearly increase with increasing spacer-arm length from allyl **6** (n = 1) to *n*-pentenyl (n = 3) glycoside **8** except in the case of *n*-undecylenyl glycoside **9**. The difficulty in copolymerization of glycoside **9** is mostly attributable to its poor solubility in water.

[9] S.-I. Nishimura, K. Matsuoka, T. Furuike, S. Ishii, K. Kurita, and K. M. Nishimura, *Macromolecules* **24,** 4236 (1991).
[10] N. K. Kochetkov, *Pure Appl. Chem.* **56,** 923 (1984).

TABLE I
POLYMERIZATIONS OF ω-ALKENYL GLYCOSIDES OF *N*-ACETYL-D-GLUCOSAMINE WITH ACRYLAMIDE

Sugar monomer	Monomer ratio[a]	Total yield (%)	Polymer composition[a]	Sugar (wt%)	$[\alpha]_D$	η_{inh}[b] (dl/g)	Molecular weight[c]
6	1:4	46.1	1:12	23.5	−6.2°	0.27	230,000
6	1:10	57.4	1:42	8.0	−4.1°	1.15	>300,000
7	1:4	61.9	1:9	29.6	−9.6°	0.43	180,000
7	1:10	59.0	1:31	11.0	−4.5°	1.26	>300,000
8	1:4	65.8	1:8	34.6	−11.4°	0.34	190,000
8	1:10	75.6	1:28	12.5	−7.6°	1.03	>300,000
9[d]	—	—	—	—	—	—	—

[a] Ratio of carbohydrate monomer to acrylamide.
[b] In water at 25°.
[c] Molecular weights were determined by the GPC method with an Asahipack (Asahi Chemical Co. Ltd.) GS-510 column [pullulans (5,800, 12,200, 23,700, 48,000, 100,000, 186,000, and 380,000; Shodex Standard P-82) were used as standards.
[d] Insoluble in water.

Because copolymerization of *n*-pentenyl glycoside **8** with acrylamide gives a polymer containing 34.6 wt% of the Glc*p*NAc, a commercially available 4-penten-1-ol seemed to be an excellent and convenient polymerizable aglycon among this type of simple alcohols having C═C bond at the ω-position. In addition to remarkably chemospecific properties of *n*-pentenyl glycosides in organic chemistry,[11] we have found here an extended applicability of *n*-pentenyl glycosides in the chemistry for the molecular design of artificial glycoconjugates.

Although the sugar contents in the macromolecules could be easily adjusted by the feed ratio of carbohydrate monomers and acrylamide as needed, homopolymers from the glycosides could not be obtained under these reaction conditions. This might be caused mainly by the innate low chemical reactivity of simple olefin-type aglycons.

*Copolymerization of GlcpNAc Monomer **8** with Acrylamide*. A polymerization mixture contains 99.2 mg of monomer **8** and 245.5 mg (10 equivalents) of acrylamide in 1 ml of deionized water, to which are added 5.2 μl of *N,N,N′,N′*-tetramethylethylenediamine (TEMED) and 2.8 mg of ammonium peroxodisulfate (APS). The mixture is stirred for 2 hr at room temperature, diluted with 1.5 ml of 0.1 *M* pyridine–acetic acid buffer (pH 5.1), dialyzed against deionized water, and freeze-dried to give a water-soluble polymer as an amorphous powder (260.5 mg). Some physical

[11] B. Fraser-Reid, U. E. Udodong, Z. Wu, H. Ottosson, J. R. Merritt, C. S. Rao, C. Roberts, and R. Madsen, *Synlett.* 927 (1992).

data are listed in Table I. The carbohydrate content of the copolymer is determined by integration of the signals arising from methine (2.2 ppm) and *N*-acetyl protons (2.0 ppm) in the ^1H NMR spectrum and also by the modified Park–Johnson colorimetric method. ^{13}C NMR: δ (D_2O) 25.0 ($COCH_3$), 37.5, and 73.4 (CH_2), 58.2 (C-2), 63.5 (C-6), 72.8 (C-4), 76.4 (C-3), 78.5 (C-5), 104.3 (C-1), 176.9 (C=O), and 182.1 ($CONH_2$).

Polymers Having Pendant-Type Oligosaccharides from n-Pentenyl Glycosides

The above procedure is also applicable to the syntheses of more complicated and longer oligosaccharide chains. First, we have planned the syntheses of polymers having branches of aminodisaccharides such as *N,N'*-diacetylchitobiose and *N*-acetyllactosamine in the same manner as for the preparation of *N*-acetyl-D-glucosamine-containing polyacrylamide. Because the *N,N'*-diacetylchitobiose structure is known as the partial component of a core structure of asparagine-linked type oligosaccharides,[12] a simple polymer having this disaccharide sequence seems to be one of the available models or ligands for lectins which recognize this disaccharide structure. In addition, much attention has been paid to the biological and structural significance of carbohydrates containing *N*-acetyllactosamine both as an important repeating unit and as the core structure of a number of glycoconjugates.[13] Polymers bearing *N*-acetyllactosamine-type side chains may become not only a powerful ligand for *N*-acetyllactosamine binding molecules but a convenient polymeric substrate for a variety of glycosyltransferases.[14]

n-Pentenyl glycoside (**12**) is successfully prepared in a high yield from chitobiose octaacetate through the reaction of oxazoline derivative **10**[15] with 4-penten-1-ol. Intermediate **11** is de-O-acetylated in the same manner as described before to give the monomer **12**. Similarly, *N*-acetyllactosamine peracetate (**13**), prepared from lactose in good yield by the conventional method,[16,17] is converted to oxazoline derivative **14** by treating with TMSOTf. Compound **14** is allowed to react with 4-penten-1-ol in the

[12] Y. C. Lee and J. R. Scoca, *J. Biol. Chem.* **247,** 5753 (1972); R. Kornfeld and S. Kornfeld, *Annu. Rev. Biochem.* **54,** 631 (1985).

[13] K. Maemura and M. Fukuda, *J. Biol. Chem.* **267,** 24379 (1992).

[14] For example, see Y. Ichikawa, J. L.-C. Liu, G.-J. Shen, and C.-H. Wong, *J. Am. Chem. Soc.* **113,** 6300 (1991).

[15] S.-I. Nishimura, H. Kuzuhara, Y. Takiguchi, and K. Shimahira, *Carbohydr. Res.* **194,** 223 (1989).

[16] J. Arnarp and J. Lonngren, *J. Chem. Soc., Perkin Trans. 1,* 2070 (1981).

[17] R. U. Lemiuex, S. Z. Abbas, M. H. Burzynska, and R. M. Ratcliffe, *Can. J. Chem.* **60,** 63 (1981).

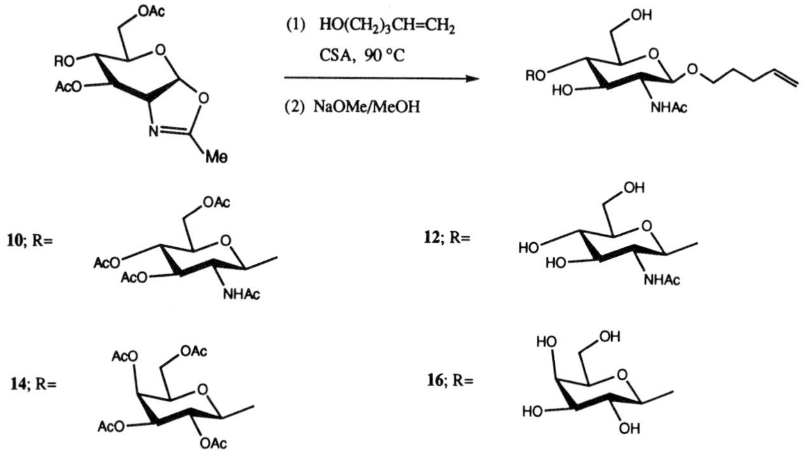

SCHEME 3. Oligosaccharides from *n*-pentenyl glycosides: Zemplen method. CSA, 10-Camphorsulfonic acid.

presence of CSA as a promoter, giving the corresponding β-glycoside **15** in high yield. Consequently, *n*-pentenylated *N*-acetyllactosamine **16** is obtained by O-deacetylation through the Zemplen method (Scheme 3).

As anticipated, the radical copolymerization of monomers **12** or **16** with acrylamide proceeds smoothly in a similar way as described for the preparation of the polymer having Glc*p*NAc branches (Table II). The

TABLE II
POLYMERIZATIONS OF ω-ACRYLAMIDOGLYCOSIDES OF *N*-ACETYL-D-GLUCOSAMINE

Sugar monomer	Monomer ratio[a]	Total yield (%)	Polymer composition[a]	Sugar (wt%)	$[\alpha]_D$	η_{inh}[b] (dl/g)	Molecular weight[c]
17	1 : 0[d]	100	1 : 0	100	−31.0°	0.41	220,000
17	1 : 4	91.8	1 : 3	61.1	−21.9°	1.97	>300,000
18	1 : 0[d]	35.0[e]	1 : 0	100	−18.6°	0.12	9000
18	1 : 4	90.0	1 : 5	51.0	−7.1°	0.35	70,000

[a] Ratio of carbohydrate monomer to acrylamide.
[b] In water at 25°.
[c] Molecular weights were determined by the GPC method with an Asahipack GS-510 column [pullulans (5,800, 12,200, 23,700, 48,000, 100,000, 186,000, and 380,000; Shodex Standard P-82) were used as standards.
[d] Homopolymerization of glycoside (polymerization without acrylamide as a comonomer.
[e] Ethanol–water (1 : 1, v/v) was used for the polymerization solvent owing to a poor solubility in water.

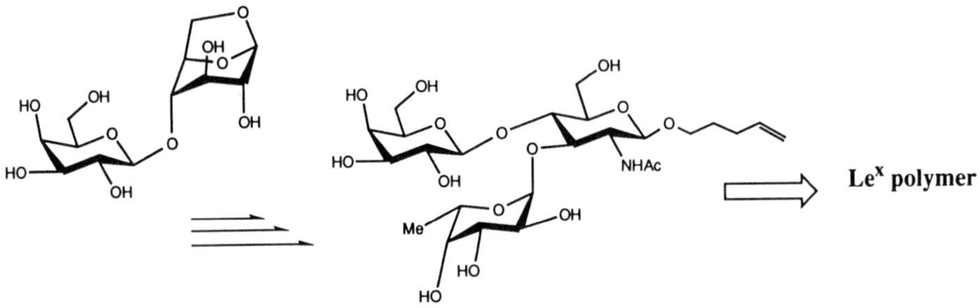

SCHEME 4. Preparation of Le^x polymer.

structures of the polymers are supported by the spectroscopic data in addition to chemical analyses. Furthermore, as one of the functionalization studies of polymers of this type, we have demonstrated that rat hepatocytes clearly exhibit much higher affinity for the surface of polyacrylamide matrices containing *N*-acetyllactosamine compared with that containing *N,N'*-diacetylchitobiose.[9] Successful preparations of the polymer carrying pendant Le^x-type trisaccharides, one of the most important sequences related to the tumor-associated oligosaccharide antigens,[18] by employing the *n*-pentenyl group as a polymerizable aglycon (Scheme 4) have been reported.[19,20]

*n-Pentenyl O-(β-D-Galactopyranosyl)-(1 → 4)-2-acetamido-2-deoxy-β-D-glucopyranoside (**16**).*[9] A solution of peracetate **13** (1.7 g, 2.51 mmol) in dry 1,2-dichloroethane (10 ml) is treated with TMSOTf (0.5 ml, 2.63 mmol) under a nitrogen atmosphere, and the mixture is stirred for 15 hr at 50°. Triethylamine (1.8 ml) is added, and the solution is applied directly to a column of silica gel and eluted with 100 : 200 : 1 (v/v/v) toluene–ethyl acetate–triethylamine, giving the syrupy oxazoline derivative **14** (1.55 g): R_f 0.6 on TLC in 5 : 4 : 1 (v/v/v) chloroform–ethyl acetate–methanol.

To a solution of freshly prepared **14** (1.0 g, 1.61 mmol) in dry 1,2-dichloroethane (10 ml) are added 4-penten-1-ol (0.94 ml, 4.84 mmol) and CSA (30 mg, 0.026 mmol), and the solution is stirred under a nitrogen atmosphere for 2 hr at 90°, cooled, and poured into ice–water. The mixture is washed successively with aqueous sodium hydrogen carbonate and water, dried, and evaporated. The residue is chromatographed on silica gel with 50 : 1 (v/v) chloroform–methanol as the eluant to afford amorphous **15** (980 mg, 86%): $[\alpha]_D$ −19.6° (*c* 0.121, chloroform).

[18] S.-I. Hakomori, *Annu. Rev. Biochem.* **50,** 733 (1981).

[19] S.-I. Nishimura, S. Murayama, K. Kurita, and H. Kuzuhara, *Chem. Lett.*, 1413 (1992).

[20] S.-I. Nishimura, K. Matsuoka, T. Furuike, N. Nishi, S. Tokura, K. Nagami, S. Murayama, and K. Kurita, *Macromolecules* **27,** 157 (1994).

Compound **15** (388.9 mg, 0.553 mmol) is de-O-acetylated by the method described for the preparation of **8**, yielding the corresponding *N*-acetyllactosamine derivative **16** (217.7 mg, 87%): mp 234° (dec.); $[\alpha]_D$ −19.0° (*c* 0.26, water); ^1H NMR: δ [$(CD_3)_2SO$ plus one drop of D_2O] 1.52 (m, 2 H, OCH_2CH_2), 1.78 (s, 3H, $NHCOCH_3$), 2.04 (m, 2H, $CH_2CH{=}CH_2$), 4.21 (d, 1 H, J = 8.3 Hz, H-1), 4.30 (d, 1 H, J = 8.3 Hz, H-1′), 5.00 (m, 2 H, $CH{=}CH_2$), 5.80 (m, 1 H, $CH{=}CH_2$), and 7.76 (d, 1 H, J = 8.5 Hz, $NHCOCH_3$).

Analysis: Calculated for $C_{19}H_{33}O_{11}N \cdot 1.5H_2O$: C, 47.69; H, 7.58; N, 2.93. Found: C, 47.57; H, 7.19; N, 2.93.

*Copolymerization of LacNAc Monomer **16** with Acrylamide.* To a solution of carbohydrate monomer **16** (100 mg) and 160 mg (10 equivalents) of acrylamide in 1 ml of deionized water are added 3.3 μl of TEMED and 2.0 mg of APS. The mixture is stirred for 2 hr at room temperature, diluted with 1.5 ml of 0.1 *M* pyridine–acetic acid buffer (pH 5.1), dialyzed against deionized water, and freeze-dried to give the polymer as an amorphous powder (98.1 mg, 77%): $[\alpha]_D$ −5.6° (*c* 0.163, water); sugar content is estimated to be 23 wt% from the ^1H NMR data; molecular weight > 300,000 [gel-permeation chromatography (GPC) method]; ^{13}C NMR: δ (D_2O): 25.0 ($COCH_3$), 37.5 and 73.2 (CH_2), 44.5 (CH), 57.8 (C-2), 62.9 (C-6), 63.6 (C-6′), 71.2 (C-4′), 73.6 (C-2′), 74.9 (C-3), 75.0 (C-5′), 77.4 (C-3′), 78.0 (C-5), 81.5 (C-4), 103.6 (C-1), 105.6 (C-1′), 176.8 (C=O), and 182.0 ($CONH_2$).

Homopolymers from Glycosides: A Novel Type of Cluster Glycosides

Lee *et al.* have proposed and demonstrated the significance of sugar density on the glycoprotein when some cells or specific ligand molecules interact with carbohydrates.[21] Chemically designed "cluster glycosides," first reported in 1978,[22] have been found to provide valuable information on these recognition processes. For example, they have elegantly demonstrated for Gal*p*/Gal*p*NAc–hepatocyte binding systems that a "cluster" or "multipoint" carbohydrate–ligand interaction may be involved in the successful binding process.[23,24] Moreover, work on the eventual design of inhibitors for influenza virus has also revealed that the cluster or multivalent sialoside inhibitors showed markedly amplified inhibitory effects on

[21] Y. C. Lee, *FASEB J.* **6,** 3193 (1992).
[22] Y. C. Lee, *Carbohydr. Res.* **67,** 509 (1978).
[23] R. T. Lee and Y. C. Lee, this series, Vol. 138, p. 424.
[24] R. T. Lee, P. Lin, and Y. C. Lee, *Biochemistry* **23,** 4255 (1984).

the hemagglutination induced by influenza virus as compared to monovalent sialosides.[25]

As one of the most effective methods to increase the density of sugar moieties, preparing "homopolymers" from simple polymerizable glycosides by a facile polymerization reaction will be desirable. For this purpose, we must design and prepare some new polymerizable glycosides with much higher reactivity besides spacing functions, since the desired homopolymers could not be obtained on the basis of *n*-pentenyl glycosides described in the preceding section.

Preparation and Polymerization of N-Acetyl-D-glucosamine Derivative Containing Terminal Acrylamide Structure

Examples with *N*-acetyl-D-glucosamine derivative **17**,[26] in addition to the known *N*-acetyl-D-glucosamine derivative **18**,[27] has been tested as tentative candidates with an appropriate spacer-arm structure for the polymerization of a variety of oligosaccharides containing Glc*p*NAc residues at the reducing end. Scheme 5 shows the synthetic procedure for acrylamide-type monomers **17** and **18**. As shown in Scheme 5, N-protected simply amino alcohols, 3-(benzyloxycarbonylamino)-1-propanol (*N*-Cbz-propanol) and 6-(benzyloxycarbonylamino)-1-hexanol,[28] are used as precursors of polymerizable aglycons.

Coupling reactions of oxazoline derivative **1** with N-protected amino alcohols proceed smoothly and give the corresponding intermediates **19** and **20** in moderate yield. General N-deprotection by hydrogeneration and the following N-acryloylation afford compounds **21** and **22** as peracetates.[29] Finally, O-deacetylation by the usual Zemplen procedure gives monomers **17** and **18**, respectively. As anticipated, these compounds show an excellent polymerizability comparable to those of the glycosides having simple ω-olefin type aglycons (Table II).

3-(N-Benzyloxycarbonylamino)-1-propanol. To a solution of 3-amino-1-propanol (7.6 ml, 0.1 mol) and sodium hydrogen carbonate (21.8 g, 0.26 mol) in water are added dropwise a mixture of benzyloxycarbonyl chloride

[25] A. Spaltenstein and G. M. Whiteside, *J. Am. Chem. Soc.* **113,** 686 (1991).

[26] T. Furuike, N. Nishi, S. Tokura, K. Maruyama, K. Kurita, K. Matsuoka, and S.-I. Nishimura, *Macromolecules* **27,** in press (1994).

[27] S. R. Sarfati, S. Pochet, J.-M. Neumann, and J. Igolen, *J. Chem. Soc., Perkin Trans. 1,* 1065 (1990).

[28] S. Chipowsky and Y. C. Lee, *Carbohydr. Res.* **31,** 339 (1973).

[29] Acryloyl chloride is toxic and causes lachrimation. It should be handled with great care under ventilation. For an alternative procedure for the acryloylation of aminoalkyl glycosides, see, for example, P. H. Weigel, R. L. Schnaar, S. Roseman, and Y. C. Lee, *Methods in Carbohydr. Chem.* **9,** 187 (1993).

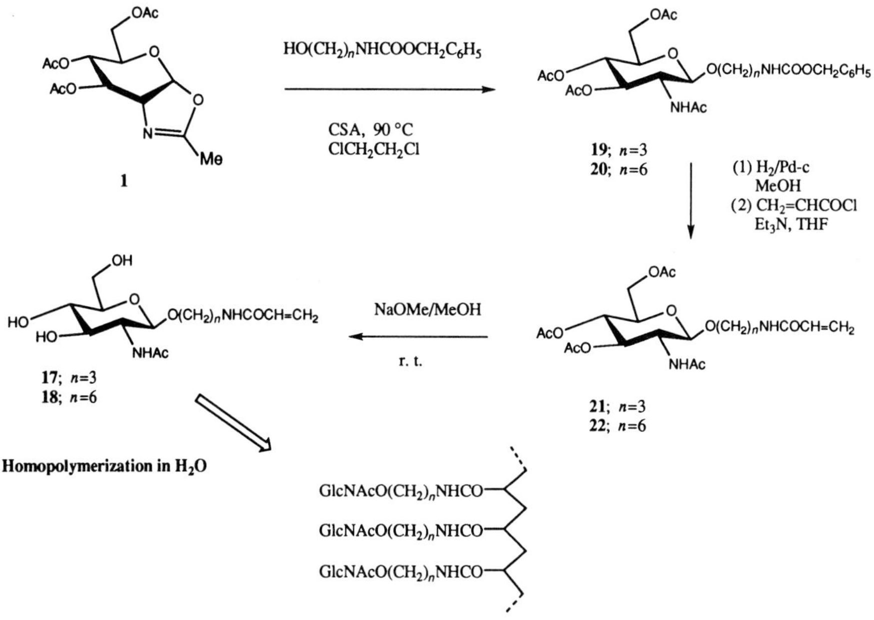

SCHEME 5. Synthesis of acrylamide-type monomers. THF, Tetrahydrofuran; r.t., room temperature.

(20.8 ml, 0.13 mol) and ether (20 ml), and the mixture is stirred for 3 hr at room temperature. The mixture is filtered with Celite, extracted with ether (200 ml), washed with water, dried over magnesium sulfate, and evaporated. The residual syrup is chromatographed on silica gel eluted first with 20 : 1 then with 2 : 1 (v/v) toluene–ethyl acetate as an eluant to give the *N*-Cbz-propanol (15.7 g): mp 43°.

Analysis: Calculated for $C_{11}H_{15}O_3N$: C, 63.07; H, 7.16; N, 6.69. Found: C, 63.27; H, 7.25; N, 6.80.

*(3-Benzyloxycarbonylamino)propyl 2-Acetamido-3,4,6-tri-O-acetyl-2-deoxy-β-D-glucopyranoside (**19**).* A solution of oxazoline derivative **1** (3.9 g, 11.8 mmol) and *N*-Cbz-propanol (5.0 g, 23.7 mmol) in 1,2-dichloroethane (15 ml) is stirred under a nitrogen atmosphere for 2 hr in the presence of CSA (50 mg). The solution is cooled to room temperature, diluted with chloroform, and poured into ice–water. The extract with chloroform is washed successively with aqueous sodium hydrogen carbonate and water, dried over magnesium sulfate, filtered, and evaporated. The residue is purified by chromatography on silica gel with 20 : 1 (v/v) toluene–ethyl acetate to afford compound **19** (5.7 g, 89%): mp 145°–146°; $[\alpha]_D$ −7.4° (*c* 0.229, chloroform).

Analysis: Calculated for $C_{25}H_{34}O_{11}N_2$; C, 55.73; H, 6.31; N, 5.20. Found: C, 55.51; H, 6.32; N, 5.26.

*3-(N-Acryloylamino)propyl 2-Acetamido-3,4,6-tri-O-acetyl-β-D-glucopyranoside (**20**).* Compound **19** (4.7 g, 8.72 mmol) is hydrogenated in the presence of 10% palladium-on-carbon (0.4 g) in methanol (100 ml) for 2 hr at room temperature. The reaction is monitored by TLC in 65 : 25 : 4 (v/v/v) chloroform–ethyl acetate–methanol. The mixture is filtered and evaporated to give the crude 3-aminopropyl glycoside.

To a solution of the crude 3-aminopropyl glycoside in tetrahydrofuran (THF) (40 ml) are added triethylamine (1.57 ml) and freshly distilled acryloylchloride (0.90 ml) at 0°, and the mixture is stirred for 24 hr at room temperature. The mixture is poured into ice–water and extracted with chloroform. The organic layer is washed with brine, dried, and evaporated. The residue is subjected to silica gel chromatography with 1 : 2 (v/v) toluene–ethyl acetate as eluant to yield **20** (3.2 g, 80%): mp 149°–150°; $[\alpha]_D$ −51.4° (*c* 0.23, chloroform).

Analysis: Calculated for $C_{20}H_{30}O_{10}N_2$; C, 51.79; H, 6.65; N, 6.04. Found: C, 51.79; H, 6.76; N, 6.14.

*3-(N-Acryloylamino)propyl 2-Acetamido-2-deoxy-β-D-glucopyranoside (**17**).* Treatment of compound **20** (0.8 g, 1.75 mmol) with sodium methoxide (28 mg) in dry methanol (30 ml), as described for the preparation of **8**, gives a quantitative yield of **17** (0.5 g): mp 168°–169°; $[\alpha]_D$ −29.3° (*c* 0.23, water).

Analysis: Calculated for $C_{14}H_{24}O_7N_2$; C, 49.79; H, 7.34; N, 8.29. Found: C, 49.69; H, 7.31; N, 8.32.

*Homopolymerization of Glycoside **17**.* A solution of monomer **17** (99.5 mg, 0.30 mmol) in deionized water (1.0 ml) is deaerated by a water aspirator for 20 min, and to the solution are added TEMED (4.5 μl, 30 μmol) and APS (2.44 mg, 10.7 μmol). The mixture is stirred for 17 hr at room temperature, diluted with 1.9 ml of 0.1 *M* pyridine–acetic acid buffer (pH 5.1), dialyzed against deionized water for 24 hr, and freeze-dried to give a water-soluble polymer in quantitative yield: $[\alpha]_D$ −31.0° (*c* 0.25, water); molecular weight 220,000 (GPC method); ^{13}C NMR: δ (D_2O) 25.1 (CH_3), 31.5 (—CH_2—), 38.0 (—CH_2—, main chain), 39.1 (NCH_2), 45.0 (—CH—), 58.2 (C-2), 63.6 (C-6), 70.4 (OCH_2), 72.8 (C-4), 76.6 (C-3), 78.6 (C-5), 103.8 (C-1), 176. 8 (C=O, aglycon), and 178.7 (C=O, GlcNAc).

[23] Synthesis of Branched Polysaccharide by Chemical and Enzymatic Reactions and Its Hypoglycemic Activity

By KENICHI HATANAKA

Introduction

Glycosylation can be carried out using chemical or enzymatic methods. Glycosyltransferases have disadvantages for preparative-scale synthesis, because of limited availability, lability, and expensive substrates (sugar nucleotide). When the glycosylation is carried out with glycosidases in the presence of high substrate concentrations and/or organic solvents, useful yields of glycosides can be obtained stereoselectively. For example, stereoselective glycosylation with glycosidases using *o*-nitrophenyl[1] and *p*-nitrophenyl glycosides[2] as glycosyl donors has been reported. Kobayashi *et al.* synthesized cellulose enzymatically using cellulase.[3] However, the syntheses of glycosides using glycosidases mostly show low yields and regioselectivity. On the other hand, chemical synthesis of glycosides, in general, gives a mixture of α and β anomers, requiring separation of the anomers by crystallization or chromatography. However, when the acceptor is a polymer such as a polysaccharide, it is impossible to separate the anomeric mixture because both anomeric units are contained in the same polymer. This chapter describes the synthesis of a branched (1→6)-α-D-glucopyranan by ring-opening polymerization of anhydro sugar derivative, chemical glycosylation, and enzymatic hydrolysis with cellulase.[4] The hypoglycemic activity of the synthetic branched polysaccharide[5] is also described.

[1] K. G. I. Nilsson, *Trends Biotechnol.* **6,** 256 (1988).

[2] K. Ajisaka and M. Shirakabe, *Carbohydr. Res.* **224,** 291 (1992).

[3] S. Kobayashi, K. Kashiwa, T. Kawasaki, and S. Shoda, *J. Am. Chem. Soc.* **113,** 3079 (1991).

[4] K. Hatanaka, S. C. Song, A. Maruyama, T. Akaike, A. Kobayashi, and H. Kuzuhara, *J. Carbohydr. Chem.* **11,** 1027 (1992).

[5] K. Hatanaka, S. C. Song, A. Maruyama, A. Kobayashi, H. Kuzuhara, and T. Akaike, *Biochem. Biophys. Res. Commun.* **188,** 16 (1992).

Copyright © 1994 by Academic Press, Inc.
All rights of reproduction in any form reserved.

Synthesis of (1→6)-α-D-Glucopyranan Having α-D-Glucopyranosyl Branch at C-3 Position

1,6-Anhydro-3-O-benzoyl-2,4-di-O-benzyl-β-D-glucopyranose

1,6-Anhydro-β-D-glucopyranose can be synthesized from glucose or cellulose by three methods as follows.

Method 1. To 4.0 liters of 2,3 *N* potassium hydroxide solution contained in a 5-liter round-bottomed flask is added 150 g of phenyl tetra-*O*-acetyl-β-D-glucopyranoside or 210 g of pentachlorophenyl tetra-*O*-acetyl-β-D-glucopyranoside, which have been prepared by the reaction of penta-*O*-acetyl-β-D-glucopyranose with phenol or tetra-*O*-acetyl-β-D-glucopyranosyl bromide with pentachlorophenol, respectively. Gentle reflux is maintained for 20 hr. At the end of the reflux period, the solution is cooled to the room temperature and carefully neutralized with 3 *N* sulfuric acid. After filtration and evaporation, the product is dried under reduced pressure at 40°. The residue is thoroughly extracted with boiling ethanol, and the extract is filtered and treated with activated carbon. The filtered solution is concentrated to dryness.

The crude 1,6-anhydro-β-D-glucopyranose is O-acetylated by mixing with 350 ml of acetic anhydride and 450 ml of pyridine for 12 hr. It is then concentrated *in vacuo* and the residue dissolved in chloroform. The chloroform solution is neutralized with aqueous sodium hydrogen carbonate, washed with water, and dried over anhydrous sodium sulfate. After evaporation, the syrupy product is dissolved in ethanol and crystallized. Recrystallizations are carried out from ethanol several times. The 1,6-anhydro-β-D-glucopyranose pentaacetate (70 g) is then dissolved in 700 ml of methanol and treated with 70 ml of 1% sodium methoxide solution in methanol. After decationization with Amberlite IR-120 (H^+), the solution is concentrated to give a crystalline 1,6-anhydro-β-D-glucopyranose (yield 34 g, 59% overall from phenyl glucopyranoside derivative).

Method 2. Treatment of 6-*O*-tosyl-β-D-glucopyranose with sodium hydroxide also gives 1,6-anhydro-β-D-glucopyranose.[6] To a stirred suspension of D-glucose (100 g, 560 mmol) in dried pyridine (1.0 liter) is added dropwise *p*-toluenesulfonyl chloride (159 g, 830 mmol) in 300 ml of dried pyridine, with the temperature of the reaction mixture being maintained at approximately 20°. The yellowish solution is stirred at room temperature for 1.5 hr and then carefully adjusted to pH 9.0 by the addition of 6 *N* sodium hydroxide. The alkaline solution is kept at room

[6] M. A. Zottola, R. Alonso, G. D. Vite, and B. Fraser-Reid, *J. Org. Chem.* **54,** 6123 (1989).

temperature for 1 hr, then neutralized to pH 7 by the addition of 3 *N* hydrochloric acid. The solution is evaporated to dryness, and the residue is triturated with methanol and filtered through a sintered glass filter. Evaporation of the filtrate gives 1,6-anhydro-β-D-glucopyranose. Purification of the 1,6-anhydro-β-D-glucopyranose by acetylation followed by deacetylation is carried out as described for Method 1 (yield 25 g, 28% from D-glucose).

Method 3. 1,6-Anhydro-β-D-glucopyranose can also be prepared by the pyrolysis of microcrystalline cellulose under reduced pressure. Microcrystalline cellulose (120 g) in a 1-liter round-bottomed short-necked Pyrex flask is heated with the luminous flame of a Bunsen burner. The volatile products are led through an outlet tube, the diameter of which is of sufficient size to prevent clogging, into a 2-liter three-necked flask cooled in an ice bath. The outlet of the system is connected to a vacuum pump, and the distillation requires about 1 hr. The combined dark colored distillate from four such runs is dissolved, except for some tar, in methanol, and the solution is filtered. The dark brown colored filtrate is concentrated under reduced pressure to a thick syrup. Purification of 1,6-anhydro-β-D-glucopyranose by acetylation followed by deacetylation is carried out as described for Method 1 (yield 29 g, 6% from cellulose).

Regioselective benzylation of 1,6-anhydro-β-D-glucopyranose is carried out with benzyl bromide and barium oxide.[7] To a solution of 1,6-anhydro-β-D-glucopyranose (7.0 g, 43 mmol) in *N,N*-dimethylformamide (80 ml) is added 28 g (183 mmol) of barium oxide. The suspension is stirred at 85° for 30 min. Then 12 ml (101 mmol) of benzyl bromide is added dropwise to the reaction mixture, which is stirred at 85° for 2 hr. Next, 90 ml of methanol is added to the reaction mixture at 60° to destroy the excess benzyl bromide. After 30 min, chloroform is added, and the mixture is centrifuged to remove barium oxide. The supernatant is washed with water, dried on sodium sulfate, and evaporated to give syrupy 1,6-anhydro-2,4-di-*O*-benzyl-β-D-glucopyranose, which is then purified by column chromatography on silica gel, with benzene–ethyl acetate (4 : 1, v/v) as eluent. The purified product is crystallized from absolute ethanol.

1,6-Anhydro-2,4-di-*O*-benzyl-β-D-glucopyranose thus obtained is benzoylated to give 1,6-anhydro-3-*O*-benzoyl-2,4-di-*O*-benzyl-β-D-glucopyranose. 1,6-Anhydro-2,4-di-*O*-benzyl-β-D-glucopyranose (5.0 g) and benzoyl chloride (3.7 ml) are dissolved in 92 ml of dried pyridine. After stirring at room temperature for 20 min, the reaction mixture is poured into ice–water and then extracted with chloroform. The extract is concentrated *in vacuo*. Column chromatography of the syrupy product on silica gel,

[7] T. Iverson and D. R. Bundle, *Can. J. Chem.* **60**, 299 (1982).

with benzene–ethyl acetate (20 : 1, v/v) as eluent, gives a polymerizable 1,6-anhydro-3-*O*-benzoyl-2,4-di-*O*-benzyl-β-D-glucopyranose.

Polymerization

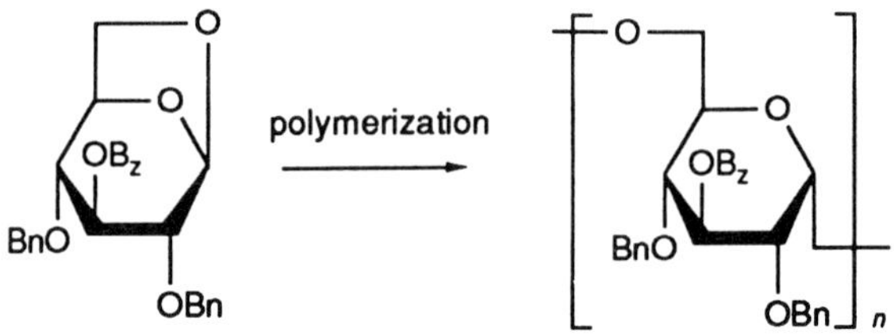

Polymerization is carried out using a high vacuum technique. After 1.0 ml of dichloromethane is distilled into a polymerization ampoule, 0.5 g of 1,6-anhydro-3-*O*-benzoyl-2,4-di-*O*-benzyl-β-D-glucopyranose is dissolved completely. Then the gaseous catalyst (phosphorus pentafluoride) is condensed in the polymerization ampoule by cooling in a liquid nitrogen bath. After the ampoule is sealed off from the vacuum line, polymerization is started by shaking the ampoule for 2 min in a temperature-controlled bath (−40° or −60°), followed by standing for a fixed time (19 hr). After the polymerization is terminated by adding a small amount of methanol, the polymer solution in chloroform is neutralized with sodium hydrogen carbonate and washed with water. Purification of the polymer is performed three times by dissolution–reprecipitation using the chloroform–petroleum ether system. The 3-*O*-benzoyl-2,4-di-*O*-benzyl-(1→6)-D-glucopyranan thus obtained is finally freeze-dried.

Polymerization of 1,6-anhydro-3-*O*-benzoyl-2,4-di-*O*-benzyl-β-D-glucopyranose gives a polymer having number-average molecular weights of 1.0×10^4 to 3.2×10^4. Polymerization with a higher concentration of PF_5 (i.e., 7–15 mol%) gives the polymer in high yield, whereas polymerization does not occur at lower concentrations of catalyst. The yields and the molecular weights of polymers increase with decreasing polymerization temperature. The ^{13}C nuclear magnetic resonance (NMR) spectrum of the polymer obtained at lower temperature (below −40°) shows a single peak for each carbon, indicating high stereoregularity, whereas two kinds of anomeric carbon signals are shown in the spectrum of the polymer obtained at −20°. The high positive specific rotation of polymers prepared at lower temperatures suggests that they have primarily the α configuration. Therefore, the resultant polysaccharide derivative is stereoregular 3-*O*-benzoyl-2,4-di-*O*-benzyl-(1→6)-α-D-glucopyranan.

Debenzoylation

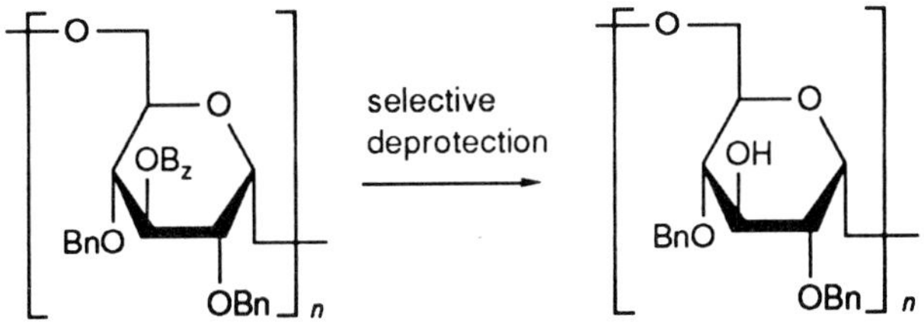

Benzoyl groups of the polymers are removed using sodium methoxide in dichloromethane–*N*,*N*-dimethylformamide–methanol. To a polymeric 3-*O*-benzoyl-2,4-di-*O*-benzyl-(1→6)-α-D-glucopyranan (0.73 g) solution in 14 ml of dichloromethane is added 30 ml of *N*,*N*-dimethylformamide and 25 ml of methanol containing 1.6 g of sodium methoxide. The mixture is stirred at room temperature for 1 day and treated with acetic acid (2.5%). The chloroform extract is washed with water, dried with sodium sulfate, and concentrated to a small volume. The polymer is precipitated with petroleum ether and purified by reprecipitation with chloroform–petroleum ether. The polymer is freeze-dried. Removal of benzoyl groups can be confirmed by ^{1}H NMR spectroscopy. The H-3 proton signal at 5.78 ppm in the spectrum of 3-*O*-benzoyl-2,4-di-*O*-benzyl-(1→6)-α-D-glucopyranan is shifted to 4.05 ppm after debenzoylation, and the new proton signal assignable to the free hydroxyl group appears at 2.35 ppm, indicating that removal of benzoyl groups is successful. The number average degree of polymerization of the debenzoylated polymer is almost the same as that of the starting polymer, indicating that no significant chain scission of the polymer backbone takes place during the debenzoylation reaction.

Glucosylation with Glucopyranosyl Imidate

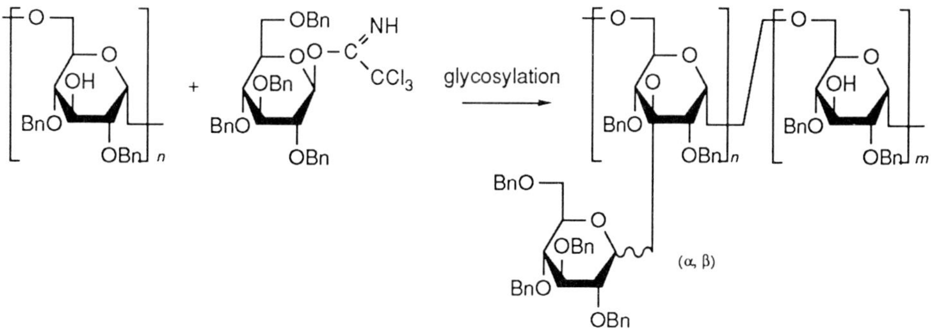

Regioselectively deprotected 2,4-di-*O*-benzyl-(1→6)-α-D-glucopyranan is glucosylated with 2,3,4,6-tetra-*O*-benzyl-β-D-glucopyranosyl trichloroacetimidate to give a branched polysaccharide. 2,4-Di-*O*-benzyl-(1→6)-α-D-glucopyranan (0.45 g, 1.32 mmol) and 2,3,4,6-tetra-*O*-benzyl-β-D-glucopyranosyl trichloroacetimidate (1.36 g, 1.98 mmol) are dissolved in 50 ml of dichloromethane (dried over calcium hydride) at room temperature. After addition of *tert*-butyldimethylsilyl triflate (0.44 mmol), the solution is stirred at room temperature for 5 hr. The reaction mixture is treated with solid sodium hydrogen carbonate and then neutralized with aqueous sodium hydrogen carbonate. After extraction with chloroform, the extract is washed with water, dried on sodium sulfate, and then concentrated to dryness. Purification of the polymer is performed by dissolution–reprecipitation using the chloroform–petroleum ether system. The polymer is finally freeze-dried.

The degree of branching can be calculated from ^{13}C NMR spectra by measuring the integration ratio of the C-6 carbon signal of branching residues to that of the main chain. The branched polymers obtained show branch contents in the range of 0.48–0.87 depending on the molar ratio of reactants. The molecular weight of the polymer decreases with increasing branch content, suggesting that a few chain cleavages occur under the glycosylation reaction conditions.

Debenzylation of Branched Polysaccharide

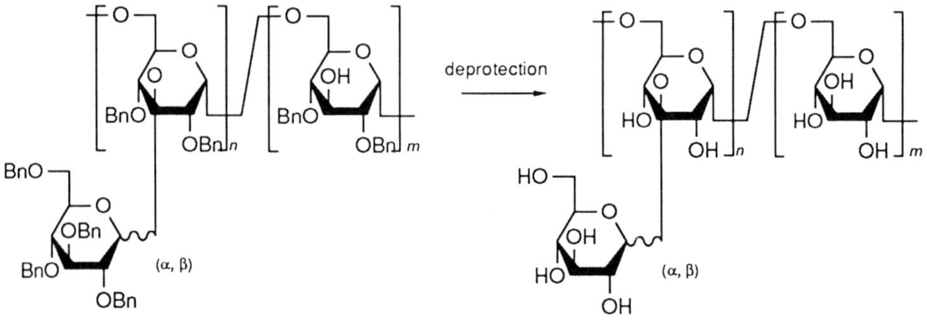

The reduction of the branched polysaccharide with sodium metal in liquid ammonia gives a debenzylated polysaccharide, that is, OH-free polysaccharide. The branched polymer (0.45 g) is dissolved in 10 ml of dichloromethane, and the solution is added dropwise to 50 ml of liquid ammonia containing 0.6 g of sodium at −78° under nitrogen which is dried by passage through successive columns of soda lime and potassium hydroxide. After 1.5 hr of stirring at −78° under nitrogen, about 10 g of solid ammonium chloride is added until the blue color of the solution disappears, at which time a small amount of distilled water is added. The reaction mixture is air-dried in fume chamber to remove ammonia. Then, the suspension of the deprotected polymer in water is washed five times with methylene chloride. The water-insoluble part is filtered off, and the remaining aqueous solution is dialyzed against distilled water for 3 days with daily changes of water. The branched polymer with free hydroxyl groups is freeze-dried.

Complete removal of benzyl groups is confirmed by the disappearance of the absorption in the aromatic region of the ^{1}H or ^{13}C NMR spectrum of the polymer solution in deuterium oxide. In the ^{13}C NMR spectrum of the polymer, three kinds of peaks with different chemical shifts in the region of the anomeric carbon can be observed at 105.5, 101.9 ppm, and 100.4 ppm, which can be assigned to anomeric carbons of the β-D-glucopyranosyl branching unit, α-D-glucopyranosyl branching unit, and α-D-glucopyranosyl main chain unit, respectively. The degree of branching and the ratio of α-D-glucopyranosyl branching units to β-D-glucopyranosyl branching units can be calculated from the integration ratios of the C-1 carbon signals in the ^{13}C NMR spectra.

Enzymatic Hydrolysis

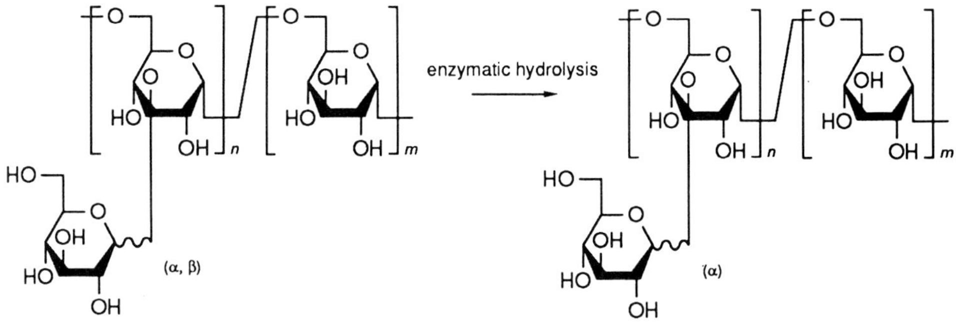

Enzymatic hydrolysis of the branched polysaccharide, which has α- and β-D-glucopyranosyl branching units, is carried out with cellulase ("Onozuka" 3S, Yakult Co. Ltd., Tokyo, Japan) to remove specifically β-D-glucopyranosyl branching units in order to complete the "stereospecific branching reaction." That the cellulase has β-glucosidase activity must be confirmed by using p-nitrophenyl β-D-glucopyranoside as a substrate, and that it does not contain the α-glucosidase activity must be tested by using p-nitrophenyl α-D-glucopyranoside as a substrate.

To the solution of 60 mg of debenzylated branched polymer in 2 ml of 50 m*M* acetate buffer (pH 4.8) is added 2 ml of cellulase (30 mg, 90 units) in 50 m*M* acetate buffer (pH 4.8). The reaction mixture is shaken at 45° in a water bath incubator for a few days. The progress of the reaction can be checked by measuring the concentration of released glucose by the glucose oxidase method. After the release of glucose is stopped, the reaction is terminated by the addition of 1.5 ml of ice-cold phosphotungstic acid (5% in 0.5 *N* HCl). The precipitated protein (enzyme) is removed by centrifugation and decantation, and then the supernatant is neutralized by the addition of 0.5 *N* NaOH. After dialysis against water for 3 days, the α-D-glucopyranose branched polysaccharide is obtained by freeze-drying from the aqueous solution. The complete removal of β-D-glucopyranosyl branching units can be confirmed by ^{13}C NMR spectroscopy of the obtained polysaccharide. In the spectrum of the branched polysaccharide having only α-D-glucopyranosyl branching units, the peak at 105.5 ppm, which can be assigned to the anomeric carbon of β-D-glucopyranoside branching unit, disappears.

The anomeric ratios of the branching units in the polysaccharides are affected by the glycosylation method and the glycosylation reaction conditions. However, treatment with cellulase affords branched polysaccharides having only α-D-glucopyranosyl branching units. Regardless of the content of β-D-glucopyranosyl residues, β-D-glucopyranosyl branches can be completely removed by enzymatic hydrolysis using cellulase as confirmed by ^{13}C NMR spectroscopy. In each branched polysaccharide, the content of α-D-glucopyranosyl branches in the polysaccharide before enzymatic reaction, determined by the branching content and anomeric ratio of the branch, must be close to the content of α-D-glucopyranosyl branches in the polysaccharide after enzymatic reaction, because the α-D-glucopyranosyl branch is hardly hydrolyzed by cellulase. Moreover, the degree of polymerization of the branched polysaccharide must be the same before and after the enzymatic reaction, because the main chain is made of α-glucopyranosidic linkages which should be resistant to cellulase.

Control of Branching Content by Copolymerization

Both the branching content and branching distribution can be controlled with the copolymerization technique.[8] Copolymerization of 1,6-anhydro-3-*O*-benzoyl-2,4-di-*O*-benzyl-β-D-glucopyranose with 1,6-anhydro-2,3,4-tri-*O*-benzyl-β-D-glucopyranose carried out in the same manner

[8] K. Hatanaka, S. C. Song, A. Maruyama, A. Kobayashi, H. Kuzuhara, and T. Akaike, *Polym. J.* **25,** 373 (1993).

as homopolymerization of 1,6-anhydro-3-*O*-benzoyl-2,4-di-*O*-benzyl-β-D-glucopyranose as detailed above affords copolymers with a different spacing of the sugar branching separated by 3-*O*-benzoyl-2,4-di-*O*-benzyl-β-D-glucopyranose residues in the copolymer.

The copolymer composition, which is equal to the content of branching points, in the initial stage of the copolymerization can be calculated from the values of monomer reactivity ratios[9] and the monomer composition in the feed. The monomer reactivity ratio of 1,6-anhydro-3-*O*-benzoyl-2,4-di-*O*-benzyl-β-D-glucopyranose is 0.27 and that of 1,6-anhydro-2,3,4-tri-*O*-benzyl-β-D-glucopyranose is 2.5, indicating that the initial monomer concentration of 1,6-anhydro-3-*O*-benzoyl-2,4-di-*O*-benzyl-β-D-glucopyranose must be three times as high as the monomer concentration of 1,6-anhydro-2,3,4-tri-*O*-benzyl-β-D-glucopyranose in order to obtain a copolymer containing 50% branching points. The number average sequence length of each monomeric unit, which indicates the distribution of the branching point, can be also calculated from the values of monomer reactivity ratios and the monomer composition in the feed.

Synthesis of (1→6)-α-D-Glucopyranan Having α-D-Mannopyranosyl Branch at C-3 Position

α-D-Mannopyranose-branching (1→6)-α-D-glucopyranan is synthesized by mannosylation of the 2,4-di-*O*-benzyl-(1→6)-α-D-glucopyranan prepared as above. In the orthoester method for mannosylation, chlorobenzene is used as the solvent and 2,6-lutidinium perchlorate as catalyst. 2,4-Di-*O*-benzyl-(1→6)-α-D-glucopyranan (300 mg, 0.88 mmol) and 3,4,6-tri-*O*-acetyl-β-D-mannopyranose-1,2-(methyl orthoacetate) (1.05 g, 2.60 mmol) are dissolved in chlorobenzene (25 ml), and the solution is refluxed for 1 hr using a Dean–Stark trap. After distillation of a few milliliters of solvent, 2,6-lutidinium perchlorate (1.8 mg) is added, and the mixture is refluxed for 40 min. The mixture is evaporated, and the product is purified by dissolution–reprecipitation using chloroform–methanol system three times and freeze-dried from benzene. ^{13}C NMR is used to confirm that all the mannopyranose branching units in the obtained mannose-branching polysaccharide are of the α configuration. Debenzylation of the mannopyranose-branching polysaccharide is carried out in the same manner as the deprotection of glucopyranose-branching polysaccharide by using sodium metal in liquid ammonia as above.

[9] J. P. Kennedy, T. Kelen, and F. Tüdös, *J. Polym. Sci. Polym. Chem. Ed.* **13,** 2277 (1975).

Hypoglycemic Activity of Synthetic (1→6)-α-D-Glucopyranan Having α-D-Glucopyranosyl Branch at C-3 Position

The hypoglycemic effect of the monosaccharide-branching (1→6)-α-D-glucopyranan is evaluated by intraperitoneal injection of the polysaccharide in mice. Male mice (Std : ddy, closed colony, National Institute of Health, Japan, 8 weeks old) are used in groups of five. The mice are given drinking water freely and are not fed for 2 hr immediately before the injection and measurement of the blood glucose level. The synthetic polysaccharides are dissolved in physiological saline and injected (10, 30, mg/kg i.p.) into the mice. As a control experiment, saline solution without polysaccharide is also injected. After 5, 10, and 24 hr from the administration, blood is drawn from the tail. The glucose level of the drawn blood is determined by a glucose level analyzer (Glucoboy, Eiken Chemical Co. Ltd., Tokyo, Japan; glucose oxidase method).

The effect of the synthetic polysaccharide on blood glucose levels in mice is shown in Fig. 1. α-D-Glucose-branched polysaccharide (**1**) Fig. 1 shows a remarkable hypoglycemic activity at a dosage of 10 and 30 mg/kg. After 5 and 10 hr of administration of **1**, the blood glucose level is in the range of 65 to 81% of the value before administration. After 24 hr, the blood glucose level is restored to normal values. The hypoglycemic activity exists in α,β-D-glucose-branched polysaccharide (**2**, Fig. 1) as

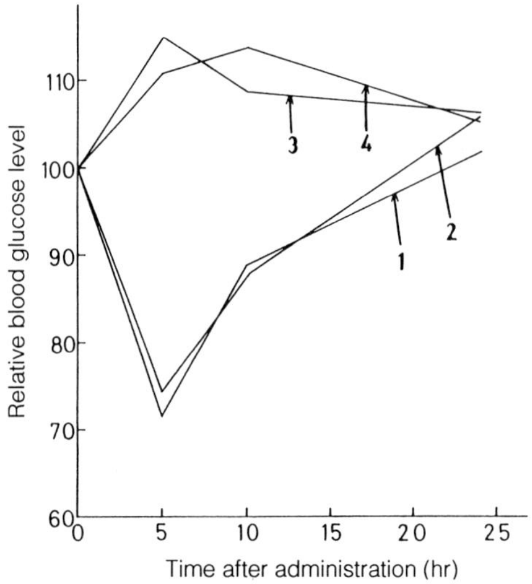

FIG. 1. Hypoglycemic activity of synthetic polysaccharides [Reprinted from K. Hatanaka, S. C. Song, A. Maruyama, A. Kobayashi, H. Kuzuhara, and T. Akaike, *Biochem. Biophys. Res. Commun.* **188,** 16 (1992), with permission].

well, indicating that a β-D-glucopyranose branch at least does not inhibit the biological functions. On the other hand, the linear (1→6)-α-D-glucopyranan (**3**, Fig. 1) and mannopyranose-branched polysaccharide (**4**, Fig. 1) does not exhibit hypoglycemic action. From these data, it is apparent that the α-D-glucosyl branch is essential to the hypoglycemic activity.

[24] Tailor-made Glycopolymer Syntheses

By F. D. TROPPER, A. ROMANOWSKA, and R. ROY

Introduction

Synthetic glycoconjugates are a family of carbohydrate derivatives which includes (neo)glycoproteins, (neo)glycolipids, and glycopolymers. Glycopolymers are gaining increasing interest owing to the many advantageous physico- and immunochemical properties with which they can be designed.[1-5] Moreover, the well-recognized "cluster effect" of carbohydrate ligands for their specific receptors can be fully appreciated by virtue of the intrinsic multivalent structures of the polymers.[6] To expand further the usefulness of glycopolymers in various bio- and immunochemical assays, three-component copolymerizations (terpolymerizations) are described.[7] The custom-designed syntheses of such glycopolymers allows for the introduction of defined properties.

The strategy is very simple and relies on readily available N-acryloylated precursors: carbohydrate derivatives,[8] probes or structures of desired properties, and monomers to be used as the core in the glycopolymer backbone. In the present work, the N-acryloylated carbohydrate derivatives serve as ligands for lectins and antibodies. Acrylamide or methacrylamide is used to confer the polymers with water-soluble properties having

[1] V. Hořejší and J. Kocourek, this series, Vol. 34, p. 361.

[2] R. T. Lee and Y. C. Lee, this series, Vol. 83, p. 299.

[3] A. Y. Chernyak, A. B. Levinsky, B. A. Dmitriev, and N. K. Kochetkov, *Carbohydr. Res.* **128,** 269 (1984).

[4] P. Kosma, P. Waldstätten, L. Daoud, G. Schulz, and F. M. Unger, *Carbohydr. Res.* **194,** 145 (1989).

[5] R. Roy, F. D. Tropper, and A. Romanowska, *Bioconjugate Chem.* **3,** 256 (1992).

[6] Y. C. Lee, R. R. Townsend, M. R. Hardy, J. Lönngren, J. Arnap, M. Haraldsson, and H. Lönn, *J. Biol. Chem.* **258,** 199 (1983).

[7] R. Roy, F. D. Tropper, and A. Romanowska, *J. Chem. Soc., Chem. Commun.*, 1611 (1992).

[8] R. Roy and F. D. Tropper, *Glycoconjugate J.* **5,** 203 (1988).

Copyright © 1994 by Academic Press, Inc.
All rights of reproduction in any form reserved.

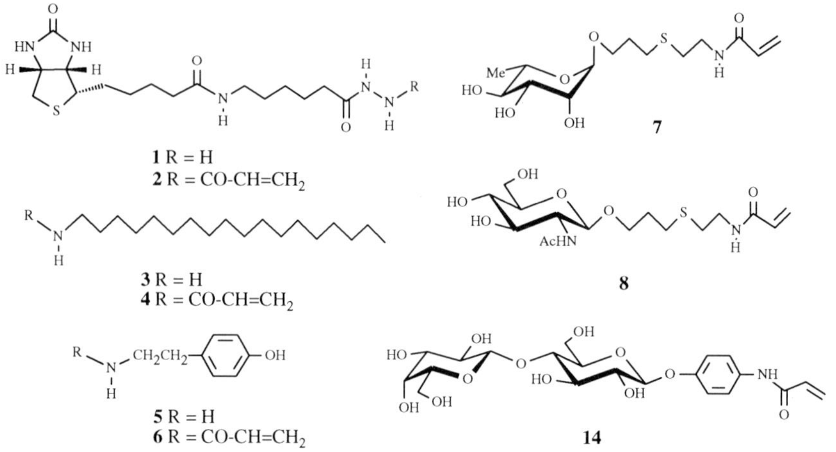

SCHEME 1

balanced hydrophilic components. The third element of structures added confers on the terpolymers certain specific properties which can be exploited. Three examples are given below.

Biotin is an obvious comonomer since many avidin- or streptavidin-labeled conjugates are commercially available.[9] In the second example, N-acryloylated 1-aminooctadecane (stearylamine) is incorporated to provide improved lipohilic properties which make these polymers more suitable as coating antigens in enzyme-linked immunosorbent assays (ELISA) or enzyme-linked lectin assays (ELLA). In the third example, an N-acryloylated tyramine moiety is also introduced because it can enable the glycopolymers to be radiolabeled, thus mimicking tyrosine residues on proteins. Another example illustrates the effect of introducing a different carbohydrate residue as an effector molecule. The dual antigenicity of the resulting heterobifunctional poly(acrylamide-*co*-*N*-acetylglucosamine-*co*-rhamnose) polymer has been evaluated with a plant lectin (wheat germ agglutinin, WGA) and rabbit antibodies raised against the capsular polysaccharide from *Streptococcus pneumoniae* serotype 23 having α-L-rhamnose residues as a branch in the repeating unit.[10] The structures of the carbohydrate and effector monomers are illustrated in Scheme 1. The terpolymers structures are presented in Schemes 2 and 3.

[9] M. Wilchek and E. A. Bayer, *Anal. Biochem.* **171,** 1 (1988).

[10] C. Jones, *Carbohydr. Res.* **139,** 75 (1985).

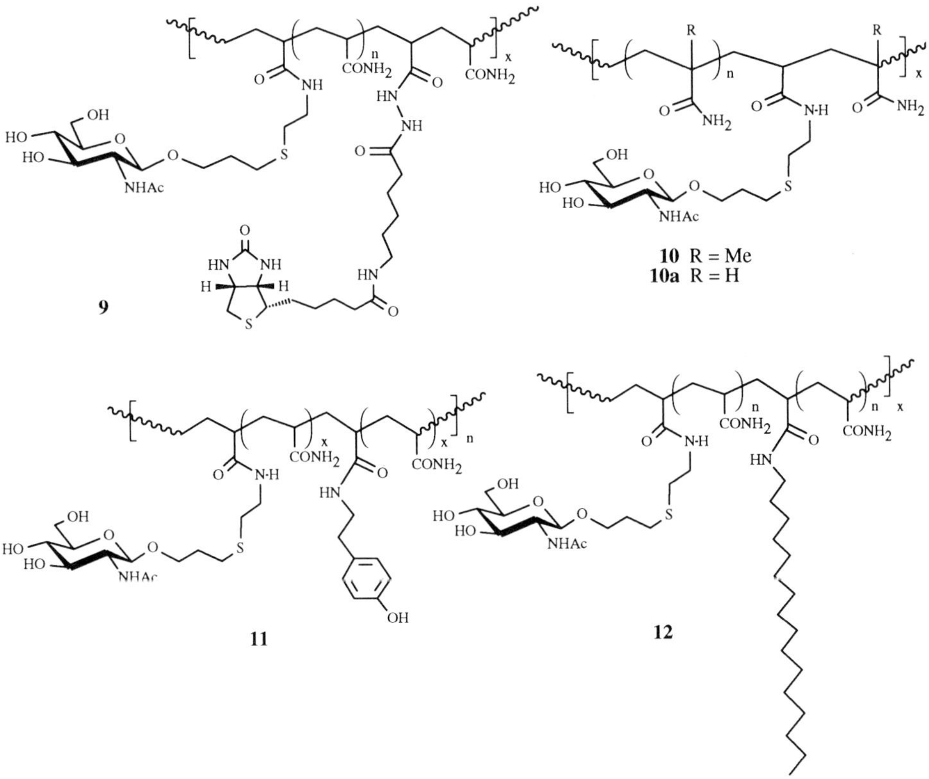

SCHEME 2

Materials and General Methods

Materials and general methods are essentially the same as those described in the accompanying paper.[11]

Acryloylation of Probes

*Acryloylation of Biotin Amidocaproylhydrazide (**1**).* Biotin amidocaproylhydrazide (**1**) [45.0 mg, 0.121 mmol, Sigma, St. Louis, MO, ~95% pure by thin-layer chromatography (TLC)] is dissolved in warm methanol (85 ml). Anion-exchange resin (HO^-) is added and the solution placed in an ice bath. To the stirring solution is added, in a slow dropwise fashion, a solution of 11 μl (0.135 mmol) acryloyl chloride in 500 μl of 1,4-dioxane.

[11] C. A Laferrière, F. O. Andersson, and R. Roy, this volume [25].

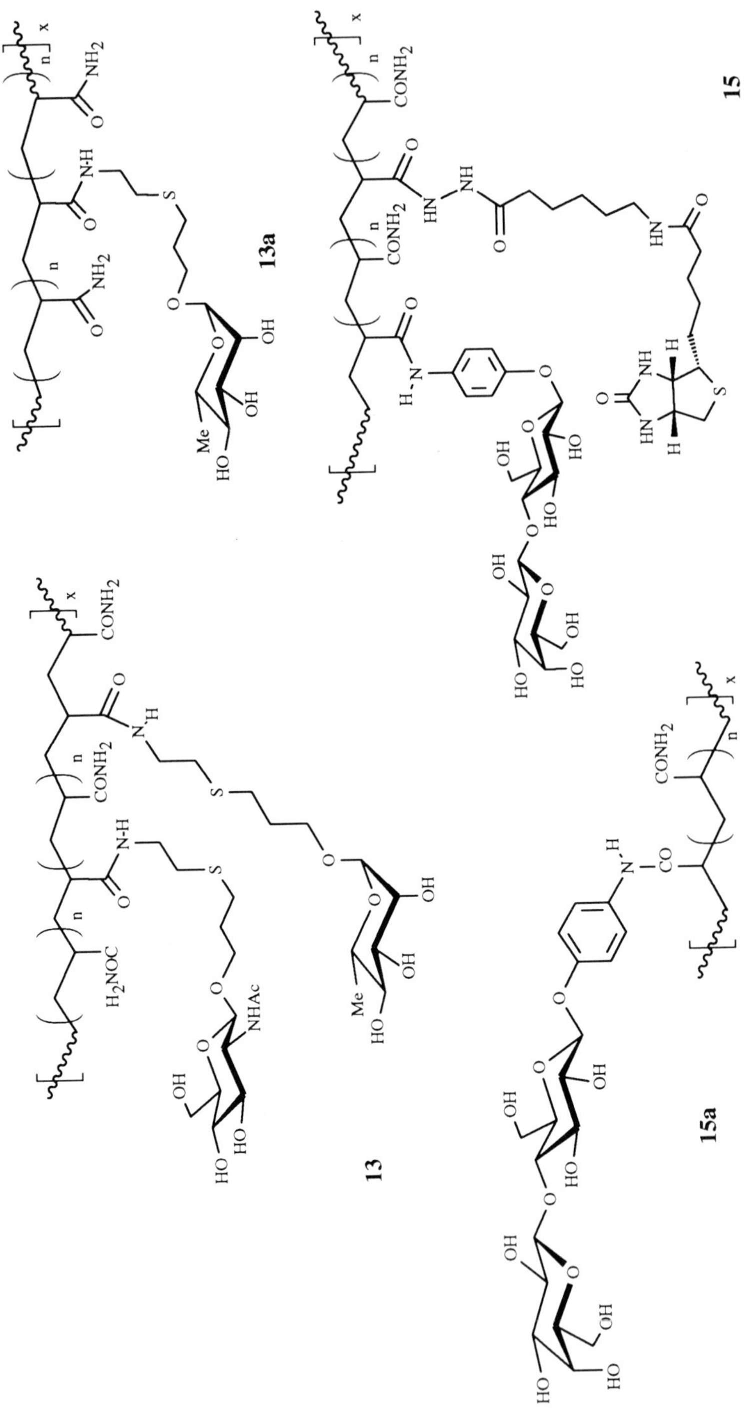

13a

15

13

15a

SCHEME 3

The reaction is monitored by TLC ($CHCl_3$/methanol, 4 : 1, v/v) which shows a complete and clean conversion of **1** (R_f 0.20) to **2** (R_f 0.40) after 3 hr. The impurity (R_f 0.50, presumably biotin amidocaproyl methyl ester) in the starting material is not affected. The reaction is then warmed to near 50°, and the resin is filtered and washed with warm methanol. Evaporation of the methanol gives a white solid which is redissolved in a minimum amount of hot methanol (~20 ml), allowed to cool at room temperature, then stored in the freezer compartment of a refrigerator (−5°). After a few days, a white, cottony mass is filtered, washed with cold ethanol, then dried under vacuum to yield 22 mg (43%) of pure **2** by TLC and nuclear magnetic resonance (NMR). The product has mp 185.7°–187.1° and $[\alpha]_D$ +35.4° [*c* 0.82, dimethyl sulfoxide (DMSO)]. The mother liquor is concentrated, and 200 μl of DMSO is added to solubilize the remaining crude material completely. The solution is applied to a half-size preparatory TLC plate. The resulting wide and double application band is concentrated to single thin band by multiple short elutions with pure ethanol. The TLC plate is then eluted with $CHCl_3$/methanol (4 : 1, v/v). The band corresponding to the desired product **2** is scraped from the glass and then extracted with ethyl acetate/ethanol (4 : 1, v/v). The silica is filtered and the solvent evaporated. An additional 18 mg (35%) of pure **2** is thus obtained, giving a total yield of 78%.

*N-Acryloylated Stearylamine (**4**).* 1-Aminooctadecane (stearylamine, **3**) (250 mg, 0.928 mmol) is dissolved in 25 ml of CH_2Cl_2 containing 400 μl of triethylamine. The solution is cooled to 0° and a solution of acryloyl chloride 100 μl in CH_2Cl_2 (8 ml) is added dropwise over a 40-min period. The precipitated triethylammonium hydrochloride is filtered and washed with ice-cold CH_2Cl_2. The CH_2Cl_2 solution is evaporated and the resulting solid redissolved in hot ethanol. The product (R_f 0.58 in ethyl acetate/hexane containing 0.5% 2-propanol) is left to crystallize at room temperature. Three crops give 289.5 mg (96%) of **4** as a white microcrystalline powder which has mp 74.5°–75.1°.

*N-Acryloylated 2-(p-Hydroxyphenyl)ethylamine (**6**).* Tyramine hydrochloride (**5**) (0.32 g, 1.84 mmol, Aldrich, Milwaukee, WI) is dissolved in 10 ml of methanol and 1 ml of triethylamine. The solution is cooled to 0° and a slow, dropwise addition of acryloyl chloride (200 μl in 5 ml 1,4-dioxane) is made until TLC ($CHCl_3$/methanol, 10 : 1, v/v) indicates a complete consumption of tyramine **5** (R_f 0) to give as the major product **6** (R_f 0.52). The reaction is worked up by evaporating the solution to a crystalline mass and redissolving the crude mixture in 30 ml of ethyl acetate. The organic phase is washed with 0.5 *M* HCl to remove excess triethylamine followed by saturated $NaHCO_3$, water, and brine (NaCl). The organic extracts are dried (Na_2SO_4), filtered, and evaporated to an oil. Attempts at crystallizing the product from various solvent systems proved fruitless. The product **6** is isolated by silica gel chromatography using $CHCl_3$/methanol (15 : 1, v/v) as eluent. The desired product is isolated in a 225 mg (64%) yield as a glassy solid which cannot be crystallized.

Synthesis of N-Acryloylated Sugar Precursors

The 3-(2-*N*-acrylamidoethylthio)propyl glycosides are all prepared by the same general procedure outlined below. The reactions are performed equally well on scales of 50 mg to 4 g to give almost quantitative yields of product.

Pure 3-(2-aminoethylthio)propyl glycoside precursors[8,12,13] are dissolved in methanol. An excess of strong anion-exchange resin [Amberlite IRA-400 (HO^-) form], wet from prior washing with distilled water, is then added and the solutions cooled to 0° with an ice bath. While the solution is stirring at a low rate, acryloyl chloride (1.05 equivalents) as a 5% (v/v) solution in either chloroform or 1,4-dioxane is slowly added in a dropwise fashion. Although acryloylation of amines is essentially instantaneous, rapid addition of acryloyl chloride leads to an incomplete consumption of the amine and the formation of unknown secondary products. Slow addition of acryloyl chloride is essential for a clean quantitative transformation to the desired products. The reactions are monitored by the disappearance of the amine by applying a small drop of the reaction solution on a piece of TLC plate and then dipping it in a solution of 0.2% (w/v) ninhydrin in 2% (v/v) ethanolic pyridine. After mild heating, the absence of a purple-colored spot indicates complete consumption of the amine. Completeness of reaction and product purity are confirmed by TLC using $CHCl_3$/methanol (5 : 1, v/v) as eluent.

Once the reaction is complete, the anion-exchange resin is filtered and washed with methanol. The filtrate is then treated for few minutes with cation-exchange resin [Amberlite IR-120 or Dowex 50W (H^+ form)]. Removal of the resin followed by evaporation of methanol under reduced pressure below 25° gives essentially quantitative yields of the desired glycosides, pure by TLC and NMR.

3-(2-*N*-Acrylamidoethylthio)propyl α-L-rhamnopyranoside (**7**) is obtained as an oil, $[\alpha]_D$ −27.9° (*c* 1.7, DMSO)[13]; ^{1}H NMR δ (DMSO-d_6): 8.36 (t, 1H, $J_{e,NH}$ = 5.6 Hz, NH), 6.12 (dd, 1H, J_{cis} = 10.0, J_{trans} = 17.1 Hz, —CH═), 5.95 (dd, 1H, J_{gem} = 2.2 Hz, ═CH_2 *trans*), 5.46 (dd, 1H, ═CH_2 *cis*), 4.41 (d, 1H, $J_{1,2}$ = 1.2 Hz, H_1), 3.03–3.59 (mult, 8H, H_2–H_5, H_a, H_e), 2.49 (mult, 4H, H_c + H_d), 1.69 (mult, 2H, H_b), 1.16 (d, 3H, J = 6.1

[12] Y. C. Lee and R. T. Lee, *Carbohydr. Res.* **37,** 193 (1974).

[13] R. Roy and F. D. Tropper, *J. Chem. Soc., Chem. Commun.*, 1058 (1988).

Hz, CH_3). The H_{a-e} protons refer to the spacer (2-*N*-acrylamidoethylthio)-propyl group, with H_a being the closest to the sugar.

3-(2-*N*-Acrylamidoethylthio)propyl 2-acetamido-2-deoxy-β-D-glucopyranoside (**8**) is obtained as crystals from warm acetonitrile, mp 135.5°–137.1°, $[\alpha]_D$ −17.2° (*c* 0.92, DMSO).[8] Analytical data: Percentage calculated for $C_{16}H_{28}N_2O_2S \cdot 1.5H_2O$: C, 46.82; H, 7.37; N, 6.82; S, 7.81. Found: C, 46.66; H, 7.54; N, 6.75; S, 8.02; ^{1}H NMR δ (DMSO-d_6): 6.21 (dd, 1H, J_{cis} = 9.9, J_{trans} = 17.1 Hz, —CH═), 6.08 (dd, 1H, J_{gem} = 2.4 Hz, ═CH_2 *trans*), 5.59 (dd, 1H, ═CH_2 *cis*), 4.25 (d, 1H, $J_{1,2}$ = 8.0 Hz, H_1), 3.76 (mult, 1H), 3.67 (mult, 1H), 3.45 (mult, 2H), 3.31 (mult, 4H), 3.06 (mult, 2H), 2.56 (t, 2H, $J_{b,c}$ = 7.3 Hz, H_c), 2.50 (t, 2H, $J_{d,e}$ = 7.2 Hz, H_d), 1.80 (s, 3H, CH_3), 1.70 (mult, 2H, H_b).

Copolymerizations of Sugars and Probes with Acrylamide

*Poly(acrylamide-co-biotin-co-β-D-GlcNAc) (**9**).* N-Acryloylated biotin amidocaproylhydrazide (**2**) (4 mg, 10 μmol), acrylamide (71.1 mg, 1.00 mmol), 3-(2-*N*-acrylamidoethylthio)propyl β-D-*N*-acetylglucosaminide (**8**) (39 mg, 0.10 mmol), and a small magnetic stir bar are placed in a test tube. The test tube is sealed with a septum, evacuated, and filled with a nitrogen atmosphere. Deoxygenated water (800 μl) and DMSO (300 μl) are then added using a syringe followed by 15 μl of *N,N,N′,N′*-tetramethylethylenediamine (TMEDA or TEMED). The solution is stirred at 65° for a few minutes, then ammonium persulfate (50 μl, 100 mg/ml) is injected into the solution. The polymerization is allowed to proceed for 2 hr at room temperature. The resulting clear viscous solution is diluted with 10 ml of hot water and then dialyzed once against 5 liters of 0.5 m*M* acetic acid followed by two 5-liter volumes of distilled water. Lyophilization gives 118.8 mg (97%) of **9** as a white solid. ^{1}H NMR (D_2O) indicates a 1 : 5 : 50 molar incorporation ratio of **2 : 8** : acrylamide. The biotin content in the glyco(ter)polymer is confirmed by treating a small sample of **9** with 0.2% (w/v) ethanolic 4-dimethylaminocinnamaldehyde, which results in a pink coloration.

*Poly(methacrylamide-co-β-D-GlcNAc) (**10**).* Compound **8** (22.1 mg, 56.3 μmol) and methacrylamide (48 mg, 0.564 mmol) are dissolved in deoxygenated water (600 μl). Ammonium persulfate (25 μl, 50 mg/ml, 0.9 mol%) is then injected and the solution stirred vigorously at 100° for 12 min. Dialysis and lyophilization give 21 mg (30%) of **10** as a white solid. ^{1}H NMR (D_2O) analysis gives a 1/10.1 molar incorporation ratio of β-D-GlcNAc/methacrylamide. A reference poly(acrylamide-*co*-β-D-GlcNAc) polymer (**10a**) having a sugar content (by moles) of one residue per nine acrylamide residues is obtained as previously described.[8] This polymer is used as a refereence standard in enzyme-linked lectin assays (ELLA) with horseradish peroxidase-labeled wheat germ agglutinin (HRP–WGA).

Poly(acrylamide-co-tyramide-co-β-D-GlcNAc) ***(11)***. 2-*N*-*p*-Hydroxyphenylethyl acrylamide (**6**) (100 μl of 20 mg/ml 0.1 *M* NaOH, 10 μmol), acrylamide (71.1 mg, 1.00 mmol), and the β-D-acetylglucosaminide (**8**) (39.2 mg, 0.1 mmol) are boiled in 1.0 ml of deoxygenated water under a nitrogen atmosphere for a few minutes before ammonium persulfate (3.5 mg, 1.4 mol%) is added. After 15 min of vigorous stirring at 100°, the clear viscous solution is diluted with 12 ml of water and dialyzed against three 5-liter volumes of distilled water. Lyophilization gives 59.4 mg (53%) of **11** as a white solid. ^{1}H NMR confirms the 1 : 10 : 100 ratio of **6** : **8** : acrylamide in the polymer.

Poly(acrylamide-co-stearylamine-co-β-D-GlcNAc) ***(12)***. N-Acryloylated stearylamine (**4**) (3.2 mg, 10 μmol), acrylamide (71.1 mg, 1.00 mmol), and the N-acryloylated β-D-*N*-acetylglucosaminide (**8**) (39.2 mg, 0.1 mmol) are vigorously stirred in 1.1 ml of deoxygenated water at 100°, under a nitrogen atmosphere. After 10 min, 5 mg (2% by moles) of solid ammonium persulfate is added; the test tube is refilled with a nitrogen atmosphere and fitted with a septum. The polymerization mixture is left stirring at high speed for 12 min at 100°. The resulting viscous solution is diluted with 12 ml of water and dialyzed against 5 volumes of water (5 liters) over the course of 3 days. The slightly foggy solution is centrifuged at 14,000 *g* for 15 min. No sedimentation or change in the appearance of the solution is observed. The dialyzed solution is then lyophilized to yield 66.5 mg (59%) of **12** as a white solid. ^{1}H NMR (D_2O) analysis of **12** reveals a structural composition of 1 : 10 GlcNAc : acrylamide, whereas a 1 : 23 estimate of stearylamine : GlcNAc is made.

Poly(acrylamide-co-α-L-rhamnose-co-β-D-GlcNAc) ***(13)***. Acrylamide (110 mg, 1.55 mmol), α-L-rhamnosyl monomer (**7**) (174 mg, 0.537 mmol), and the β-D-*N*-acetylglucosaminide (**8**) (117 mg, 0.298 mmol) are dissolved in 8.0 ml of deoxygenated water in a flask filled with nitrogen. The flask is immersed in a boiling water bath, and after a few minutes polymerization is initiated with 40 μl of $(NH_4)_2S_2O_8$ (100 mg/ml, 0.4% by moles). The polymerization reaction is stirred vigorously at 100° for 15 min before being diluted and exhaustively dialyzed against distilled water. Lyophilization gives 401 mg (59%) of **13** as a white solid. ^{1}H NMR analysis of **13** reveals a monomer incorporation ratio of α-L-Rha : β-D-GlcNAc : acrylamide of 1 : 1 : 6. A reference poly(acrylamide-*co*-α-L-Rha) (**13a**) with a ratio 3 : 1 is similarly prepared.

Poly(acrylamide-co-biotin-co-β-D-lactose) ***(15)***. Acrylamide (27.4 mg, 386 μmol), biotin monomer (**2**) (8.2 mg, 19.3 μmol), and the *p*-acrylamidophenyl lactoside (**14**) (18.8 mg, 36.8 μmol) are dissolved in 400 μl of deoxygenated water and 200 μl of DMSO in a small stoppered test tube, under a nitrogen atmosphere. The solution is warmed to 65° and stirred vigorously to dissolve all the monomers properly. Once the solution is homogeneous, 5 μl of TEMED followed by 50 μl of ammonium persulfate (50 mg/ml) are injected. After the mixture is stirred at room temperature for 3 hr, the clear, viscous solution is diluted with water and dialyzed. The initial dialysis is performed in 5 liters of 0.5 m*M* acetic acid followed by two other dialyses in 5 liters of distilled water. Lyophilization gives 31.0 mg (57%) of **15** as a white solid. The biotin incorporation is verified by treating a small sample of **15** with 0.2% (w/v) 4-dimethylaminocinnamaldehyde in acidified ethanol (Sigma, a reagent specific for detecting biotin) and by observing the generation of a pink to red coloration. An analog deprived of biotin does not react with 4-dimethylaminocinnamaldehyde. ^{1}H NMR (D_2O) analysis of **15** confirms the presence of biotin from the characteristic resonance of **14** found unobstructed between 2.6 and 3.4 ppm. The aromatic resonances of the lactoside (7.0–7.5 ppm) as well as the anomeric signals at 5.14 (br, H-1 Glc) and 4.50 (d, $J_{1,2}$ = 7.5 Hz, H-1 Gal) are clear indications of the lactose content in **15**. Comparison of intensities for various reporter groups of the biotin and the lactoside monomers to the characteristic methine and methylene resonances of the polymer backbone at 2.3 and 1.7 ppm indicated a 1 : 2 : 10 molar incorporation ratio for biotin : lactose : acrylamide in terpolymer **15**. Corrections for the eight methylene resonances of biotin, found under the backbone methylene signal, are made for this determination. A control copolymer composed of only lactose and acrylamide residues (**15a**) is also used as reference.

Quantitative Immunoprecipitation with Rabbit Antiserum against *Streptococcus pneumoniae* Type 23

The rabbit antiserum against *Streptococcus pneumoniae* type 23 (Statens Serum Institut, Copenhagen, Denmark) is diluted 3 times with phosphate-buffered saline (PBS), and 100 μl of the solution is added to 1.5-ml Eppendorf microcentrifuge tubes (Brinkmann Instruments Co., Westbury, NY). Aliquots of 1.0 to 100 μl of a stock solution of the copolymer (**13**) having an acrylamide-GlcNAc-Rha ratio of 5.2 : 1.0 : 1.8 (1.0 mg/ml) in PBS are added to the tubes. The final volume of all the tubes are then adjusted to 300 μl with PBS. The reactions are run in triplicate. The solutions are gently agitated and allowed to incubate for 2 hr at 37°, then 3 days at 4°. The tubes are centrifuged at 25° for 15 min at 14,000 *g* in a Fisher (Ottawa, Ontario, Canada) microcentrifuge (Model 235B). The precipitin pellets are washed three times with 300 μl of cold PBS with

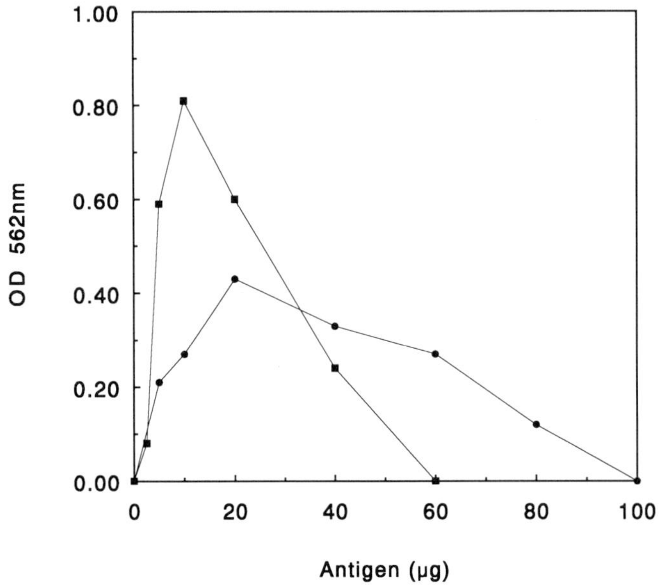

FIG. 1. Quantitative precipitin curves of *Streptococcus pneumoniae* type 23 rabbit antiserum (●) or wheat germ agglutinin (WGA) (■) with heterobifunctional terpolymer **13** made of acrylamide, *N*-acetylglucosamine, and L-rhamnose. The two precipitin curves show that the terpolymer conserves its independent bifunctional antigenicity.

centrifugation between washes. The protein content of the pellets is determined using the bicinchoninic acid (BCA) protein assay reagents of Pierce (Rockford, IL) following the enhanced protocol described in the instruction manual.[14] The absorbances are measured at 562 nm.

Lectin Quantitative Precipitation

Wheat germ agglutinin (WGA, Sigma, L-9640) is diluted to 200 μg/ml in PBS, and 100-μl aliquots of the solution are placed in 1.5-ml Eppendorf microcentrifuge tubes. Aliquots of 1.0 to 100 μl of 1.0 mg/ml of copolymer **13** in PBS are added to give final volumes of 300 μl. The reactions are run in triplicate and processed as described above for the rabbit antiserum. The results of the quantitative precipitation experiments of **13** with WGA and the rabbit antiserum are illustrated in Fig. 1. The results clearly

[14] P. K. Smith, R. I. Krohn, G. T. Hermanson, A. K. Mallia, F. H. Gartner, M. D. Provenzano, E. K. Fukimoto, N. M. Goeke, B. J. Olson, and D. C. Klench, *Anal. Biochem.* **150,** 76 (1985).

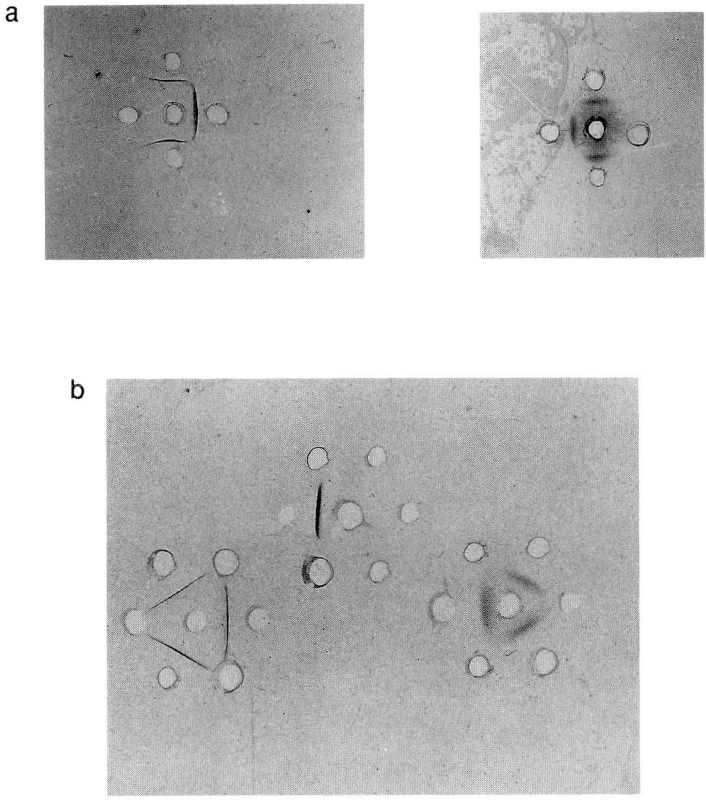

FIG. 2. Agar double immunodiffusions showing that the terpolymers conserved their bifunctional properties. (a) *Left*: WGA (middle well) with polymer **10a** clockwise (well 2), **13a** (well 4), and heterobifunctional polymer **13** (wells 1 and 3). *Right*: *Streptococcus pneumoniae* type 23 rabbit antiserum (middle well) and copolymers **10a**, **13a**, and **13** placed as above. (b) Clockwise from the 1 o'clock position: lactose polymer **15a** (well 1), GlcNAc polymer **10a** (well 2), lactose–biotin (1 : 0.2) copolymer **15** (well 3), GlcNAc–biotin (1 : 0.2) copolymer **9** (well 4), lactose–biotin (1 : 1) copolymer **15** (well 5), GlcNAc–biotin (1 : 0.2) copolymer **9** (well 6). Middle wells contained wheat germ agglutinin (left), streptavidin (center), and peanut lectin (right).

indicate that the heterobifunctional copolymer possesses binding activity for both the lectin and the antibodies.

Agar Gel Double Radial Diffusion

Double radial diffusion is performed on 1% (w/v) agarose (BDH, Toronto, Ontario, Canada) containing 2% (w/v) polyethylene glycol

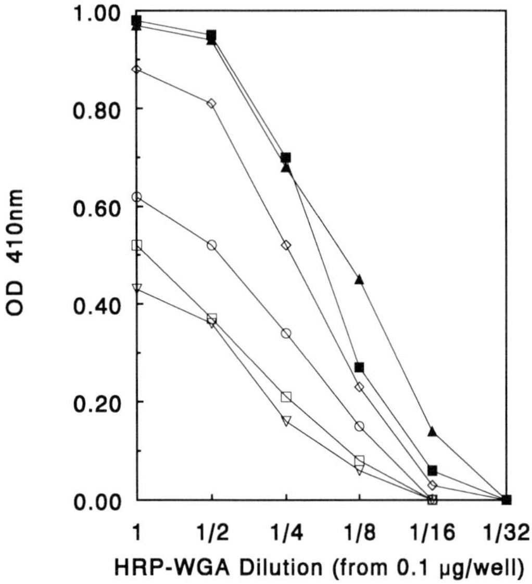

FIG. 3. Enzyme-linked lectin assay (ELLA) showing the coating properties of the lipophilic terpolymer (**12**) used at different coating concentrations with HRP–WGA at 2-fold dilution using ABTS/H_2O_2 as peroxidase substrates: ▽, 0.01 ng; □, 0.10 ng; ○, 1 ng; ◇, 10 ng; ▲, 100 ng; ■, 1 μg.

(PEG, molecular weight 8000, Sigma) in 0.1 *M* PBS, pH 7.2, according to Ouchterlony and Nilsson.[15] Both glycopolymers and lectins are used at a concentration of 1 mg/ml PBS unless indicated otherwise. The rabbit antiserum against *Streptococcus pneumoniae* type 23 is used as received. Aliquots (20 μl) of respective antigens and lectins are used to fill the wells perforated in the agarose gel slabs (~1 mm thickness). The precipitin bands are allowed to form overnight at 4°. Figure 2 clearly illustrates that each carbohydrate hapten on the terpolymer can be recognized as such by the respective lectin, thus demonstrating that the terpolymers behave as bifunctional antigens.

Enzyme-Linked Lectin Assay with Horseradish Peroxidase-Labeled Wheat Germ Agglutinin and N-Acetylglucosamine Copolymers

The wells of Linbro enzyme immunoassay (EIA) microtiter plates (Flow Laboratories, Mississauga, ON, Canada) are coated at room temperature overnight by using 100 μl of 10-fold dilutions of the glycopolymers

[15] O. Ouchterlony and L. A. Nilsson *in* "Handbook of Experimental Immunology" (D. M. Weir, ed.), [19]. Blackwell Scientific, Oxford, 1978.

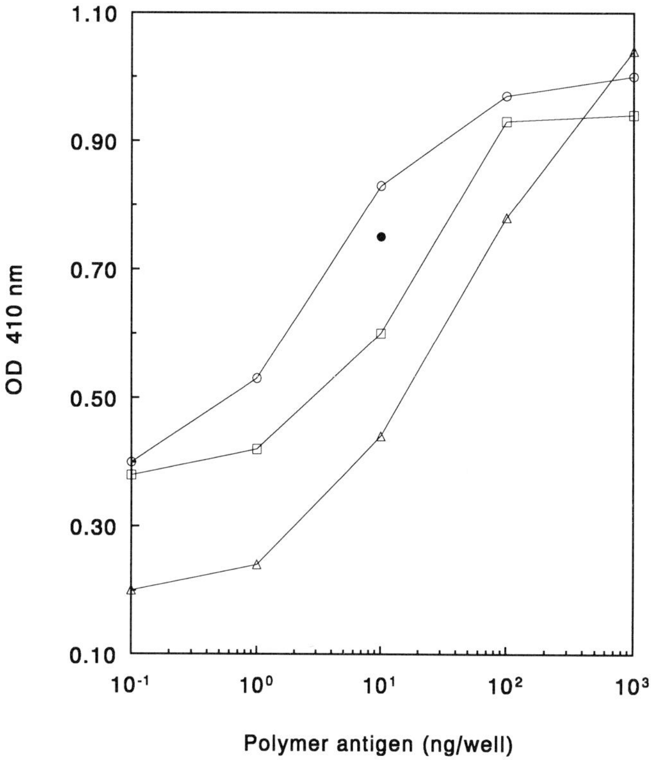

FIG. 4. Comparative enzyme-linked lectin assays (ELLA) of the terpolymers **10–12** used as coating antigen with HRP-labeled WGA and ABTS/H_2O_2 as peroxidase substrates: ● (one point only), **10**; △, **10a**; ○, **11**; □, **12**. The assays show that the more lipophilic terpolymer (**12**) at 10 ng/well was slightly better than the other copolymers including the *co*-methacrylamide polymer (**10**).

10, **10a**, **11**, and **12** (10 μg/ml). The plates are blocked with 300 μl of 0.2% (w/v) bovine serum albumin (BSA) in PBS for 1 hr at room temperature. After washing (four times with 300 μl PBS), the wells are filled with 100 μl of horseradish peroxidase-labeled wheat germ agglutinin (HRP–WGA, Sigma, L3892, 1 μg/100 μl). The plates are incubated at room temperature for 3 hr and then washed five times with PBS. A 50-μl volume of the enzyme substrate 2,2′-azinobis(3-ethylbenzothiazoline-6-sulfonic acid) (ABTS, 250 μg/ml in 0.2 *M* citrate–phosphate buffer, pH 4.0, containing 0.01% (v/v) H_2O_2) is then added. After 15 min, the optical density is measured at 410 nm on a Dynatech (Alexandria, VA) MR 600 spectrophotometer. Figures 3 and 4 illustrate the coating properties of the terpolymers and indicate that as little as 1–10 ng/well of the polymers is

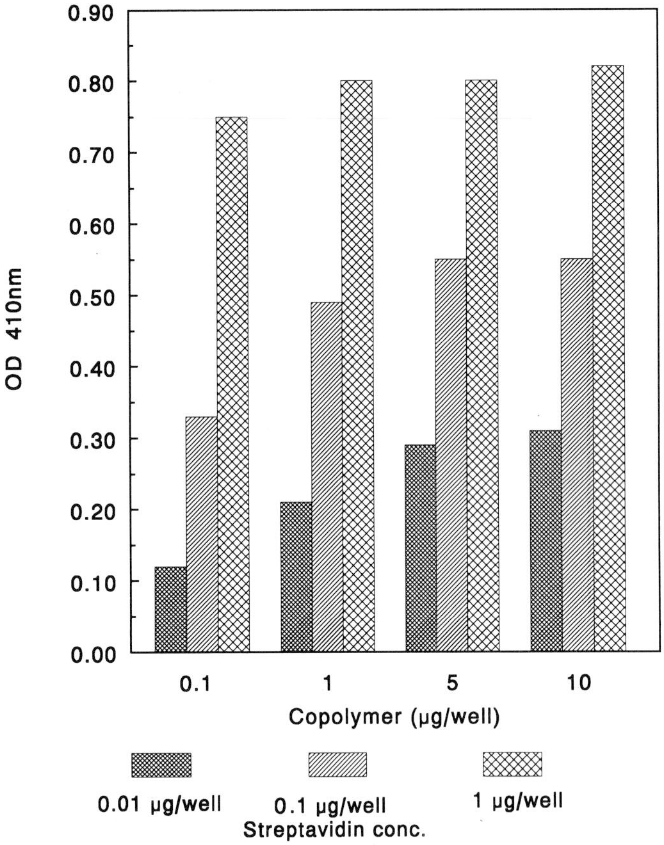

FIG. 5. Bar diagram of the combined results of sandwich enzyme-linked lectin assays (ELLA) using streptavidin as coating antigen for the capture of the biotin–lactose copolymer (**15**). The detection of **15** was done with HRP–peanut lectin using ABTS/H_2O_2 as peroxidase substrates.

sufficient to capture the lectins. Figure 4 also demonstrates that the more lipophilic properties of the *co*-stearylamine terpolymer **12** confer better coating properties. However, it should be kept in mind that the other terpolymers (*co*-biotin and *co*-tyramine) are designed for different immunoassays.

Enzyme-Linked Lectin Assay with Streptavidin

The wells of Linbro EIA microtiter plates are coated at room temperature overnight with 0.01, 0.1 and 1.0 μg (100 μl/well) of streptavidin (Boehringer Mannheim GmbH, Mannheim, Germany, Cat. No. 973 190) used as capture receptor for the biotin-containing lactose-*co*-polymer (**15**). The plates are blocked with 300 μl of 1% (w/v) BSA in PBS for 1 hr at room temperature. After washing (5 times), the wells are filled with 100 μl/well of the poly(acrylamide-*co*-biotin-*co*-lactose) polymer (**15**) (10 : 1 : 1) containing 10.0, 5.0, 1.0, and 0.1 μg/well respectively. The plates are incubated at room temperature for 90 min. After washing, HRP-labeled peanut lectin (0.5 μg/well) is added and the mixture allowed to incubate for 3 hr at room temperature. The plates are washed and 50 μl/well of ABTS (1 mg/4 ml) in 0.2 *M* citrate–phosphate buffer, pH 4.0, containing 0.01% H_2O_2 is added. After 15 min the absorbance is measured at 410 nm on a Dynatech MR 600 spectrophotometer. Figure 5 shows the results of the sandwich ELLA assays. Maximum sensitivity is obtained with 1 μg/well of streptavidin. At that level of streptavidin coating, as little as 0.1 μg/well of copolymer is sufficient to saturate the plates.

[25] Syntheses of Water-Soluble Polyacrylamide-Containing Sialic Acid

By C. A. LAFERRIÈRE, F. O. ANDERSSON, and R. ROY

Introduction

Glycopolymers still constitute a relatively unexploited family of carbohydrate-containing macromolecules with demonstrated biological and immunochemical activities. Of particular interest is the application of sialic acid-containing water-soluble polymers as inhibitors of influenza A and C virus hemagglutinins.[1–7] In general, glycopolymers have excellent antigenic properties since they can be used in double immunodiffusion experiments, in quantitative immunoprecipitations, and, most importantly, as

[1] R. Roy, F. O. Andersson, G. Harms, S. Kelm, and R. Schauer, *Angew. Chem., Int. Ed. Engl.* **31,** 1478 (1992).

[2] R. Roy and C. A. Laferrière, *Carbohydr. Res.* **177,** C1 (1988).

[3] R. Roy, C. A. Laferrière, A. Gamian, and H. J. Jennings, *J. Carbohydr. Chem.* **6,** 161 (1987).

[4] A. Spaltenstein and G. M. Whitesides, *J. Am. Chem. Soc.* **113,** 686 (1991).

[5] N. E. Byramova, L. V. Mochalova, J. M. Belyanchikov, M. N. Matrosovich, and N. V. Bovin, *J. Carbohydr. Chem.* **10,** 691 (1991).

[6] W. Spevak, J. O. Nagy, D. H. Charych, M. E. Schaefer, J. H. Gilbert, and M. Bednarski, *J. Am. Chem. Soc.* **115,** 1146 (1993).

[7] J. O. Nagy, P. Wang, J. H. Gilbert, M. E. Schaefer, T. G. Hill, M. R. Callstrom, and M. Bednarski, *J. Med. Chem.* **35,** 4501 (1992).

Copyright © 1994 by Academic Press, Inc.
All rights of reproduction in any form reserved.

coating antigens in microtiter plate enzyme-linked immunosorbent assays (ELISA).[8] They possess distinct advantages over their neoglycoprotein counterpart: cost, stability, ease of characterization, and control of chemical, physical, and biological properties.

Previous syntheses of sialic acid-containing glycopolymers relied on the direct copolymerization of allyl sialosides with acrylamide. This chapter describes the chemical synthesis of three types of sialyl glycopolymers which differ by the length and hydrophobicity of the spacer arm introduced between the sialic acid residues and the polyacrylamide backbone. Some of these glycopolymers have been successfully used as inhibitors of viral hemagglutinins.[1,9]

Materials

Silver salicylate is prepared according to a published procedure.[10] All common reagents and solvents are reagent grade and are used as received (Aldrich Chemical Co., Milwaukee, WI). 4-Nitrothiophenol (80% w/w, No. N2,270-9) is also from Aldrich and is used as such. Ammonium persulfate and acrylamide are of electrophoretic grade from Bio-Rad Laboratories (Richmond, CA). Size-exclusion chromatography is done with the supports from Pharmacia Fine Chemicals (Piscataway, NJ). Polyacrylamides used as high-performance liquid chromatography (HPLC) standards are from American Polymer Standards Corp. (Mentor, OH) and have M_w/M_n ranging from 1.2 (15 kDa) to 2.1 (400 kDa). Small-scale ultrafiltrations are done on Minicon concentrators from Amicon Canada Ltd. (Oakville, Ontario, Canada, No. A25 and B125).

General Methods

Melting points are determined on a Gallenkamp (London, England) apparatus and are uncorrected. The ^{1}H nuclear magnetic resonance (NMR) and ^{13}C NMR spectra are recorded on a Varian (Palo Alto, CA) XL-300 at 300 and at 75.4 MHz, respectively. The proton chemical shifts (δ) are given relative to internal chloroform (7.24 ppm) for $CDCl_3$ solutions and to the HOD signal (4.75 ppm) for D_2O solutions. The analyses are done as a first-order approximation, and assignments are based on correlated spectroscopy (COSY) and heteronuclear correlated spectroscopy (HETCOR) experiments. Optical rotations are measured on a Perkin-

[8] R. Roy, C. A. Laferrière, R. A. Pon, and A. Gamian, this series, Vol. 247 [24].

[9] A. Gamian, M. Chomik, C. A. Laferrière, and R. Roy, *Can. J. Microbiol.* **37,** 233 (1991).

[10] D. J. M. van der Vleugel, W. A. R. van Heeswijk, and J. F. G. Vliegenthart, *Carbohydr. Res.* **102,** 121 (1982).

Elmer (Norwalk, CT) 241 polarimeter and are run in chloroform at 23° for 1% (w/v) solutions unless stated otherwise. Thin-layer chromatography (TLC) is performed using silica gel 60F-254 and column chromatography on silica gel 60 (230–400 mesh, E. Merck, Darmstadt, Germany, No. 9385). Molecular weight determination is effected on a Waters (Milford, MA) Model 991/625 LC HPLC system equipped with a diode-array detector. The calibration is done with polyacrylamide standards (American Polymer Standards Corp.) in 0.1 *M* phosphate buffer (pH 7.0) on a TSK G-4000 SWXL column (7.5 × 300 mm) at a flow rate of 1 ml/min. Absorbances are measured at 205 nm.[11]

Membrane-Exclusion Ultrafiltration

Minicon concentrators (A25, molecular weight cutoff 25,000 and B125, molecular weight cutoff 125,000) equipped with an ultrafiltration membrane with a specified molecular weight cutoff are filled with aqueous solutions of the allyl-based polymer [16% (w/w) NeuAc] **16**. After concentration (5 times), water is added and the solutions are again allowed to concentrate. Water is added once again, and the solutions are recovered and tested for sialic acid content by the colorimetric resorcinol method.[12] The polymer is totally recovered from the A25 miniconcentrator (molecular weight >25,000), and only 20% is recovered from the B125 miniconcentrator (molecular weight <125,000), indicating that the bulk of the molecular weight of the polymers range around 80,000–100,000.

Gel-Permeation Chromatography

Following the results of Kosma *et al.*,[13] it is found that all the copolyacrylamide polymers are retained on a Sepharose CL-4B gel-filtration column (1.5 × 80 cm, 30 m*M* NH_4HCO_3) having a molecular weight cutoff of 5,000,000 and that the copolymers appear in the void volume of other size exclusion gels with lower molecular weight cutoffs (Sephadex G-100 and Sephacryl S-300). HPLC of the polymers using a standardized TSK G-4000 SWXL column (7.5 × 300 mm) is done as described above.[11]

Syntheses of Copolymer Precursors

*Allyl α-Sialoside Methyl Ester (**3**).* The chloride **1** (Scheme 1) (0.75 g, 1.4 mmol)[14] is dissolved in dry allyl alcohol (5 ml) (4 Å molecular sieves)

[11] R. Roy, F. D. Tropper, and A. Romanowska, *Bioconjugate Chem.* **3,** 256 (1992).

[12] L. Svennerholm, *Biochim. Biophys. Acta* **24,** 604 (1957).

[13] P. Kosma, J. Gass, G. Schulz, R. Christian, and F. M. Unger, *Carbohydr. Res.* **167,** 39 (1987).

[14] R. Kuhn, P. Lutz, and D. L. McDonald, *Chem. Ber.* **99,** 611 (1966).

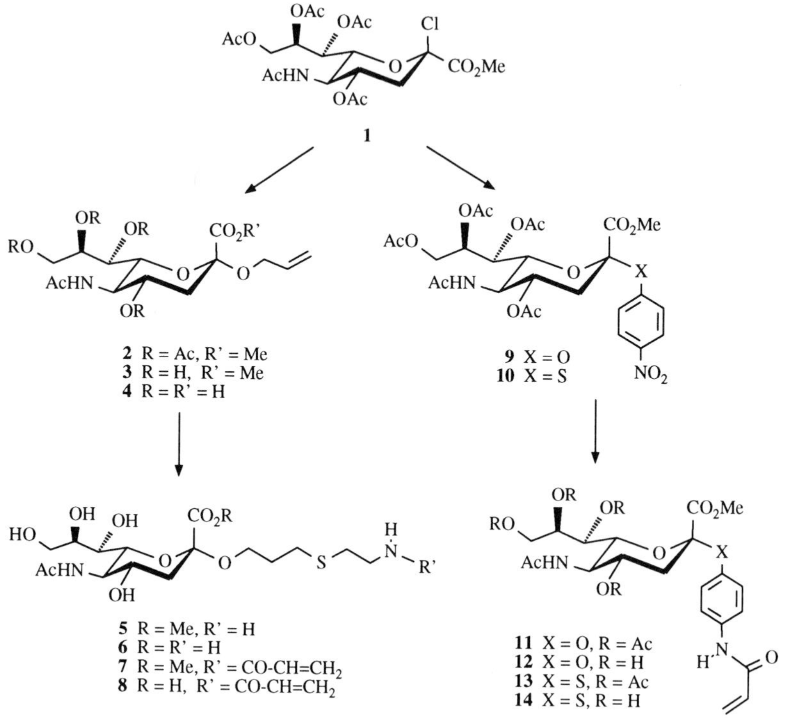

SCHEME 1

containing freshly prepared silver salicylate (0.35 g)[10] and 4 Å molecular sieves (0.5 g). The reaction is stirred for 2 hr at room temperature in the dark. The slurry is filtered over Celite and washed with dichloromethane. The resulting clear solution is evaporated to dryness, and the residue is dissolved in dichloromethane (25 ml). The organic solution is successively washed with cold 5% (w/v) aqueous sodium bicarbonate, 5% (w/v) aqueous sodium thiosulfate, and finally with water. The organic layer is dried over sodium sulfate and evaporated to a clear oil that is homogeneous by TLC (80% yield). An analytical sample is obtained by crystallization from methanol–ether (0.59 g, 1.1 mmol, 73%). The glycoside **2** has mp 153.6°–157.7°; $[\alpha]_D$ −14.2°.[15]

Compound **2** (1.00 g, 1.9 mmol) is de-O-acetylated in a dry methanol (20 ml) containing a few drops of 1 *M* sodium methoxide. The reaction is complete after 2 hr at room temperature. The solution is neutralized with

[15] R. Roy and C. A. Laferrière, *Can. J. Chem.* **68,** 2045 (1990).

Amberlite 120 (H^+) resin, filtered, and evaporated to approximately 5 ml. Compound **3** crystallizes on addition of ether to afford 0.64 g of pure material (93%) having mp 143°–144°; $[\alpha]_D$ −9.1° (*c* 0.66, H_2O).

The sodium salt of the acid **4** is obtained by treating the methyl ester **3** (0.28 g, 0.77 mmol) at room temperature for 1 hr with 0.1 *M* NaOH (5 ml). The reaction mixture is cooled to 0° and neutralized with Amberlite 120 (H^+) resin (pH reached 2). After filtration of the resin, the pH is quickly adjusted to 7 with 0.1 *M* NaOH. The aqueous solution is lyophilized to yield **4** as a white powder (0.25 g, 0.72 mmol, 93%) having mp 245°–250° (dec.); $[\alpha]_D$ −9.1° (*c* 0.66, water); negative ion fast atom bombardment-mass spectrometry (FAB-MS) for $C_{14}H_{23}NO_9$: 348 (M − 1).

*3-(2-N-Acrylamidoethylthio)propyl α-Sialoside (**8**) and Methyl Ester (**7**).* The methyl ester **3** (75.3 mg, 0.21 mmol) is dissolved in water (2 ml). Cysteamine hydrochloride (35.2 mg, 0.31 mmol, Sigma, St. Louis, MO), is added, and the reaction is stirred under a UV lamp for 3 days. The solution is loaded on a 1.5 × 60 cm column of Sephadex G-10 and eluted with water. The product **5** is identified and lyophilized to a white powder; yield 71.1 mg, 0.16 mmol (77%); $[\alpha]_D$ −9.2° (*c* 1.0, water).

The free acid **4** (0.148 g, 0.425 mmol) is dissolved in water (2 ml), cysteamine hydrochloride (0.104 g, 0.912 mmol) is added, and the solution is placed under vacuum for 2 min, then purged with nitrogen. The solution is stirred for 18 hr under a UV lamp, then loaded directly on a 3 × 60 cm column of Sephadex G-10 eluted with water. Compound **6** is obtained as a white powder after lyophilization; yield 0.152 g, 0.357 mmol, 84%; $[\alpha]_D$ −8.3° (*c* 1.0, water, pH 5.0).

Compound **5** (120 mg, 0.28 mmol) is dissolved in methanol (10 ml). Amberlite IRA 400 anion-exchange resin (OH^- form) (1 g) is added, and the solution is stirred at 0°. A 10% (v/v) solution of acryloyl chloride in dioxane, approximately 1 ml, is added dropwise until a ninhydrin test of an aliquot (100 μl) shows no presence of amine. The solution is filtered and evaporated. Compound **7** is obtained as an oil after silica gel chromatography with chloroform–methanol (8 : 2, v/v); yield 89 mg, 0.18 mmol, 64%; $[\alpha]_D$ −11.3° (*c* 1.4, water, pH 5.0).

Compound **6** (0.102 g, 0.24 mmol) is dissolved in water (1 ml), and Na_2CO_3 (0.2 g) is added followed by methanol (1 ml). The slurry is cooled to 0°, and acryloyl chloride in dioxane [20% (v/v), 0.5 ml] is added with stirring over a 3-min period, after which time TLC shows the reaction to be complete. The solution is evaporated to dryness, and the residues are dissolved in water (1 ml). The mixture is desalted on a 3 × 60 cm column of Sephadex G-10, and the product is eluted with water. The product **8** is obtained as an oil after lyophilization; yield 0.108 g, 0.22 mmol, 93%; $[\alpha]_D$ −11° (*c* 1.0, water, pH 5.0).

*p-N-Acrylamidophenyl α-Sialoside Methyl Ester (**12**).* To a solution of **1** (420 mg, 0.82 mmol, 1 equivalent), 4-nitrophenol (229 mg, 1.65 mmol, 2 equivalents), and tetrabutylammonium hydrogen sulfate (279 mg, 0.82 mmol, 1 equivalent) in dichloromethane (5 ml) is added 1 *M* sodium hydroxide (2.5 ml), and the reaction mixture is stirred vigorously at room temperature for 20 min. The mixture is diluted with dichloromethane (100 ml) and washed successively with saturated sodium hydrogen carbonate (2 times, 50 ml), water (50 ml), and saturated sodium chloride (50 ml). The organic phase is dried with sodium sulfate, filtered, and evaporated. The residue is chromatographed on a short silica gel column [toluene–ethyl acetate, 1 : 1 (v/v)], and the collected product **9** is crystallized from chloroform–carbon tetrachloride (yield 438 mg, 87%); mp 87°–88°, $[\alpha]_D$ +26.6° (*c* 1.28, $CHCl_3$).

Compound **9** (200 mg, 0.327 mmol, 1 equivalent) is mixed with ammonium formate (309 mg, 4.91 mmol, 15 equivalents) in methanol (10 ml). The reaction mixture is heated to boiling, then stirred at 40° for 20 min and filtered through a Celite plug. The reaction mixture is concentrated and the residue dissolved in dichloromethane (8 ml) and triethylamine (0.2 ml). The mixture is cooled in an ice bath, and acryloyl chloride (106 μl, 1.308 mmol, 4 equivalents) in dichloromethane (2 ml) is added dropwise. The solution is stirred for 10 min and then brought to room temperature. It is then diluted in dichloromethane (100 ml) and washed successively with 1 *M* HCl (50 ml), saturated sodium hydrogen carbonate (50 ml), water (50 ml), and saturated sodium chloride (50 ml). The organic phase is dried with sodium sulfate, filtered, and evaporated. The residue is chromatographed on a silica gel column (ethyl acetate) to give **11** as a clear foam (yield 179 mg, 86%); $[\alpha]_D$ +18.6° (*c* 0.86, $CHCl_3$).

Compound **11** (54 mg, 85 μmol) dissolved in anhydrous methanol (2 ml) is cooled in an ice bath, treated with sodium methoxide in methanol (0.2 *M*, 100 μl), and stirred overnight. Neutralization with resin (H^+), filtration, and concentration afford crude **12**. The residue is chromatographed on a silica gel column [ethyl acetate–methanol, 6 : 1 (v/v)] to give a white foam (yield 37 mg, 93%); $[\alpha]_D$ +33.1° (*c* 1.6, methanol).

*p-N-Acrylamidophenylthio α-Sialoside Methyl Ester (**14**).* 4-Nitrothiophenol (80%, Aldrich) (396 mg, 2.04 mmol, 1.1 equivalents) and tetrabutylammonium hydrogen sulfate (692 mg, 2.04 mmol, 1.1 equivalents) are added to 1 *M* NaOH (12 ml) and stirred for 30 min at room temperature. Compound **1** (964 mg, 1.857 mmol, 1 equivalent) dissolved in dichloromethane (12 ml) is added, and the reaction mixture is stirred for 5 min. The mixture is diluted with ethyl acetate (150 ml) and washed successively with saturated sodium hydrogen carbonate (2 times, 70 ml), water (70 ml), and saturated sodium chloride (70 ml). The organic phase is dried with sodium sulfate, filtered, and evaporated. The residue is chromatographed on a short silica gel column (ethyl acetate), and the collected product **10** is crystallized from chloroform–carbon tetrachloride (yield 942 mg, 81%); mp 170°–171°, $[\alpha]_D$ +18.2° (*c* 1.01, $CHCl_3$).

To a solution of compound **10** (400 mg, 0.637 mmol, 1 equivalent) in ethanol (2 ml) is added tin(II) chloride dihydrate (730 mg, 3.236 mmol, 5 equivalents), and the mixture is heated to 70° and stirred for 30 min. The solution is cooled, poured into ice–water (100 ml), and adjusted to pH 7–8 with a saturated sodium hydrogen carbonate solution. Ethyl acetate (100 ml) is added, and the organic phase is separated and washed with water (2 times, 50 ml). The dried organic phase (sodium sulfate) is filtered and evaporated. The residue is dissolved in dichloromethane (20 ml) containing triethylamine (0.5 ml), the solution is cooled in an ice bath, and acryloyl chloride (210 μl, 2.59 mmol, 4 equivalents) in dichloromethane (5 ml) is added dropwise. The solution is stirred for 10 min and then brought to room temperature. The reaction mixture is diluted with dichloromethane (150 ml) and washed successively with 1 *M* HCl (50 ml), saturated sodium hydrogen carbonate (80 ml), water (80 ml), and saturated sodium chloride (80 ml). The organic phase is dried with sodium sulfate, filtered, and evaporated. The residue is chromatographed on a silica gel column [ethyl acetate–toluene, 20 : 1 (v/v)] to give **13** as a clear foam (yield 365 mg, 88%); $[\alpha]_D$ +45.6° (*c* 1.13, $CHCl_3$).

Compound **13** (92 mg, 141 μmol) dissolved in anhydrous methanol (5 ml) is cooled in an ice bath, treated with sodium methoxide in methanol (0.2 *M*, 150 μl), and stirred overnight. The solution is neutralized with cationic resin (H^+), filtered, and concentrated to give a white foam (yield 65.5 mg, 98%); $[\alpha]_D$ +89.5° (*c* 0.63, methanol). The product (**14**) is used as its methyl ester in the polymerization reactions since it has proved difficult to hydrolyze the methyl ester without modifying the *N*-acrylamido residues.

Syntheses of Glycopolymers

The following is a typical procedure for the nonaromatic monomers **3**, **4**, **7**, and **8** (Scheme 2). The α-sialoside monomers and acrylamide in the proportions indicated in Table I are dissolved in deoxygenated water obtained by multiple freeze–thaw cycles and bubbling with nitrogen. Alternatively, water can be deaerated by simply boiling it vigorously over a flame for 10–15 min followed by bubbling a steady stream of nitrogen through it via a fritted glass filter rod as it cools. Bubbling with helium or argon offers no distinct advantage. Ammonium persulfate (1 mg/50 mg acrylamide), preferably from a fresh stock solution (50 mg/ml), and

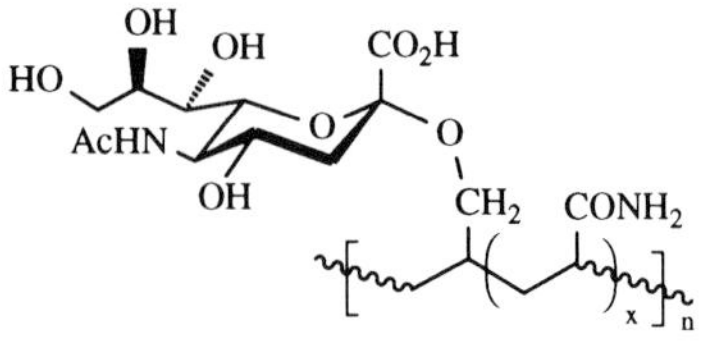

15 R = Me
16 R = H

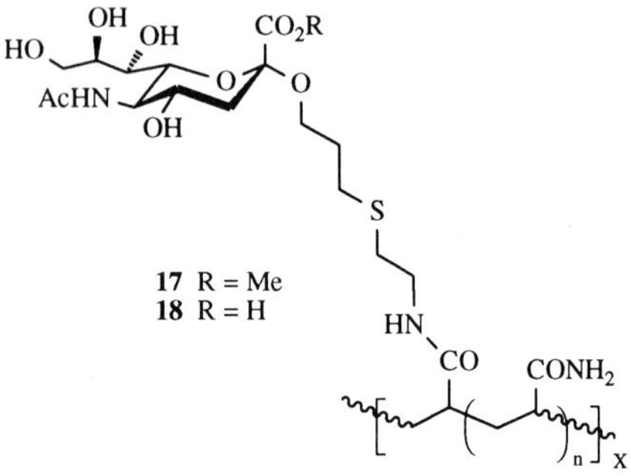

17 R = Me
18 R = H

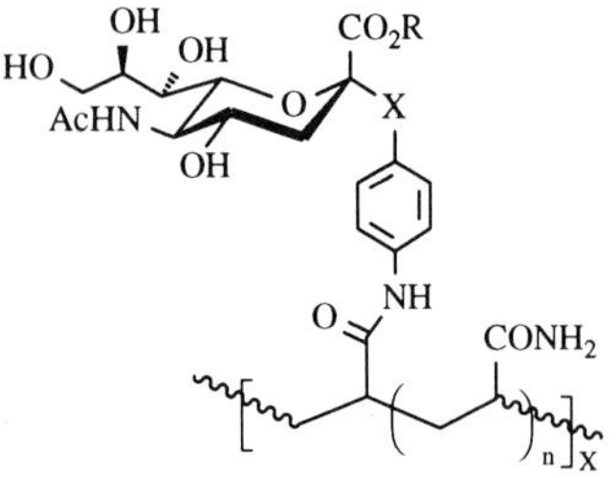

19 X = O, R = Me
20 X = O, R = H
21 X = S, R = Me
22 X = S, R = H

SCHEME 2

TABLE I
INCORPORATION OF SIALOSIDE MONOMERS INTO ACRYLAMIDE COPOLYMERS

Sialoside monomer	Molar ratio acrylamide/sialoside		Yield (%)[c]
	Reactions[a]	Products[b]	
3	9.4	52	36
	4.7	27	26
	1.2	6.8	17
4	13	70	—
	5.1	22	—
	3.3	20	—
7	13	12	64
	6.5	5.3	70
	1.6	1.8	47
8	6.7	25.3	—
	1.8	22.2	—
12	5.1	5.2	78
14	4.5	5	60

[a] Polymerizations by heat.
[b] Based on ^{1}H NMR data.
[c] Based on total mass of product recovered.

N,N,N',N'-tetramethylethylenediamine (TMEDA or TEMED) (5 μl) are then added. The reaction mixture is stirred under nitrogen at 90° for a few minutes or alternatively kept overnight at room temperature. The starting materials must be pure products; otherwise, the polymerization may not proceed to completion. If this happens, more initiator [$(NH_4)_2S_2O_8$/heat or $(NH_4)_2S_2O_8$/TEMED/25°] may be added. The reaction mixtures are checked by TLC for residual α-sialoside monomers. The cooled reaction mixtures are diluted with water and dialyzed (10,000 molecular weight cutoff) exhaustively against distilled water (6 times, 5 liters). Aqueous solutions of the polymers are freeze-dried to afford white spongy solids. The α-sialoside content within the polymers can be determined by the resorcinol method[12] and by ^{1}H NMR spectroscopy.

The extent of sialoside incorporation within the copolymers is highly dependent on the quality of the starting materials. Allyl α-sialosides tend to give much lower yields, and the content of sialyl residue is hardly predictable. In contrast, *N*-acrylamido-terminated sialosides give a highly predictable sugar content. Usually, the molar ratio of *N*-acrylamido-terminated sialoside to acrylamide in the reaction mixtures is well reflected in the copolymers.

The following is a typical procedure for aryl(thio) α-sialoside methyl esters **12** and **14** having low water solubility. To a solution of **14** (10.3 mg, 22 μmol) and acrylamide (7.7 mg, 108 μmol, 5 equivalents) in a deoxygenated water–dimethyl sulfoxide (DMSO) mixture (2 : 1 v/v, 300 μl) is added ammonium persulfate (15 μl, 100 mg/ml). The solution is stirred at 100°, and after 15 and 25 min, ammonium persulfate is added (15 μl each time). Thirty-five minutes after the last ammonium persulfate addition, the reaction mixture is cooled, 0.2 *M* NaOH (5 ml) is added, and the solution is stirred at room temperature overnight in order to hydrolyze all the methyl ester groups on the polymer (**21**), which can be isolated as such if so desired. The reaction mixture is finally dialyzed against distilled water (3 times, 5 liters). The aqueous solution is lyophilized to give **22** as a white powder (10.8 mg, 60%). The molar ratio of α-sialoside to acrylamide is found to be 1 : 5 by ^{1}H NMR spectroscopy.

To a solution of the methyl ester **12** (5.3 mg, 11 μmol), acrylamide (4.3 mg, 61 μmol, 5 equivalents), and ammonium persulfate (15 μl, 100 mg/ml) in deoxygenated water (150 μl) is added TEMED (15 μl). The mixture is stirred overnight at room temperature to form the polymer **19** which is treated with 0.2 *M* NaOH (5 ml) for an additional 24 hr at 75°. After dialysis and lyophilization, the copolymer **20** is obtained as a white spongy solid (7.6 mg, 78%). The ratio of α-sialoside to acrylamide is calculated to be 1 : 5.2 from the relative intensities of the bulk of the signals attributable solely to the sialoside residues (aromatic or *N*-acetyl signal at δ 2.01 ppm) to those of the polymer backbone (methine —CH— at 2.1–2.5 and methylene —CH_2— at 1.2–1.8 ppm).

The amount of ammonium persulfate or TEMED used in the copolymerization reactions is not critical to the success of the reaction. However, too large a quantity will favor polymers with lower molecular weight.

[26] Polymer-Supported Solution Synthesis of Oligosaccharides

By JIRI J. KREPINSKY, STEPHEN P. DOUGLAS, and DENNIS M. WHITFIELD

Introduction

Oligosaccharides [usually conjugated to proteins (glycoproteins) and lipids (glycolipids)] are finally being recognized as biomolecules of paramount importance in life processes[1] with significant promise as future therapeutics.[2] Consequently, the efficient preparation of oligosaccharides is of central importance for further studies of the compounds in biological sciences and for application in medicine. Preparation of neoglycoconjugates in particular depends on the availability of oligosaccharides. Naturally occurring oligosaccharides can be made enzymatically employing glycosyltransferases.[3] Any oligosaccharide, whether natural or not, can be synthesized by chemical synthesis using solution methodology,[4a–d]

[1] R. A. Dwek and F. A.. Quiocho, *Curr. Opin. Struct. Biol.* **1,** 709 (1991); and following articles; H. J. Allen and E. C. Kisalius, (eds.), "Glycoconjugates—Composition, Structure, and Function." Dekker, New York, 1992; L. A. Lasky, *Science* **258,** 964 (1992); A. Varki, *Glycobiology* **3,** 97 (1993).

[2] B. N. Cronstein and G. Weissman, *Arthritis Rheum.* **36** (2), 147 (1992); J. C. Bystryn, S. Ferrone, and P. Livingston (eds.), *Ann. N.Y. Acad. Sci.* **690,** (1993).

[3] S. Roth, U.S. Patent 5,180,674 (1993).

[4a] H. Paulsen, *Angew. Chem., Int. Ed. Engl.* **21,** 155 (1982).

[4b] H. Paulsen, *Angew. Chem., Int. Ed. Engl.* **29,** 823 (1990).

[4c] R. R. Schmidt, *Angew. Chem., Int. Ed. Engl.* **25,** 212 (1986).

[4d] K. Toshima and K. Tatsuta, *Chem. Rev.* **93,** 1503 (1993).

METHODS IN ENZYMOLOGY, VOL. 242

Copyright © 1994 by Academic Press, Inc.
All rights of reproduction in any form reserved.

which has made dramatic advancements.[5] However, the desired products invariably have to be purified from complex reaction mixtures by time-consuming and labor-intensive chromatography. This major obstruction to greater efficiency of synthesis is removed by polymer-supported synthetic methodology.[6] Solid-state procedures eliminate laborious workup and purification by chromatography, and they potentially reduce the time required for the synthesis of an oligosaccharide from months to days.

Polymer-supported liquid synthesis is another polymer-supported strategy also utilized for the synthesis of oligomers of peptides and nucleotides,[7] which when applied to oligosaccharide synthesis requires the polymer–carbohydrate synthon to be soluble under conditions of glycosylation, and insoluble during the workup. It allows syntheses of smaller oligomers in gram or even larger quantities. Poly(ethylene glycol) monomethyl ether (PEG, molecular weight 5000) is one suitable polymer which can be linked to different carbohydrate hydroxyl groups through ester linkages of succinic acid (PEG-Su). The PEG can be considered just another protecting group, and the usual manipulations of protecting groups and glycosylation chemistry can be exploited. As any other protecting group, PEG is incompatible with some reagents and reaction conditions, and it may also influence the reactivity in the glycosylation reaction. For instance, PEG works best under the Schmidt glycosylation conditions.[4c] Using PEG-bound saccharide as an acceptor, glycosylations can be driven to virtual completion by repeated additions of the glycosylating agent

[5] K. C. Nicolaou, J. L. Randall, and G. T. Furst, *J. Am. Chem. Soc.* **107,** 5556 (1985); P. Fügedi, P. J. Garegg, H. Lönn, and T. Norberg, *Glycoconjugate J.* **4,** 97 (1987); D. R. Mootoo, V. Date, and B. Fraser-Reid, *J. Am. Chem. Soc.* **110,** 2662 (1988); P. Konradsson and B. Fraser-Reid, *J. Chem. Soc. Chem. Commun.*, 1124 (1989); B. Fraser-Reid, Z. Wu, U. K. Udodong, and H. Ottoson, *J. Org. Chem.* **55,** 6068 (1990); G. H. Veeneman, S. H. van Leeuwen, H. M. Zuurmond, and J. H. van Boom, *J. Carbohydr. Chem.* **6,** 783 (1990); O. Kanie, M. Kiso, and A. Hasegawa, *J. Carbohydr. Chem.* **7,** 501 (1988); G. V. Reddy, V. R. Kulkarni, and H. B. Mereyala, *Tetrahedron Lett.* **30,** 4283 (1989); R. W. Friesen and S. J. Danishefsky, *J. Am. Chem. Soc.* **111,** 6656 (1989); R. W. Friesen and S. J. Danishefsky, *Tetrahedron* **46,** 103 (1990); H. M. Zuurmond, P. A. M. van der Klein, G. A. van der Marel, and J. H. van Boom, *Tetrahedron Lett.* **33,** 2063 (1992); M. Mori, Y. Ito, J. Uzawa, and T. Ogawa, *Tetrahedron Lett.* **31,** 3191 (1990).

[6] S. P. Douglas, D. M. Whitfield, and J. J. Krepinsky, *J. Am. Chem. Soc.* **113,** 5095 (1991); J. J. Krepinsky, S. P. Douglas, and D. M. Whitfield, U.S. Patent 5,278,303 (1994); S. J. Danishefsky, K. F. McClure, J. T. Randolph, and R. B.. Ruggeri, *Science* **260,** 1307 (1993).

[7] G. M. Bonora, C. L. Scremin, F. P. Colonna, and A. Garbesi, *Nucleic Acids Res.* **18,** 3155 (1990); G. M. Bonora, G. Biancotto, M. Maffini, and C. L. Scremin, *Nucleic Acids Res.* **21,** 1213 (1993); K. Kamaike, Y. Hasegawa, and Y. Ishido, *Tetrahedron Lett.* **29,** 647 (1988); E. Bayer and M. Mutter, *Nature (London)* **237,** 512 (1972); E. Bayer and M. Mutter *in* "The Peptides" (E. Gross and J. Meienhofer, eds.), Vol. 2, p. 286. Academic Press, New York, 1980.

since any excess of reactants not bound to PEG is washed off the precipitated PEG-bound product. The PEG-bound products can be further purified by crystallization from ethanol. The progress of reactions is monitored by ^{1}H nuclear magnetic resonance (NMR) spectroscopy using the signal of a single *O*-CH_3 group (δ 3.380 ppm) in PEG as the internal standard.

We describe in detail examples of a few protection/deprotection and glycosylation reactions of several PEG-bound substrates.[8a–c] These reactions are portrayed in Schemes 1–7. Glycosylating agents are added several times, if required, in order to complete the glycosylation monitored by NMR spectroscopy. After completion of the reaction, the reaction mixture is filtered to remove any solids present (e.g., molecular sieves) and concentrated to 5–10 ml per gram of PEG. The PEG–saccharide is precipitated from the solution after the addition of a 10-fold excess of diethyl ether or *tert*-butylmethyl ether at 0°–5° with vigorous stirring. This precipitate is filtered under an inert gas pressure and can be further purified by recrystallization from absolute ethanol: the precipitate is dissolved in warm absolute ethanol (50 ml/g PEG), and solids are filtered off under pressure. The solution is cooled, and the solid precipitated is collected, dried *in vacuo*, and used in the following step of the synthetic sequence. The PEG-Su is eventually easily cleaved from the saccharide by 1,8-diazabicyclo[5.4.0]undec-7-ene (DBU)-catalyzed methanolysis in dichloromethane or by hydrazinolysis if a phthalimido group is to be removed. Peracetylated oligosaccharides for final purification are obtained from dried residues after methanolysis by acetylation with acetic anhydride in pyridine. The expected anomer is formed in each glycosylation; the other anomer is not detected.

Although many oligosaccharides can be efficiently synthesized using PEG-supported methodology with a succinoyl diester linker,[8a–c] the sensitivity of the linker toward basic conditions will restrict the use of base-labile protecting groups. This drawback is eliminated by employing the α,α'-dioxyxylyl diether, —$OCH_2C_6H_4CH_2O$— (DOX), linker, which is substantially more stable than the succinoyl diester linker.[9] The DOX linker makes it possible to bind the supporting polymer to all hydroxyls including the anomeric hydroxyl (anomeric esters, such as succinoyl, are unstable). The entire PEG–DOX tether is removed from the completed oligosaccharide by hydrogenolysis under slightly acidic conditions. Otherwise PEG alone can be removed from the PEG–DOX–oligosaccharide by

[8a] O. T. Leung, D. M. Whitfield, S. P. Douglas, H. Y. S. Pang, and J. J. Krepinsky, *New J. Chem.* **18,** 349 (1994).

[8b] R. Verduyn, P. A. M. van der Klein, M. Douwes, G. A. van der Marel, and J. H. van Boom, *Recl. Trav. Chim. Pays-Bas* **112,** 464 (1993).

[8c] A. A. Kandil, N. Chan, P. Chong, and M. Klein, *Synlett*, 555 (1992).

[9] S. P. Douglas, D. M. Whitfield, and J. J. Krepinsky, *J. Am. Chem. Soc.*, submitted.

hydrogenolysis under neutral conditions. Thus, PEG–DOX is transformed into a protecting group in which a *p*-methylbenzyl group (*p*-tolylmethyl, TM) protects the carbohydrate hydroxyl to which PEG–DOX was originally bound. Such derivatives may be used for further synthetic purposes.

Materials and Methods

The starting carbohydrate derivatives are available from a number of commercial sources (e.g., Pfanstiehl, Toronto Research Chemicals). The PEG polymer and derivatives can be obtained from Hoechst through Fluka (Roukonkowa, NY) or from Shearwater Polymers (Huntsville, AL). Poly (ethylene glycol) monomethyl ether from Fluka is recrystallized from CH_2Cl_2–diethyl ether. It is ground to a fine powder and dried overnight in high vacuum. Because dry PEG is highly hygroscopic, due precautions must be taken. *tert*-Butylmethyl ether is obtained from ARCO Chemicals in bulk; it is of excellent quality and sufficiently dry and could be used without further purification. Dry diethyl ether is obtained from Mallinckrodt Specialty Chemicals, Inc. (Paris, KY). Ethanol (99%) is used for recrystallization. Other reagents are obtainable from a number of commercial sources; the highest purity available is used. All starting materials are dried overnight *in vacuo* (10^{-3} mm Hg) over KOH or P_2O_5 prior to use, and the solvents are distilled from appropriate drying agents. Solutions are concentrated at 10 mm Hg pressure in a rotary evaporator using a Teflon vacuum pump. Thin-layer chromatography (TLC) is performed on silica gel $60F_{254}$ (Merck, Darmstadt, Germany) plates and visualized by spraying with 50% aqueous sulfuric acid and heating at 200°. Silica gel (230–400 mesh, Toronto Research Chemicals) is used for flash chromatography. Sephadex LH-20 is obtained from Pharmacia (Piscataway, NJ). ^1H NMR spectra at 500 MHz are obtained at 19° ± 2° either in $CDCl_3$ containing a trace of trimethylsilyl (TMS) (0 ppm, ^1H and ^{13}C) as the internal standard or in D_2O (99.98%, Aldrich, Milwaukee, WI) containing a trace of acetone [2.225 ppm relative to internal 2,2-dimethyl-2-silapentane-5-sulfonic acid (DDS), ^1H] as the internal standard using δ (ppm) scale. Peak suppression is at δ 3.640, using 3.0 sec irradiation.

Experimental Procedures

Linking Polyethylene Glycol to Carbohydrate Derivative

*Example I: Preparation of **1c** by Method 1*

*Step 1: Saccharide–hemisuccinate **1b**.* Methyl 4,6-benzylidene-2-deoxy-2-*N*-phthalimido-β-D-glucopyranoside (**1a**) (0.44 g, 1.07 mmol), succinic anhydride (0.54 g, 5.3 mmol), and 4,4′-dimethylaminopyridine (DMAP) (50 mg) are stirred in dry pyridine (50 ml) at room temperature (Scheme 1). After completion of the reaction [monitored by TLC using ethyl acetate–hexane, 2 : 1 (v/v)], pyridine is removed by evaporation *in vacuo*, and the residue is subjected to flash chromatography in ethyl acetate to give 3-*O*-hemisuccinate **1b** (0.4 g, 70%).

Step 2: PEG-Su–saccharide **1c**. Poly(ethylene glycol) monomethyl ether (3.2 g, 0.8 equivalents), mixed with the 3-*O*-hemisuccinate, is dried overnight at high vacuum over P_2O_5. The mixture is dissolved in anhydrous CH_2Cl_2 (25 ml), and a catalytic amount of DMAP, followed by 1,3-dicyclohexylcarbodiimide (DCC) (0.16 g, 0.77 mmol), is added. The solution becomes cloudy in 15 min and is stirred overnight at room temperature. The precipitated urea is removed by filtration, washed with dry CH_2Cl_2, and the combined filtrates are reduced to the original volume. The solution is cooled to 0°, anhydrous ether (250 ml) is added with vigorous stirring, and **1b** precipitates out. After filtration, the solid is dissolved in hot absolute ethanol (50 ml), the solution is filtered and cooled to 4°, and the recrystallized **1c** is filtered, washed with diethyl ether, and dried. ^{1}H NMR: phthalimido (Phth), 7.860 and 7.726 (m, 4H); PhCH=, 5.550 (s, 1H); H-1, 5.337 (d, $J_{1,2}$ = 8.32 Hz, 1H); sugar-OCH_3, 3.444 (s, 3H); PEG-OCH_3, 3.378 (s, 3H); Su-CH_2, 2.35–2.50 (m, 4H).

Example I: Preparation of **1c** *by Method II, General Procedure*

Step 1: PEG–hemisuccinate (**2**; *PEG-Su*). Poly(ethylene glycol) monomethyl ether (20 g) is dried overnight at high vacuum with succinic anhydride (2 g, 5 equivalents) and DMAP (200 mg). To this mixture are added

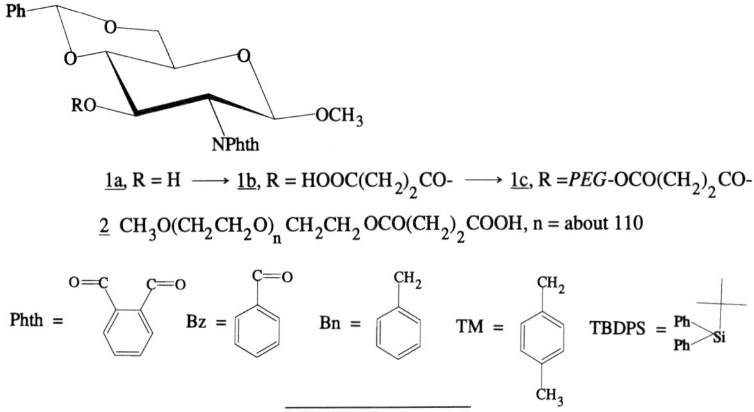

SCHEME 1. Ph, Phenyl.

CH_2Cl_2 (140 ml) and pyridine (30 ml). After stirring overnight the mixture is concentrated to 75 ml, cooled to 0° in ice, and diluted with stirring to 1.0 liter with cold diethyl ether. The mixture is allowed to stand for 1 hr on ice, and the solid is filtered off by suction, washed with diethyl ether, and air-dried for 1 hr. Further purification by recrystallization from hot ethanol (700 ml) as above gives **2**. ^{1}H NMR: PEG-CH_2—O-Su, 4.259 (brdd, 2H); PEG-OCH_3, 3.380 (s, 3H); Su-CH_2, 2.631 (m, 4H).

Step 2: PEG-Su–saccharide. To a portion of the solid PEG-Su (**2**) (5 g) are added a monosaccharide with one free OH (e.g., **1a**) (1.5 equivalents), and DMAP (100 mg), and the mixture is dried at high vacuum overnight. Under argon are added CH_2Cl_2 (25 ml), CH_3CN (25 ml), and DCC (1.5 ml of a 1 *M* solution in CH_2Cl_2, 1.5 equivalents), and the reaction mixture is left to stir at room temperature overnight. After the workup as for Method 1, sugar attached to PEG is obtained. The unreacted sugar is recovered from the combined filtrates.

Example II: Preparation of ***3b***. For the preparation of **3b** (Scheme 2),[8a] a mixture of benzyl 2,3-di-*O*-benzoyl-β-D-galactopyranoside (**3a**) (0.55 g, 1.15 mmol, 1.5 equivalents), DCC (0.17 g, 0.805 mmol, 1.05 equivalents), PEG-Su (**2**) (3.83 g, 0.77 mmol, 1.0 equivalents), and DMAP (a small crystal) is dried in a desiccator under vacuum overnight. After cooling to 0° under dry argon, CH_2Cl_2 (30 ml) is added. The stirred reaction mixture is allowed to warm to room temperature, and stirring is continued for 20 hr. The resulting cloudy solution is filtered and washed with dry CH_2Cl_2 (2 times, 5 ml). The combined filtrate and washings are concentrated to approximately 5 ml, *tert*-butyl methyl ether (20 ml) is added at 0°, and a white solid precipitates out. The solid is filtered, recrystallized from hot 99% ethanol, filtered, washed with *tert*-butyl methyl ether (2 times, 10

3a, $R_1 = R_2 = H$ ⟶

3b, R_1 = *PEG*-OOC(CH_2)$_2$CO-, $R_2 = H$ ⟶

3c, R_1 = *PEG*-OOC(CH_2)$_2$CO-, R_2 = 6b ⟶

3d, PERACETYLATED DISACCHARIDE

6a, R_1 = OC(=NH)CCl_3, $R_2 = H$

6b, R_1 = O-, $R_2 = H$

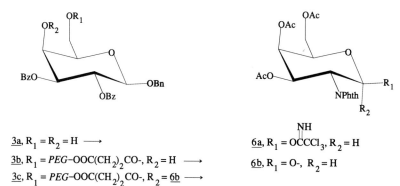

SCHEME 2. Bz, Benzoyl; Bn, benzyl; Ac, acetyl.

ml), and dried *in vacuo* overnight to give benzyl 2,3-di-*O*-benzoyl-6-*O*-Su-PEG-yl-β-D-galactopyranoside (**3b**) (3.54 g; 84%). ^{1}H NMR: benzoate(*o*), 7.979 (dd, 2H) and 7.918 (dd, 2H, J = 8.2 Hz, 1.3); benzoate(*p*), 7.51 (m, 2H); benzoate(*m*), 7.379 (brt, 2H, J = 7.8 Hz); C_6H_5-CH_2, 7.19 (m, 5H); C_6H_5-CH_2, 4.906 (d, 1H) and 4.712 (d, 1H, J = −12.6 Hz); Su-CH_2, 2.66 (m, 4H); PEG-OCH_3, 3.379 (s, 3H).

Glycosylation

*Preparation of Trisaccharide **4b***. Diol **4a** with attached Su-PEG (312 mg, 0.057 mmol) is mixed with bromide **5a** (56 mg, 2 equivalents), silver trifluoromethane sulfonate (AgOTf) (28 mg, 2 equivalents), 2,6-di-*tert*-butyl-4-methylpyridine (DBMP) (11 mg, 1 equivalent), and a small portion of powdered 4 Å molecular sieves (4 Å-MS) and the mixture is dried at high vacuum overnight (Scheme 3). Then the flask is cooled in ice–water under argon, CH_2Cl_2 (4 ml) is added, and the reaction mixture is stirred for 2 hr. At this point more dried bromide **5a** (2 equivalents), AgOTf (2 equivalents), and DBMP (1 equivalent) are added, and an identical addition is made after another 2 hr. The stirring is continued overnight, the reaction mixture is diluted with CH_2Cl_2 (10 ml), and the molecular sieves and precipitated silver salts are filtered off. The filtrate is evaporated to dryness, the residue is redissolved in CH_2Cl_2 (4 ml), the solution is cooled to 0° in an ice bath, and the product is precipitated by the addition of diethyl ether (40 ml) with vigorous stirring. After standing for 1 hr, the precipitate is collected by filtration, washed with diethyl ether, and dried in air for at least 1 hr. The dry solid is dissolved in warm ethanol (15 ml) and filtered from undissolved solids, and the solution is allowed to crystallize at 4°. The solid is collected by filtration, washed with cold ethanol and diethyl ether, and dried *in vacuo* to give trisaccharide attached to SuPEG (**4b**). ^{1}H NMR: Phth, 7.860 (m, 4H) and 7.75 (m, 4H); Gal H-1, 4.30 (1H); GlcNPhth(β1-6) H-1, 5.425 (d, $J_{1,2}$ = 8.5 Hz, 1H); GlcNPhth

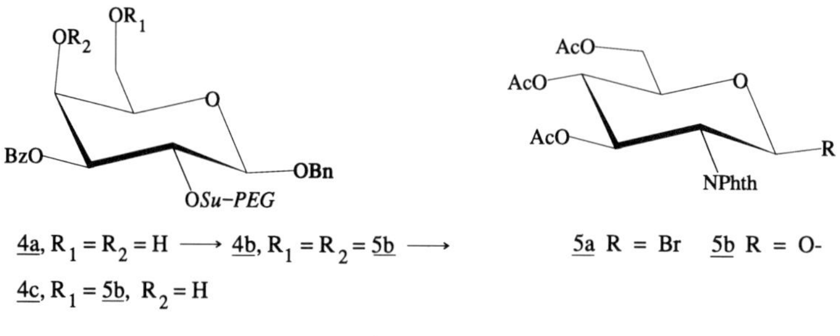

SCHEME 3

(β1-4) H-1, 5.462 (d, $J_{1,2}$ = 8.5 Hz, 1H); Bz_0, 7.999 (brd, 2H); Bz_m, 7.416 (brt, 2H); Bz_p, 7.564 (brt, 1H).

Preparation of Disaccharide ***3c***. A mixture of **3b** (0.35 g, 1.0 equivalent), trichloroacetimidate **6a** (55 mg, 1.5 equivalents), and 3 Å molecular sieves (500 mg; activated at 300° overnight) is dried in the vacuum of an oil pump overnight and cooled to 0°–5° under argon. To the mixture is added CH_2Cl_2 (4 ml) to dissolve the solid, followed after 20 min by an addition of triethylsilyl triflate (12 μl, 0.75 equivalent). After stirring for 3 hr at 0°–5° the reaction is quenched by adding 5 drops of dry diisopropylethylamine, and the solids are filtered off and rinsed with CH_2Cl_2. After concentration to the original volume, dry *tert*-butylmethyl ether (50 ml) is added, and the white precipitate that forms is filtered and recrystallized from hot 100% ethanol (35 ml). The crystallized solid is washed with *tert*-butyl methyl ether and dried under vacuum to yield disaccharide benzyl 4-*O*-(2-deoxy-2-phthalimido-3,4,6-tri-*O*-acetyl-β-D-galactopyranosyl)-2,3-di-*O*-benzoyl-6-*O*-Su-PEG-yl-β-D-galactopyranoside (**3c**) in 75% (0.33 g).

Preparation of Disaccharide ***4c***. Diol **4a** with attached Su-PEG (153 mg, 0.057 mmol) is mixed with bromide **5a** (1.1 equivalents), Ag_2CO_3 (7 equivalents), and a small portion of powdered 4 Å-MS, and the mixture is dried at high vacuum overnight. Then the flask is cooled in ice–water under argon, CH_2Cl_2 (2 ml) is added, and the reaction mixture is allowed to warm slowly to room temperature, with stirring continued for 2 days. The reaction mixture is worked up as described for **4b**. ^{1}H NMR: Gal H-1, 4.447 (d, $J_{1,2}$ = 8.0 Hz, 1H); GlcNPhth H-1, 5.425 (d, $J_{1,2}$ = 8.5 Hz, 1H); three CH_3COO, 2.117, 2.033, 1.866 (3s, 9H).

Preparation of Disaccharide ***7b***. To a cold solution (−10°) of the saccharide with attached Su-PEG (**7a**) (77 mg, 13 μmol) and trichloroacetimidate **8a** (20 mg, 40 μmol) in CH_2Cl_2 (1 ml) is added boron trifluoride etherate (80 m*M* in DCM, 6 μl, 48 μmol) (Scheme 4). The reaction mixture

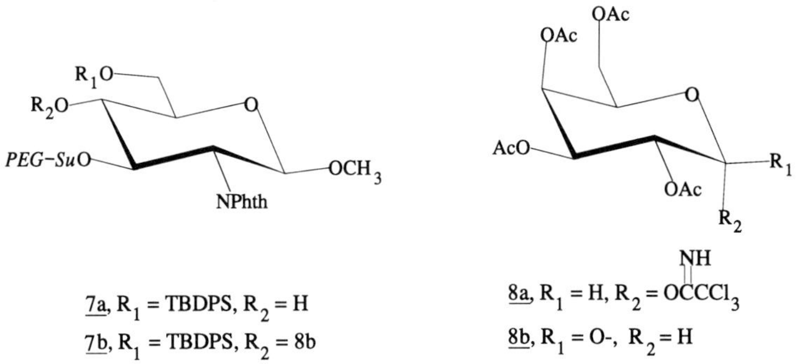

SCHEME 4

is allowed to warm slowly to room temperature, and stirring is continued overnight. The reaction mixture is worked up as described for **4b**. This procedure is repeated two more times, and complete galactosylation gives **7b**. ^{1}H NMR: Phth, 7.771 and 7.423 (m, 4H); GlcNPhth H-1, 5.231 (d, $J_{1,2}$ = 8.43 Hz, 1H); Gal H-1, 4.898 (d, $J_{1,2}$ = 8.11 Hz, 1H); GlcNPhth-OCH_3, 3.438 (s, 3H); PEG-OCH_3, 3.380 (s, 3H); Su-CH_2, 2.4–2.6 (m, 4H); four CH_3COO, 2.132, 2.086, 1.984, 1.822 (4s, 12H).

Saccharide Cleavage from Polymer Followed by Acetylation

Peracetylated oligosaccharides are usually more suitable for final purification by chromatography than completely deprotected compounds. After deprotection pure oligosaccharides are obtained.[10]

*Cleavage of Trisaccharide **4b**.* The trisaccharide moiety is removed from the polymer in **4b** (330 mg) by treatment with $N_2H_4 \cdot H_2O$ (1 ml) and ethanol (2 ml) at 70° for 2 hr. The liquids are removed by coevaporation with toluene (2 times, 10 ml), and the resulting solid is dried at high vacuum for 2 hr, then cooled on ice. Pyridine (2 ml) and acetic anhydride (1 ml) are added under argon, and the reaction mixture is stirred overnight at room temperature. The liquids are removed at oil pump vacuum, and the residue is dissolved in hot ethanol (15 ml), filtered, and allowed to precipitate at 4°. The precipitated PEG is filtered off and rinsed with cold ethanol, after which the combined filtrate and washings are evaporated to dryness and purified by chromatography on silica gel [CH_2Cl_2–methanol, 40 : 1 (v/v), followed by 10 : 1 (v/v)] to yield peracetylated trisaccharide. ^{1}H NMR: GlcNAc(β1-6) H-1, 4.28; GlcNAc(β1-4) H-1, 4.690 (d, $J_{1,2}$ = 8.4 Hz, 1H); Gal H-1, 4.387 (d, $J_{1,2}$ = 8.0 Hz, 1H).

*Cleavage of Disaccharide **4c**.* The disaccharide moiety is removed from the polymer of overnight treatment of **4c** (290 mg) dissolved in CH_2Cl_2 (2 ml) with methanol (0.5 ml) and DBU (1 drop) with stirring. The PEG and deprotected sugar are precipitated with diethyl ether as above, then removed by filtration. The precipitate containing PEG and oligosaccharide is dissolved in hot ethanol (10 ml), the PEG is allowed to crystallize out, and the solid is filtered off and washed with cold ethanol. The combined filtrate and washings are evaporated to dryness, and the residue is treated with pyridine (2 ml) and acetic anhydride (1 ml) at room temperature overnight as above. The liquids are removed by coevaporation with toluene, and the residue is purified by chromatography on silica gel to yield a peracetylated *N*-phthalimido disaccharide. ^{1}H NMR: GlcNPhth H-1, 5.435 (d, $J_{1,2}$ = 8.6 Hz, 1H); Gal H-1, 4.250 (d, $J_{1,2}$ = 7.9 Hz, 1H); Phth, 7.838 (m, 2H) and 7.730 (m, 2H).

[10] D. M. Whitfield, H. Pang, J. P. Carver, and J. J. Krepinsky, *Can. J. Chem.* **68,** 942 (1990).

A separate sample treated with hydrazine hydrate and acetylated as for **4b** gives the known peracetylated disaccharide.[11]

*Cleavage of Disaccharide **3c**.* The disaccharide moiety is removed from the polymer together with the phthalimido group by treatment of **3c** (335 mg) with $N_2H_4 \cdot H_2O$ (1 ml) and ethanol (2 ml) at 70° for 2 hr. The liquids are removed by coevaporation with toluene (2 times, 10 ml), and the resulting solid is dried at high vacuum for 2 hr and cooled on ice, after which pyridine (2 ml) and acetic anhydride (1 ml) are added under argon and the reaction mixture stirred overnight at room temperature. The liquids are removed at oil pump vacuum, and the residue is dissolved in hot 100% ethanol (15 ml), filtered, and allowed to precipitate at 4°. The precipitate is filtered off and rinsed with cold ethanol, and the combined filtrate and washings are evaporated to dryness and purified by chromatography on silica gel (40 g); the **3d**-containing fractions are eluted with ethyl acetate (70%) in hexane. After further purification through Sephadex LH-20 [$CHCl_3$–methanol, 4 : 3, v/v)], the pure disaccharide benzyl 4-*O*-(2-acetamido-2-deoxy-3,4,6-tri-*O*-acetyl-β-D-galactopyranosyl)-2,3,6-tri-*O*-acetyl-β-D-galactopyranoside (**3d**) is obtained, identical with **3d** prepared by the solution synthesis (but free from any α anomer, which was a by-product of the solution synthesis).

Examples Using α,α′-Dioxyxylyl Diether Linker

*Preparation of PEG–DOX–Cl (**9**).* Dry poly(ethylene glycol)monomethyl ether (molecular weight 5000; 5 g, 1 mmol) is dissolved with heating in anhydrous tetrahydrofuran (THF) (50 ml) with exclusion of humidity (Scheme 5). After cooling to room temperature, sodium hydride (0.12 g, 3 mmol) is added with stirring, followed after 10 min by sodium iodide (0.17 g, 1.15 mmol) and α,α′-dichloro-*p*-xylene (5.25 g, 30 mmol). The reaction mixture is stirred for 96 hr and then filtered through a bed of Celite which is subsequently washed with a small amount of CH_2Cl_2 (5 ml). The combined filtrate and washings are cooled on an ice bath, and the PEG derivative is precipitated with anhydrous diethyl ether (300 ml). The solid is filtered off, resuspended in diethyl ether (100 ml), and filtered again. The solid is redissolved in hot absolute ethanol (80 ml), then cooled to 4°. The precipitate is filtered off and dried, yielding **9** in 96% (5 g). ^{1}H NMR: aromatics, 7.321 (dd, $J = 8.33$, 14.01 Hz, 4H); $OCH_2C_6H_4$, 4.555 (s, 2H); $ClCH_2C_6H_4$, 4.536 (s, 2H); CH_3OPEG, 3.349 (s, 3H).

*Preparation of PEG–DOX–OH (**10**).* The solution of PEG–DOX–Cl (**9**) (21.95 g, 4.3 mmol) in 10% aqueous Na_2CO_3 is heated at 70° under

[11] D. M. Whitfield, C. J. Ruzicka, J. P. Carver, and J. J. Krepinsky, *Can. J. Chem.* **65**, 693 (1987).

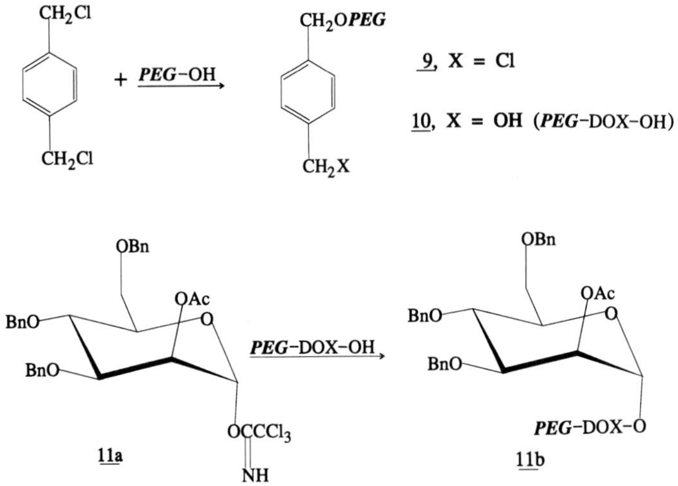

SCHEME 5

argon using an air condenser stoppered with a rubber septum for 16 hr. Water is distilled off *in vacuo*, and the residue is dried by coevaporation with toluene (2 times 250 ml) and in high vacuum. Then the residue is taken up in CH_2Cl_2, filtered, and rinsed with CH_2Cl_2 (3 times, total volume 350 ml). Toluene (100 ml) is added to the combined filtrate and washings, the solvents are evaporated, and the residue is dried *in vacuo*. The residue is redissolved in dry CH_2Cl_2 (75 ml) and filtered, and the filtrate is cooled in an ice bath and *tert*-butyl methyl ether (900 ml) added. The precipitated solid is filtered off, rinsed with *tert*-butyl methyl ether (100 ml) and diethyl ether (100 ml), and dried *in vacuo*. This residue is recrystallized from absolute ethanol (500 ml), rinsed successively with ethanol, *tert*-butyl methyl ether, and diethyl ether (100 ml each), and dried *in vacuo*, yielding **10** as a white powder (18.2 g, 83%).

*Preparation of a PEG–DOXyl Derivative **11b***. Dichloromethane (15 ml) is added with stirring to a mixture of PEG–DOX–OH (**10**) (4.0 g, 0.78 mmol) and 4 Å-MS (1 g) followed by a solution of 2-*O*-acetyl-3,4,6-tri-*O*-benzyl-α-D-mannopyranosyl trichloroacetimidate (**11a**) (875 mg, 1.36 mmol) in CH_2Cl_2 (5 ml) under an atmosphere of argon with stirring. This mixture is cooled in an ice bath, and after 30 min triethylsilyltrifluoromethane sulfonate (TESOTf) (177 μl, 0.78 mmol) is added by syringe. After 4 hr at ice bath temperature diisopropylethylamine (10 drops, 90 mg) is added to the solution, and after 5 min excess *tert*-butyl methyl ether (230 ml) is added to precipitate the polymer. The white solid is separated by filtration, and after rinsing with *tert*-butyl methyl ether (50 ml) it is recrystallized from absolute ethanol (200 ml). The white precipitate is rinsed with ethanol (50 ml), *tert*-butyl methyl ether (50 ml), and dried *in vacuo*, yielding PEG–DOXyl 2-*O*-acetyl-3,4,6-tri-*O*-benzyl-α-D-mannopyranoside (**11b**) (4.11 g, 92%). ^{1}H NMR: aromatics, 7.40–7.25 (m, 17H) and 7.150 (dd, J = 7.6, 2.0 Hz, 2H); Man H-1, 4.924 (d, 1H); Man H-2, 5.408 (dd, J_{12} = 1.8 Hz, J_{23} = 3.3 Hz, 1H); PEG-OCH_3, 3.382 (s, 3H); CH_3COO, 2.137 (s, 3H).

Saccharide Cleavage from Polyethylene Glycol–α,α'-Dioxyxylyl Ether Followed by Acetylation[10]

*Preparation of Disaccharide **12c** by Procedure A.* Raney Ni W-2 (Aldrich; 50% aqueous slurry, 5 g) is triturated with water until neutral (pH paper; 3 times) and with ethanol (3 times). The ethanolic slurry is added to PEG–DOX-bound disaccharide **12a** (1.10 g, 0.18 mmol), and the reaction mixture (total volume 50 ml) is refluxed under argon for 4 hr (Scheme 6). The hot solution is filtered through a bed of Celite and rinsed with warm ethanol (100 ml). The combined filtrate and washings, to which toluene (35 ml) is added, are concentrated to approximately 40 ml, then cooled on ice, and the precipitated PEG is filtered off. The precipitate is rinsed with ethanol (50 ml), and the combined filtrate and washings are evaporated to dryness. The residue (**12b**) is acetylated and purified as described in the following paragraph to yield 1,3,4,6-tetra-*O*-acetyl-2-*O*-(2,3,4,6-tetra-*O*-acetyl-α-D-mannopyranosyl)-α/β-D-mannopyranose (**12c**) (58 mg, 36%).

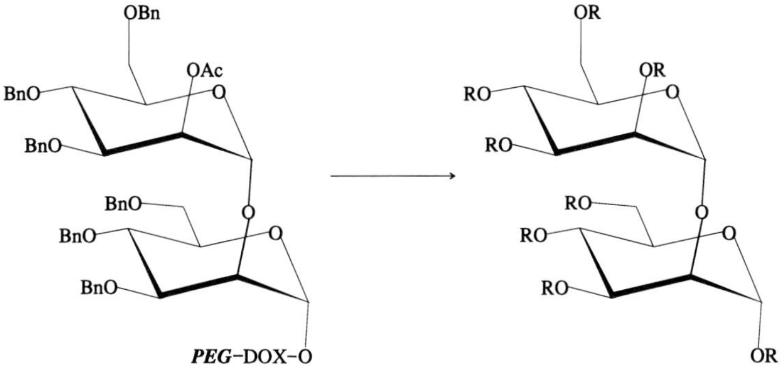

12a

12b R = H

12c R = Ac

SCHEME 6

Preparation of Disaccharide **12c** *by Procedure B.* The PEG–DOX-bound disaccharide (**12a**) (1.19 g, 0.19 mmol) dissolved in acetic acid–water [1 : 1, (v/v), 30 ml] is hydrogenated over palladium black (450 mg) at 50 psi (Parr hydrogenation apparatus) at room temperature for 16 hr. The catalyst is filtered off on a bed of Celite and rinsed well with 9 : 1 (v/v) methanol–water. The combined filtrates are evaporated at high vacuum (bath temperature 40°–45°), coevaporated with toluene (50 ml), and dried *in vacuo*. The solid residue is recrystallized from ethanol (100 ml), collected by filtration, and then rinsed with ethanol (25 ml) and *tert*-butyl methyl ether (25 ml). The combined filtrate and washings are evaporated to dryness. (If the removal of benzyl groups is incomplete, the above procedure may be repeated using Pd black in ethanol.) The residue (**12b**) is dissolved in pyridine (5 ml), then cooled to 0° under argon, acetic anhydride (2.5 ml) is added, and the mixture is stirred for 16 hr at room temperature. The solvents are evaporated at high vacuum, and the residue is chromatographed on a silica gel column. Compound **12c** (25 mg) is eluted with 2% methanol in CH_2Cl_2 (v/v). ^1H NMR: Man_A H-1α, 6.246 (d, J_{12} = 2.0 Hz, 1H); Man_B H-1, 4.950 (d, J_{12} = 1.6 Hz; 1H); CH_3COO, 2.155 (s, 6H); 2.143 (s, 3H); 2.105 (s, 3H); 2.089 (s, 3H); 2.047 (s, 6H); 2.012 (s, 3H).

Cleavage of Polyethylene Glycol from α,α'-Dioxyxylyl Diether to Tolylmethyl Group

Preparation of Disaccharide **13c***.* A mixture of the PEG–DOX-bound saccharide (**13a**) (440 mg) and 10% Pd/C (50 mg) in 95% ethanol (10 ml)

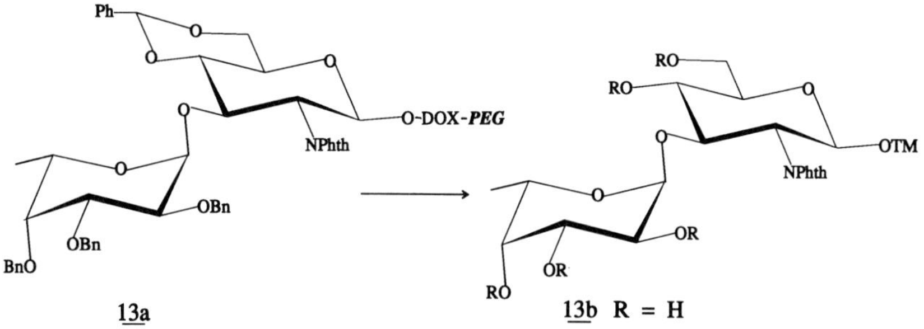

13a

13b R = H

13c R = Ac

SCHEME 7

is hydrogenated at atmospheric pressure for 24 hr (Scheme 7). The catalyst is removed by filtration through Celite. (Note that the liberated oligosaccharide may not be soluble in ethanol; subsequent rinsing of the filter cake with a suitable solvent such as CH_2Cl_2 will effect recovery.) Acetylation as described for **12c** (this step can be omitted if no protecting group removable by hydrogenation is present) furnishes the desired *p*-tolylmethyl-protected disaccharide, *p*-tolylmethyl 2-deoxy-4,6-di-*O*-acetyl-2-*O*-phthalimido-(2,3,4-tri-*O*-acetyl-β-L-fucopyranosyl)-β-D-gluco-pyranoside (**13c**) (10 mg, ~25%). ^{1}H NMR: $CH_3C_6H_4CH_2$—, 2.554 (s, 3H); GlcNPhth H-1, 5.416 (d, $J_{1,2}$ = 8.42 Hz, 1H); Fuc H-6, 0.9890 (d, J = 6.78 Hz, 3H); Phth, 7.72 (m, 2H) and 7.760 (m, 2H).

Concluding Notes

Polyethylene glycol derivatives are very hygroscopic. This property is beneficial sometimes since it helps to keep the reaction medium anhydrous. On the other hand, the more moist a PEG derivative is, the more soluble it is in ether; wet PEG derivatives resemble a wax, and they are more difficult to filter. When PEG derivatives contain too much water, they are virtually impossible to precipitate. If this happens, the solution containing a PEG derivative has to be evaporated to dryness, rendered anhydrous by coevaporation with toluene, redissolved in a solvent like dichloromethane, and precipitated again. These operations invariably reduce yields. Therefore, it is advisable to prevent prolonged exposure to the laboratory atmosphere if the air humidity exceeds 80% (air-conditioned laboratories usually are not sufficiently dry either). It is recommended to perform the precipitation under a blanket of inert gas, preferably argon because of its density, and the precipitate should not be filtered by suction. Alternatively, a closed device for filtering under pressure (such instruments are commercially available, e.g., from Aldrich) of an inert dry gas, such as nitrogen or argon, should be used.

[27] Syntheses of Clustered Lactosides by Telomerization

By WILLIAM K. C. PARK, SIVASUBRAMANIAN ARAVIND, ANNA ROMANOWSKA, JOCELYN RENAUD, and RENÉ ROY

Introduction

A number of mammalian cells have been shown to possess endogenous lectins having specific binding affinity for nonreducing terminal β-D-galactose (D-Gal) or β-D-*N*-acetylgalactosamine (D-GalNAc) residues present on glycolipids, glycoproteins, or synthetic glycoconjugates.[1–4] Hepatic asialoglycoprotein receptors (ASGP-R),[1,2] macrophages,[3] and more recently certain metastatic tumor cells[4,5] are among the most intensively studied cases of such interactions. Pioneering observations and quantitations of endogenous hepatic lectin–Gal interactions by Lee and co-workers have set the ground for the now well-recognized "cluster effect." [6] This effect suggests that both the number of sugar residues together with the respective propinquity confer to the glycosylated clusters their improved overall binding affinity (avidity).

Carbohydrate-containing polymers (glycopolymers) have been specifically designed to access the multivalent properties of clusters.[7–9] However, such glycopolymers constitute poorly defined chemical entities and thus represent mediocre therapeutic agents. To overcome this drawback, we describe here a single step synthesis of a family of clusters derived from quenching polymer growth by chain transfer reagents (telogens).[10] This process, called telomerization,[10,11] has given access to small size polymers (telomers) of 1–6 residues depending on the concentration and

[1] G. Ashwell and J. Harford, *Annu. Rev. Biochem.* **51,** 531 (1982).

[2] C. P. Stowell, R. T. Lee, and Y. C. Lee, *Biochemistry* **19,** 4904 (1980).

[3] S. Kelm and R. Schauer, *Biol. Chem. Hoppe-Seyler* **369,** 693 (1988).

[4] A. Raz, G. Pazerini, and P. Carmini, *Cancer Res.* **49,** 3489 (1988).

[5] B. Dean, H. Oguchi, S. Cai, E. Otsuji, K. Tashiro, S.-I. Hakomori, and T. Toyokuni, *Carbohydr. Res.* **245,** 175 (1993).

[6] Y. C. Lee, R. T. Lee, K. Rice, Y. Ichikawa, and T.-C. Wong, *Pure Appl. Chem.* **63,** 499 (1991).

[7] R. Roy, F. D. Tropper, and A. Romanowska, *J. Chem. Soc., Chem. Commun.*, 1611 (1992).

[8] R. Roy, F. D. Tropper, and A. Romanowska, *Bioconjugate Chem.* **3,** 256 (1992).

[9] A. Kobayashi, T. Akaiki, K. Kobayashi, and H. Sumimoto, *Makromol. Chem., Rapid Commun.* **7,** 645 (1986).

[10] C. M. Starks, "Free Radical Telomerizations," Chap. 8, p. 193. Academic Press, New York, 1974.

[11] A. Pavia, B. Pucci, J. G. Riess, and L. Zarif, *Bioorg. Med. Chem. Lett.* **1,** 103 (1991).

Copyright © 1994 by Academic Press, Inc.
All rights of reproduction in any form reserved.

nature of the chain transfer reagent used. Lactosylated telomers with and without a spacer arm have been synthesized to provide access to carbohydrate clusters useful in the studies of carbohydrate–lectin interactions.

Thus, lactosyl azide **2** is readily prepared from peracetylated lactosyl bromide (**1**) under phase-transfer catalysis (PTC) (Scheme 1).[12] Reduction of the azide group in **2** by Raney nickel followed by N-acryloylation and Zemplén de-O-acetylation provides the required monomer (taxogen) **5** in high yield. Polymerizations of similar N-acryloylated sugar derivatives have been previously described.[13,14] Telomerization of **5** in refluxing methanol under nitrogen with 2,2′-azodi(2-methylpropanenitrile) (AIBN) as initiator and *tert*-butylmercaptan as telogen in various molar ratios provides, in one single step, a family of low molecular weight lactosylated clusters. Separation of each individual telomer is accomplished by size-exclusion chromatography on a BioGel P-2 column (Bio-Rad, Richmond, CA). The results of the telomerization are shown in Table I.

To exploit further the usefulness of N-acryloylated sugar derivatives in cluster syntheses, compound **5** is dimerized by tethering the terminal *N*-acryloyl function with 1,3-propanedithiol under Michael conditions to provide dimer **10** in excellent yield.

Clusters **6–10** have been evaluated for the relative abilities to inhibit the binding of peanut lectin against a model lactosylated polymer.[8] To this end, inhibitions of enzyme-linked lectin assays are performed in microtiter plates. Telomers **6–9**, deprived of a spacer arm between the sugar residues and the backbone, are no better inhibitors than pure lactose taken as a reference. However, dimer **10** with the slightly longer spacer shows a 4-fold increased inhibitory capacity over lactose. Thus, it has been envisaged that other telomers having a longer spacer arm would offer distinct advantages. Therefore, the telomerization process is repeated with an elongated monomer **13** readily prepared by derivatizing the lactosylamine derivative **3** with *N*-acrylamidohexanoic acid (**11**) with 2-ethoxy-1-ethoxycarbonyl-1,2-dihydroquinoline (EEDQ) (Scheme 2). Preliminary inhibition experiments with peanut lectin show that the spacered trivalent telomer **16** is twice as potent as lactose.

General Methods

Melting points are determined on a Gallenkamp apparatus and are uncorrected. The ^{1}H and ^{13}C nuclear magnetic resonance (NMR) spectra

[12] F. D. Tropper, F. O. Andersson, S. Braun, and R. Roy, *Synthesis*, 618 (1992).

[13] R. Roy and C. A. Laferrière, *J. Chem. Soc., Chem. Commun.*, 1709 (1990).

[14] E. Kallin, H. Lönn, T. Norberg, and M. Elofsson, *J. Carbohydr. Chem.* **8,** 597 (1989).

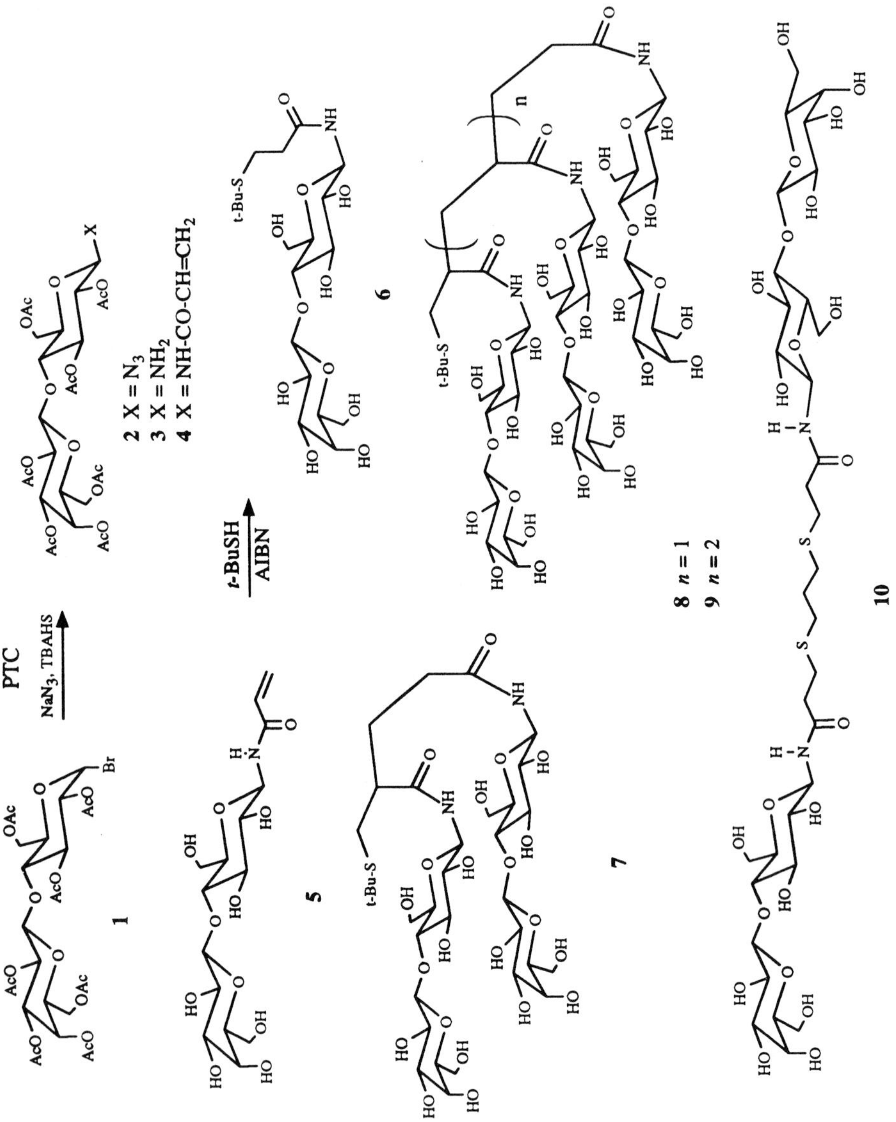
PTC
NaN3, TBAHS
t-BuSH
AIBN
1
2 X = N3
3 X = NH2
4 X = NH-CO-CH=CH2
5
6
7
8 n = 1
9 n = 2
10

SCHEME 1

TABLE I
EFFECT OF TELOGEN CONTENT ON AVERAGE TELOMER SIZE FOLLOWING TELOMERIZATION OF **5**

Telogen[a] (equivalents)	Average number of lactose residues in telomer mixture[b]
0.1	190
0.2	85
1.0	43
5	6
10	2[c]

[a] The telogen was *tert*-butylmercaptan. Based on telomerization of 50 mg (0.127 mmol) of **5**.
[b] Based on ^1H NMR spectral data.
[c] The telomer distribution was as follows: 29.4% monomer **6**, 15.8% dimer **7**, 8.4% trimer **8**, 5.8% tetramer **9**, and 41% higher telomers.

are recorded on a Bruker (Milton, Ontario, Canada) AMX-500 or a Varian (Palo Alto, CA) Gemini 200 MHz instrument. The proton chemical shifts are given relative to internal chloroform (7.24 ppm) for $CDCl_3$ solutions and to internal acetone (2.216 ppm) for D_2O solutions. Analyses of ^1H NMR data are based on a first-order approximation. Optical rotations are measured on a Perkin-Elmer (Norwalk, CT) 241 polarimeter and are run at room temperature for 0.5–1.0% (w/v) solutions in chloroform unless otherwise specified. Mass spectra are recorded on a Kratos Concept IIH spectrometer (Manchester, England). Elemental analyses are performed by M-H-W Laboratories (Phoenix, AZ). Thin-layer chromatography (TLC) is performed using silica gel 60-F254 precoated plates and column chromatography on silica gel (E. Merck, Darmstadt, Germany, No. 9385, 230–400 mesh). The developed plates are sprayed or dipped with a solution of ceric sulfate (1%, w/v) and ammonium molybdate (2.5%, w/v) in 10% (v/v) aqueous sulfuric acid and heated at 150°. All solvents and reagents used are reagent grade and obtained from Aldrich Chemical Co. (Milwaukee, WI). Optical densities in enzyme-linked lectin assays (ELLA) are measured at 410 nm on a Dynatech (Alexandria, VA) MR-600 spectrophotometer. Mixtures of telomers are separated by size-exclusion chromatography over BioGel P-2 (200–400 mesh, Bio-Rad, Mississauga, Ontario, Canada) with distilled water as eluent. Telomer separations are monitored with a differential refractometer (R401) from Millipore (Waters, Canada Ltd).

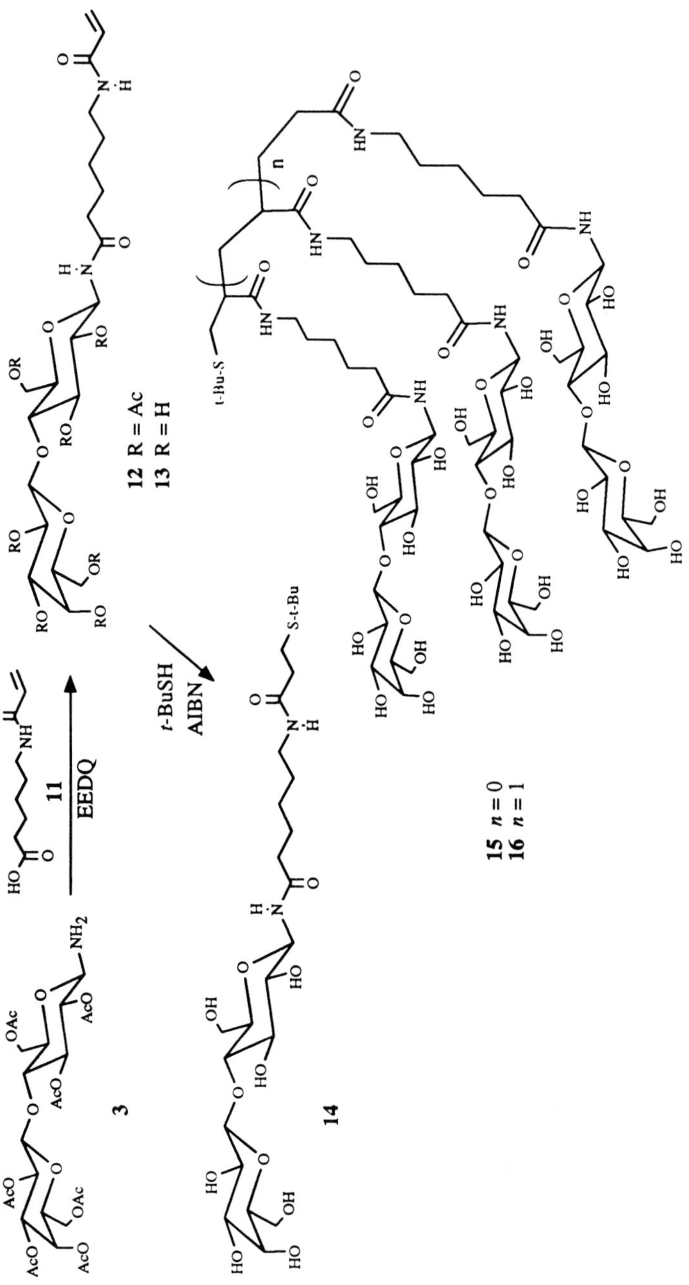

SCHEME 2

Experimental Procedures

Syntheses of Telomer Precursors

Lactosyl Azide **2**. To a solution of peracetylated lactosyl bromide (**1**) (800 mg, 1.14 mmol) and tetrabutylammonium hydrogen sulfate (TBAHS) (390 mg, 1.14 mmol) in ethyl acetate (5 ml) is added 1 *M* sodium carbonate (5 ml) and sodium azide (333 mg, 5.13 mmol). The two-phase reaction mixture is vigorously stirred at room temperature for 30 min. The completeness of the reaction is monitored by TLC using ethyl acetate as eluent; compound **1** has R_f 0.67, whereas compound **2** has R_f 0.71. The organic layer is separated and successively washed with 1 *M* $NaHCO_3$ (2 times, 20 ml), water (2 times, 20 ml), and brine (NaCl) (2 times, 20 ml). The combined organic extracts are dried with magnesium sulfate, filtered, and evaporated under reduced pressure to give crude **2**, which is purified by silica gel column chromatography using a gradient of ethyl acetate and hexane [6 : 4 to 7 : 3 (v/v)] as eluent. Pure **2** is obtained in 98% yield as an amorphous white solid. Relevant ^1H and ^{13}C NMR data ($CDCl_3$) are obtained as follows: δ (ppm) 4.45 (d, $J_{1',2'}$ = 7.8 Hz, H-1′), 4.60 (d, $J_{1,2}$ = 8.6 Hz, H-1); 101.0 (C-1′), 87.7 (C-1); mp 67.0°–70.0°; $[\alpha]_D$ −17.7° (*c* 1.0, $CHCl_3$) in agreement with literature values.[12]

Lactosylamine **3**. To a solution of compound **2** (500 mg, 0.756 mmol) dissolved in ethyl acetate (2 ml) is added activated Raney nickel (Aldrich, No. 22, 167-8, 2 g) that has been thoroughly washed with distilled water (2 times, 25 ml), ethanol (2 times, 25 ml), and ethyl acetate. The reduction is allowed to proceed at room temperature for 30 min. The progress of the reaction is monitored by TLC using ethyl acetate as eluent, in which compound **3** has R_f 0.33. Amine formation is also monitored by spraying the plate with a ninhydrin solution (1 g ninhydrin in 5 ml ethanol). Filtration of the solution over a bed of Celite and evaporation of the solvent provide compound **3** as a white solid in 94% yield. Relevant NMR and physical data are as follows: ^1H NMR ($CDCl_3$, δ ppm): 4.09 (d, 1H, $J_{1,2}$ = 6.6 Hz, H-1), 4.44 (d, 1H, $J_{1',2'}$ = 7.7 Hz, H-1′); ^{13}C NMR: 101.0 (C-1′), 84.6 (C-1); mp 81.0°–84.0°; $[\alpha]_D$ +7.40° (*c* 1.0, $CHCl_3$). Analyses: Calculated for $C_{26}H_{37}NO_{17}$: C, 49.29; H, 5.57; N, 2.21. Found: C, 49.21; H, 5.77; N, 1.98.

Peracetylated N-Acryloylated Lactosylamine **4**. To an ice-cold solution of compound **3** (100 mg, 0.157 mmol) dissolved in ethyl acetate (1 ml) is added acryloyl chloride (19 μl, 0.236 mmol, 1.5 equivalents) followed by pyridine (19 μl, 0.236 mmol, 1.5 equivalents). The reaction mixture is stirred at 0° for 45 min. A solution of 1 *M* HCl (2 ml) is then added to the reaction mixture. Usual washings of the organic phase with saturated

$NaHCO_3$ solution, water, and a brine (NaCl) solution is followed by drying the organic phase with sodium sulfate. The pure title compound **4** is obtained in 85% yield after purification by silica gel column chromatography using ethyl acetate as eluent. This treatment allows the removal of the small quantity of the α anomer of **4**. Relevant NMR and physical data are as follows. The *N*-acryloyl protons are defined as NHC(O)—CH_a=CH_b(*cis*)H_c(*trans*). ^{1}H NMR ($CDCl_3$, δ ppm): 6.37 (d, 1H, $J_{NH,H\text{-}1}$ = 9.3 Hz, NH), 6.24 (dd, 1H, $J_{a,c}$ = 17.0 Hz, $J_{b,c}$ = 1.1 Hz, H_c), 6.00 (dd, 1H, $J_{a,b}$ = 10.4 Hz, H_a), 5.68 (dd, 1H, H_b), 5.24 (dd, 1H, $J_{1,2}$ = 9.4 Hz, H-1), 4.43 (d, 1H, $J_{1',2'}$ = 7.9 Hz, H-1'); ^{13}C NMR: 129.9 (C-a), 128.4 (C-b), 100.8 (C-1'), 78.2 (C-1); mp 97.0°–99.9°; $[\alpha]_D$ +10.4° (*c* 1.0, $CHCl_3$).

N-Acryloylated Lactosylamine **5**. To a solution of **4** (100 mg, 0.145 mmol) dissolved in methanol (2 ml) is added 1 *M* sodium methoxide solution (0.15 ml) until the reaction mixture reaches pH 9.5 (pH paper strip). The reaction mixture is then stirred at room temperature for 3 hr. The evolution of the reaction is monitored by TLC using ethyl acetate, with which compound **4** has R_f 0.46. Compound **5** has R_f 0.33 in $CHCl_3$–methanol–water (10 : 10 : 0.7, by volume). The title compound precipitates on cooling the methanolic solution with an ice bath. The product is filtered, washed with cold methanol, and dried under vacuum in a desiccator. Compound **5** is obtained in 98% yield as an amorphous solid. Relevant NMR and physical data are as follows: ^{1}H NMR (D_2O, δ ppm): 6.26 (m, 2H, H_b, H_c), 5.81 (dd, 1H, $J_{a,b}$ = 4.9 Hz, $J_{a,c}$ = 6.6 Hz, H_a), 5.02 (d, 1H, $J_{1,2}$ = 9.2 Hz, H-1), 4.40 (d, 1H, $J_{1',2'}$ = 7.3 Hz, H-1'); ^{13}C NMR: 170.0 (C=O), 130.0 (CH=), 129.9 (=CH_2), 103.5 (C-1'), 80.0 (C-1); mp 241°–246° (dec); $[\alpha]_D$ −7.0° (*c* 0.5, water), +15.1° (*c* 1.0, DMSO) in agreement with literature values.[14] Fast atom bombardment-mass spectrometry in the positive ion mode [FAB-MS (pos)] yields the following: calculated for $C_{15}H_{25}NO_{11}$, 395.4; found, 396.0 (M + 1).

Telomerizations of N-Acryloylated Lactosylamine

Mixtures of the N-acryloylated lactosylamine derivative **5** (50 mg, 0.127 mmol), the telogen *tert*-butylmercaptan in various molar ratios (see Table I), and a catalytic amount of AIBN (0.2 mg) used as radical initiator are refluxed in methanol (0.57 ml) for 4 hr under a nitrogen atmosphere. The progress of the telomerizations is monitored by TLC using 8 : 2 (v/v) acetonitrile–water as eluent and is stopped by cooling the reaction mixture when all the starting monomer has been consumed. The solvent is evaporated under reduced pressure, and the resulting crude mixtures of telomers are dissolved in water and separated by size exclusion chromatography (2.0 × 90 cm, Bio-Rad) using BioGel P-2 and water as eluent. In a typical run, telomers **6–9** are obtained in yields of 29.4% (14.5 mg), 15.8% (7.9 mg), 8.4% (4.2 mg), and 5.8% (3 mg), respectively (total 59%); the remainder of the material is isolated as higher size telomers which are not processed further.

Relevant data for the telomers **6–9** are as follows. Monomer **6**: ^{1}H NMR (D_2O, δ ppm): 5.10 (d, 1H, $J_{1,2}$ = 9.3 Hz, H-1), 4.55 (d, 1H, $J_{1',2'}$ = 7.8 Hz, H-1′), 2.98 (dd, 2H, —SCH_2), 2.71 [m, 2H, —CH_2—C(O)—], 1.42 (s, 9H, ^{t}Bu); ^{13}C NMR: 99.7 (C-1′), 75.8 (C-1), 32.4 [—CH_2—C(O)—], 26.6 (^{t}Bu), 19.7 (S—CH_2); $[\alpha]_D$ +3.62° (*c* 1.0, water). Dimer **7**: ^{1}H NMR: 5.10 (d, 1H, $J_{1,2}$ = 9.3 Hz, H-1), 4.55 (d, 1H, $J_{1',2'}$ = 7.8 Hz, H-1′), 2.89 (m, 2H, —SCH_2), 2.67 [m, 1H, CH—C(O)], 2.48 [m, 2H, CH_2—C(O)], 2.03 [m, 2H, CH_2—C(O)], 1.40, 1.41 (two s, 9H, diastereomeric ^{t}Bu); ^{13}C NMR: 99.7 (C-1′), 75.8 (C-1), 42.9 [CH—C(O)], 29.6, 26.2, 24.3 (CH_2), 26.6 (^{t}Bu); $[\alpha]_D$ +9.85° (*c* 1.0, water); FAB-MS (pos): calculated for $C_{34}H_{60}NO_{22}S$, 881.0; found, 882.3 (M + 1). Trimer **8**: ^{1}H NMR: 5.10 (broad d, 1H, H-1), 4.55 (broad d, 1H, H-1), 3.75–2.27 [m, 4H, two CH, CH_2—C(O)], 2.91–2.81 (m, 2H, S—CH_2), 2.07–1.90 (m, 4H, backbone —CH_2), 1.41, 1.40, 1.39, 1.38 (four s, 9H, diastereomeric ^{t}Bu); ^{13}C NMR: 99.7 (C-1′), 75.8 (C-1), 42.9 [CH—C(O)], 31.5, 29.6, 26.2, 24.3 (CH_2), 26.6 (^{t}Bu); $[\alpha]_D$ +11.6° (*c* 1.0, water); FAB-MS (pos): calculated for $C_{49}H_{85}NO_{33}S$, 1276.3; found, 1277.8 (M + 1). Tetramer **9**: ^{1}H NMR: 5.10 (broad d, 1H, H-1), 4.55 (broad d, 1H, H-1′), 2.95–2.78, 2.66–2.44, 2.44–2.38, 2.16–1.80 (4m, 10 CH_2, 3 CH, backbone), 1.41, 1.40, 1.39 (s, 9H, diastereomeric ^{t}Bu); ^{13}C NMR: 99.7 (C-1′), 75.8 (C-1), 42.9 (CH), 31.5, 29.6, 26.2, 24.3 (CH_2), 26.6 (^{t}Bu); $[\alpha]_D$ +5.43° (*c* 1.0, water); FAB-MS (pos): calculated for $C_{64}H_{110}NO_{44}S$, 1671.6; found, 1672.4 (M + 1, 6.5%).

Tethered Dimer

To a 10-ml round-bottomed flask containing compound **5** (53.6 mg, 0.136 mmol) dissolved in methanol (2 ml), is added propanedithiol (7.36 mg, 68 μmol, 0.5 equivalent) and triethylamine (13.8 mg, 0.136 mmol, 1 equivalent). The reaction mixture is refluxed for 5 hr. The progress of the reaction is monitored by TLC using 97 : 3 (v/v) methanol–water as eluent, in which dimer **10** has R_f 0.57. The solvent and volatile reagents are evaporated under reduced pressure, and the residue is applied to a BioGel P-2 column using water as eluent. Pure compound **10** is obtained as a white and amorphous solid in 82% yield. Relevant NMR data are as follows. ^{1}H NMR (D_2O, δ ppm): 4.83 (d, 2H, $J_{1,2}$ = 9.2 Hz, H-1), 4.28 (d, 2H, $J_{1',2'}$ = 7.5 Hz, H-1′), 2.67 [t, 4H, *J* = 7.1 Hz, —C(O)—CH_2—], 2.51 (t, 4H, *J* = 7.2 Hz, —CH_2S—), 2.47 (t, 4H, *J* = 7.3 Hz, —SCH_2—), 1.70 (q, 2H, *J* = 7.1 Hz, middle CH_2); ^{13}C NMR (δ ppm): 175.5 175.4 (C=O),

102.5 (C-1′), 78.7 (C-1), 36.2, 35.4 [—C(O)—CH_2], 29.5, 29.3 (—CH_2S), 27.9, 27.5 (—SCH_2), 26.1 (middle CH_2).

Telomers with Spacer Arm

*6-N-Acrylamidohexanoic Acid (**11**)*. To an ice-cold aqueous solution (5 ml) of 6-aminohexanoic acid (1 g, 7.62 mmol) are added acryloyl chloride (0.75 g, 670 μl, 8.38 mmol), 2,6-di-*tert*-butyl-4-methylphenol (10 mg, added as a radical scavenger), and sodium hydroxide (0.33 g/5 ml water). The reaction mixture is stirred for 30 min at 0°, then at room temperature for a further 90 min. The resulting floating solid is removed by decantation, and the remaining aqueous phase is washed with ethyl acetate. The aqueous phase is cooled to 0°, the pH lowered to pH 1–2 with 6 *N* HCl, and then washed exhaustively with ethyl acetate. The organic phase is dried with sodium sulfate, filtered, and evaporated to dryness under vacuum. The resulting residue is crystallized with a mixture of ethyl acetate–hexane to provide pure 6-*N*-acrylamidohexanoic acid **11** (0.84 g) in 60% yield. Compound **11** has mp 84°–85°; ^{1}H NMR ($CDCl_3$, δ ppm): 6.15 (m, 3H), 5.58 (dd, 1H), 3.29 (q, 2H), 2.32 (t, 2H), 1.53 (m, 6H): ^{13}C NMR (δ ppm); 179.0, 166.6, 166.5, 39.9, 34.4, 29.6, 26.8, 24.8.

*Peracetylated 6-N-Acrylamidohexanoyl Lactosylamine (**12**)*. To a solution of lactosylamine derivative **3** (1.2 g, 1.89 mmol) in absolute ethanol (7 ml) is added 6-*N*-acrylamidohexanoic acid (**11**) (0.419 g, 2.26 mmol). After 5 min, EEDQ (550 mg, 2.22 mol) is added, and the reaction mixture is stirred at room temperature overnight. Compound **12** has R_f 0.25 in ethyl acetate. The solvent is evaporated under reduced pressure, and the crude residue is dissolved in chloroform. The organic phase is successively washed with 3% HCl, saturated $NaHCO_3$, and brine. After drying over Na_2SO_4 and filtration, the organic phase is evaporated to dryness. The residue is purified by silica gel column chromatography, and pure compound **12** is eluted with 8 : 2 (v/v) ethyl acetate–methanol and obtained in 73% yield (1.1 g) as an amorphous white solid which resists crystallization. Relevant ^{1}H and ^{13}C NMR data ($CDCl_3$, δ ppm) are obtained as follows: 6.22 (dd, 1H, $J_{a,c}$ = 16.5 Hz, $J_{b,c}$ = 1.0 Hz, H_c), 6.19 (d, 1H, $J_{1,NH}$ = 9.4 Hz, NH), 6.06 (dd, 1H, $J_{a,b}$ = 6.7 Hz, H_a), 5.83 (b, 1H, NH), 5.60 (dd, 1H, H_b), 5.16 (dd, 1H, $J_{1,2}$ = 9.4 Hz, H-1), 4.41 (d, 1H, H-1′), 3.85–3.69 (m, 2H, CH_2N), 3.31–3.27 (m, 2H, CH_2CO), 2.16, 1.93 (7s, m, 23H, 7CH_3CO, CH_2), 1.60–1.20 (m, 6H, three CH_2); 131.4 (CH=), 127.0 (=CH_2), 101.5 (C-1′), 78.6 (C-1), 26.7–21.1 (7 CH_2). Compound **12** has $[\alpha]_D$ +5.7.0° (*c* 1.03, water); FAB-MS (pos): calculated for $C_{35}H_{50}N_2O_{12}$, 802.3; found, 803.3 (M + 1, 16%).

TABLE II
RELATIVE POTENCIES OF LACTOSYLATED TELOMERS AS INHIBITORS OF PEANUT LECTIN BINDING TO LACTOSYLATED POLYMER

Compound	Concentration (n*M*) required for 50% inhibition	Relative potencies
Lactose	248	1.00
5	255	0.97
6	201	1.23
7	211	1.18
8	220	1.13
9	250	0.99
10	56	4.43
16	150	1.65

*6-N-Acrylamidohexanoyl Lactosylamine (**13**).* To a solution of **12** (1.0 g, 1.24 mmol) in methanol (10 ml) is added 1 *M* sodium methoxide until the solution reaches approximately pH 9.5 (pH paper strip). The reaction mixture is stirred at room temperature for 3 hr after which time it is neutralized with Amberlite IR-120 cation-exchange resin (H^+). After filtration, the solution is evaporated under vacuum to provide pure **13** as an amorphous white solid in 91% yield (576 mg) having R_f 0.42 in 8 : 2 (v/v) acetonitrile–water. ^1H and ^{13}C NMR data (D_2O, δ ppm) are as follows: 6.38–6.24 (m, 2H, H_b, H_c), 5.83 (m, 1H, H_a), 5.09 (d, 1H, $J_{1,2}$ = 9.2 Hz, H-1), 4.55 (d, 1H, $J_{1,2}$ = 7.8 Hz, H-1), 2.43 (m, 2H, CH_2), 1.69 (m, 4H, CH_2), 1.45 (m, 2H, CH_2); 179.9 (CO), 169.9 (CO), 131.5 (CH═), 128.5 (═CH_2), 104.7 (C-1′), 80.6 (C-1). Compound **13** has $[\alpha]_D$ +3.30° (*c* 1.10, water); FAB-MS (pos): calculated for $C_{21}H_{36}N_2O_{12}$, 508.2; found, 509.2 (M + 1, 0.1%).

Telomerization of 6-N-Acrylamidohexanoyl Lactosylamine

A mixture of monomer **13** (200 mg, 0.393 mmol), *tert*-butylmercaptan (195 mg, 245 μl, 2.16 mmol, 5.5 equivalents), and a catalytic amount of AIBN (10 mg) in methanol–water (3.2 ml : 0.5 ml) is refluxed for 5 hr under nitrogen. After complete conversion of the monomer **13**, the reaction mixture is evaporated to dryness under reduced pressure. The resulting mixture of telomers in the minimum volume of water is subjected to size exclusion chromatography over BioGel P-2 (2.0 × 90 cm), and each individual telomer is obtained using water as eluent. The column allows the separation of the monomer **14**, dimer **15**, and trimer **16** from the remaining bulk of higher telomers in a percent weight ratio of

26.6 : 13.9 : 1.22. Compounds **14–16** have R_f values of 0.66, 0.20, and 0.06, respectively, in 8 : 2 (v/v) acetonitrile–water. The ^{1}H and ^{13}C NMR data (D_2O, δ ppm) for compound **14** are as follows: 5.13 (d, 1H, $J_{1,2}$ = 9.2 Hz, H-1), 4.60 (d, 1H, $J_{1',2'}$ = 7.8 Hz, H-1′), 1.81–1.47 (m, s, 15 H, ^{t}Bu, CH_2); 176.0, 175.1 (two CO), 99.6 (C-1′), 75.8 (C-1), 26.7 (^{t}Bu); the compound has $[\alpha]_D$ +6.9° (*c* 0.56, water); FAB-MS (pos): calculated for $C_{25}H_{46}N_2O_{12}S$, 598.3; found, 599.2 (M + 1, 19.5%). Compound **15** has $[\alpha]_D$ +8.1° (*c* 0.48, water); FAB-MS: calculated for $C_{46}H_{82}N_4O_{24}S$, 1106.5; found, 1107.5 (M + 1, 3.5%). Compound **16** has $[\alpha]_D$ +14.9° (*c* 0.25, water); FAB-MS: calculated for $C_{67}H_{118}N_6O_{36}S$, 1614.7; found, 1615.9 (M + 1, 1.1%).

Enzyme-Linked Lectin Assay

Inhibition of Binding of Peanut Lectin to Lactose-Containing Polymer by Lactosylated Telomers. Linbro microtitration plates (Flow Laboratories, Mississauga, ON, Canada) are coated overnight with poly(acrylamide-*co*-*p*-*N*-acrylamidophenyl β-D-lactoside)[8] (2.5 μg/100 μl/well). The stock solution of polymer is prepared at 1 mg/ml in phosphate-buffered saline (PBS). After removing the excess coating capture antigen, the wells are blocked with 1% (w/v) bovine serum albumin (BSA) in PBS at 200 μl/well for 60 min at room temperature. After washing the plates five times with PBS–Tween, the inhibitors are added in serial dilutions starting from stock solutions at 2 mg/ml in PBS. Thus 50 μl/well of between 50 and 500 nmol inhibitors are used in each well. A solution of horseradish peroxidase-labeled peanut lectin from *Arachis hypogaea* (Sigma Chemical Co., St. Louis, MO, L-7759, 20 U/mg protein) in PBS (0.25 μg/100 μl/well) is then added. The inhibitors and the lectin in the wells are allowed to equilibrate for 3 hr at room temperature. The plates are then washed 5 times with PBS–Tween and 2 times with PBS alone. The peroxidase substrate, 2,2′-azinobis(3-ethylbenzothiazoline-6-sulfonic acid) (ABTS, Aldrich, 1 mg/4 ml) (50 μl/well) in 0.2 *M* citrate–phosphate buffer, pH 4.0, containing 0.015% (v/v) H_2O_2 is then added. After 15 min, the absorbances are measured at 410 nm. The concentration of inhibitors required for 50% inhibition are indicated in Table II.

Author Index

Numbers in parentheses are footnote reference numbers and indicate that an author's work is refered to although the name is not cited in the text.

A

B

C

D

E

F

G

L

M

N

O

P

S

T

U

V

W

Subject Index

A

M

N

O

P

R

S

T

W

ISBN 0-12-182143-9

90018

9 780121 821432